Applied Analysis

Mathematical Methods in Natural Science

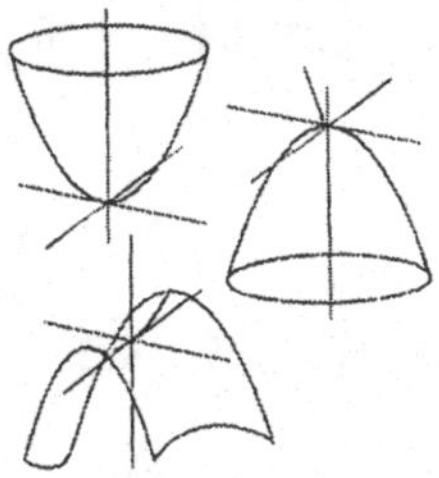

Applied Analysis

Mathematical Methods in Natural Science

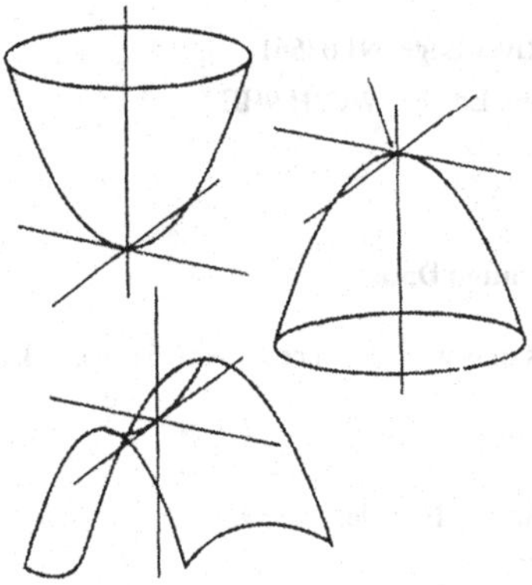

Takasi Senba
Miyazaki University, Japan

Takashi Suzuki
Osaka University, Japan

Imperial College Press

Published by

Imperial College Press
57 Shelton Street
Covent Garden
London WC2H 9HE

Distributed by

World Scientific Publishing Co. Pte. Ltd.
5 Toh Tuck Link, Singapore 596224
USA office: Suite 202, 1060 Main Street, River Edge, NJ 07661
UK office: 57 Shelton Street, Covent Garden, London WC2H 9HE

Library of Congress Cataloging-in-Publication Data
Senba, Takasi.
Applied analysis : Mathematical Methods in Natural Science / Takasi Senba, Takashi Suzuki.
p. cm.
Includes bibliographical references and index.
ISBN 1-86094-440-X (alk. paper)
1. Mathematical analysis. I. Suzuki, Takashi. II. Title.

QA300 .S3952 2004
515--dc22 200307074

British Library Cataloguing-in-Publication Data
A catalogue record for this book is available from the British Library.

Printed in Singapore by World Scientific Printers (S) Pte Ltd

Dedicated to Professor Hiroshi Fujita

Preface

This book is intended to be an introduction to mathematical science, particularly the theoretical study from the viewpoint of applied analysis. As basic materials, *vector analysis* and *calculus of variation* are taken, and then *Fourier analysis* is introduced for the eigenfunction expansion to justify. After that, *statistical method* is presented to control the mean field of many particles, and the mathematical theory to linear and nonlinear partial differential equations is accessed. *System of chemotaxis* is a special topic in this book, and well-posedness of the model is established. We summarize several mathematical theories and give some references for the advanced study. We also picked up some materials from classical mechanics, geometry, mathematical programming, numerical schemes, and so forth. Thus, this book covers some parts of undergraduate courses for mathematical study. It is also suitable for the first degree of graduate course to learn the basic ideas, mathematical techniques, systematic logic, physical and biological motivations, and so forth.

Most part of this monograph is based on the notes of the second author for undergraduate and graduate courses and seminars at several universities. We thank all our students for taking part in the project.

December 2003

Takasi Senba and Takashi Suzuki

Contents

Chapter 1

Geometric Objects

Some kind of insects and amoeba are lured by special chemical substances of their own. Such a character is called chemotaxis in biology. For its formulation, some mathematical terminologies and notions are necessary. This chapter is devoted to geometric objects.

1.1 Basic Notions of Vector Analysis

1.1.1 *Dynamical Systems*

Movement of a *mass point* is indicated by the position vector $\boldsymbol{x} = \boldsymbol{x}(t) \in \mathbf{R}^3$ depending on the time variable $t \in \mathbf{R}$. If m and $\boldsymbol{F}$ denote its mass and the force acting on it, respectively, *Newton's equation of motion* assures the relation

$$m\frac{d^2\boldsymbol{x}}{dt^2} = \boldsymbol{F}, \tag{1.1}$$

where $\frac{d^2\boldsymbol{x}}{dt^2}$ stands for the *acceleration vector*. If n points $x_i = x_i(t)$ $(i = 1, 2, \cdots, n)$ are interacting, then they are subject to the system

$$m_i\frac{d^2\boldsymbol{x}_i}{dt^2} = \boldsymbol{F}_i \qquad (i = 1, 2, \cdots, n),$$

simply written as

$$\ddot{x} = f(x, \dot{x}, t) \tag{1.2}$$

with $x = (\boldsymbol{x}_1, \boldsymbol{x}_2, \cdots, \boldsymbol{x}_n) \in \mathbf{R}^{3n}$,

$$\ddot{x} = \frac{d^2x}{dt^2}, \qquad \text{and} \qquad \dot{x} = \frac{dx}{dt}.$$

It is sometimes referred to as the *deterministic principle* of Newton, and under reasonable assumptions on f, say, continuity in all variables $(x, \dot{x}, t)$ and the Lipschitz continuity in $(x, \dot{x})$, there is a unique solution $x = x(t)$ to (1.2) locally in time with the prescribed initial position $x(0) = x_0$ and the initial velocity $\dot{x}(0) = \dot{x}_0$. At this occasion, let us recall that the initial values

$$x(0) = x_0 \qquad \text{and} \qquad \dot{x}(0) = \dot{x}_0$$

provide equation (1.2) with the *Cauchy problem.*

In some cases the degree of freedom is reduced, as $x = x(t) \in \mathbf{R}$ or $x = x(t) \in \mathbf{R}^2$. For example,

$$\ddot{x} = -k^2 x$$

with $x = x(t) \in \mathbf{R}$ is associated with the oscillatory motion of a bullet hanged by spring, and its solution is given by

$$x(t) = x_0 \cos kt + \dot{x}_0 \sin kt/k.$$

Although very few solutions to (1.2) are written explicitly even for the case of $x = x(t) \in \mathbf{R}$,

$$\ddot{x} = f(x) \tag{1.3}$$

is the simplest but general form of it. In this case

$$T = \frac{1}{2}\dot{x}^2 \qquad \text{and} \qquad U(x) = -\int^x f(\xi)d\xi$$

are referred to as the *kinetic energy* and the *potential energy*, respectively. Then, the *total energy* is given by

$$E = T + U = \frac{1}{2}\dot{x}^2 + U(x)$$

so that it is a function of $(x, \dot{x})$, denoted by $E = E(x, \dot{x})$. If $x = x(t)$ is a solution to (1.3), then it holds that

$$\frac{d}{dt}E\left(x(t), \dot{x}(t)\right) = \dot{x}\ddot{x} - f(x)\dot{x} = \dot{x}\left(\ddot{x} - f(x)\right) = 0,$$

so that $E(x(t), \dot{x}(t))$ is a constant. This fact is referred to as the *conservation law of energy.*

System (1.3) is equivalent to $\dot{x} = y$ and $\dot{y} = f(x)$, or

$$\frac{d}{dt}\begin{pmatrix} x \\ y \end{pmatrix} = \Phi(x, y) \tag{1.4}$$

with $\Phi(x, y) = {}^t(y, f(x))$. Because the right-hand side does not include the variable t explicitly, system (1.4) is said to be *autonomous*, and its solution is illustrated as a curve in $x - y$ plane. Energy conservation $E(x(t), \dot{x}(t)) = E(x_0, \dot{x}_0)$ guarantees the existence of the solution globally in time if the potential $U = U(x)$ is *coercive*, which means that $|x| \to +\infty$ implies $U(x) \to +\infty$. Then, each

$$\mathcal{O} = \{(x(t), \dot{x}(t)) \mid t \in \mathbf{R}\}$$

is called an *orbit*, which coincides with the curve $E = \frac{1}{2}y^2 + U(x)$, where $E = E(x_0, \dot{x}_0)$.

Because of the uniqueness of the solution to the Cauchy problem of (1.3), the orbit never intersects by itself. However, it may be a point, which corresponds to the zero of Φ, that is, $y = 0$ and $f(x) = -U'(x) = 0$. It is referred to as the *equilibrium point.* Each equilibrium point $(\hat{x}_0, 0)$ is *stable* or *unstable* if $\hat{x}_0$ is a local minimum or a local maximum of $U = U(x)$, respectively. This means that if the initial value $(x_0, \dot{x}_0)$ is close to $(\hat{x}_0, 0)$, then the solution to (1.3) stays near or away from it.

The solution to (1.4) may be written as

$$\begin{pmatrix} x(t) \\ y(t) \end{pmatrix} = T_t \begin{pmatrix} x(0) \\ y(0) \end{pmatrix}$$

for ${}^t(x(0), y(0)) = {}^t(x_0, \dot{x}_0)$, with the mapping $T_t : \mathbf{R}^2 \to \mathbf{R}^2$ defined for each $t \in \mathbf{R}$. Then the family $\{T_t\}_{t \in \mathbf{R}}$ induces the continuous mapping

$$T : \mathbf{R}^2 \times \mathbf{R} \quad \to \quad \mathbf{R}^2$$

by

$$T\left({}^t(x, y), t\right) = T_t \begin{pmatrix} x \\ y \end{pmatrix}.$$

This family is provided with the properties that $T_0 = Id$, the identity operator, and $T_{t+s} = T_t \circ T_s$ for $t, s \in \mathbf{R}$, with $\circ$ denoting the composition of operators. Then, we call $\{T_t\}_{t\in\mathbf{R}}$ the *dynamical system.*

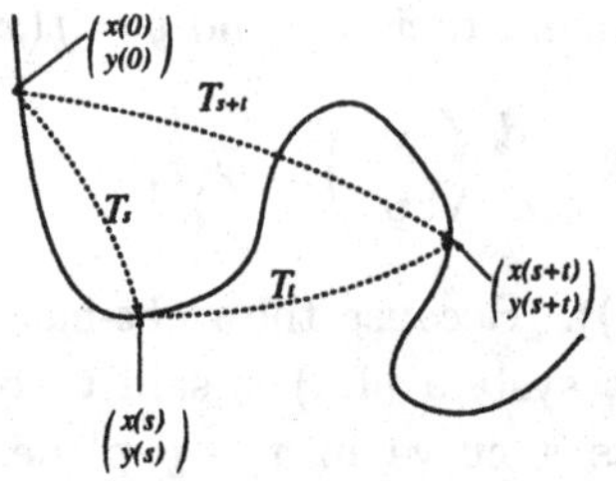

Fig. 1.1

For $x = x(t) \in \mathbf{R}^3$ in (1.3), we say that $f = f(x)$ is a *potential field* if

$$f = -\begin{pmatrix} \partial U/\partial x_1 \\ \partial U/\partial x_2 \\ \partial U/\partial x_3 \end{pmatrix} \tag{1.5}$$

holds with a scalar function $U = U(x_1, x_2, x_3)$. In use of the *gradient operator*

$$\nabla = \begin{pmatrix} \partial/\partial x_1 \\ \partial/\partial x_2 \\ \partial/\partial x_3 \end{pmatrix},$$

relation (1.5) is written as

$$f = -\nabla U.$$

Then we can define the total energy by

$$E(x, \dot{x}) = \frac{1}{2}|\dot{x}|^2 + U(x),$$

where $|\dot{x}|$ denotes the length of the velocity vector $\dot{x} \in \mathbf{R}^3$. Similarly to the one-dimensional case, this E is a quantity of conservation. In fact, if $x = x(t)$ is a solution to (1.3) it follows that

$$\frac{d}{dt}E\left(x(t), \dot{x}(t)\right) = \dot{x} \cdot \left(\ddot{x} + \nabla U(x)\right) = 0,$$

where $\cdot$ denotes the inner product in $\mathbf{R}^3$. However, this law of the conservation of energy is not sufficient to control the orbit $\mathcal{O}$ in $x - \dot{x}$ space, which is now identified with $\mathbf{R}^6$.

Exercise 1.1 Illustrate some orbits to (1.3) for $x = x(t) \in \mathbf{R}$ in $x - \dot{x}$ plane, when the potential energy is given by $U(x) = \frac{1}{4}x^4 - \frac{1}{2}x^2$. Seek all equilibrium points and judge their stability. Examine the same question for $U(x) = \pm\frac{1}{2}x^2$.

1.1.2 *Outer Product*

Here, we take the notion of vector analysis; outer product of the vector, gradient of the scalar field, and divergence and rotation of the vector field. Throughout the present chapter, three-dimensional vectors are denoted by $\boldsymbol{a}$, $\boldsymbol{b}$, $\boldsymbol{c}$, $\cdots$, while a, b, c, $\cdots$ indicate scalars. The canonical basis of $\mathbf{R}^3$ is given by

$$\boldsymbol{i} = \begin{pmatrix} 1 \\ 0 \\ 0 \end{pmatrix}, \quad \boldsymbol{j} = \begin{pmatrix} 0 \\ 1 \\ 0 \end{pmatrix}, \quad \boldsymbol{k} = \begin{pmatrix} 0 \\ 0 \\ 1 \end{pmatrix},$$

which are arranged to form a right-handed coordinate system in three dimensional space $\mathbf{R}^3$. The length of $\boldsymbol{a}$ is denoted by $|\boldsymbol{a}|$, and $\boldsymbol{a} \cdot \boldsymbol{b}$ stands for the inner product of $\boldsymbol{a}$ and $\boldsymbol{b}$. That is, $\boldsymbol{a} \cdot \boldsymbol{b} = |\boldsymbol{a}| \cdot |\boldsymbol{b}| \cos\theta$, where θ is the angle between $\boldsymbol{a}$ and $\boldsymbol{b}$. If

$$\boldsymbol{a} = \begin{pmatrix} a_1 \\ a_2 \\ a_3 \end{pmatrix} = a_1\boldsymbol{i} + a_2\boldsymbol{j} + a_3\boldsymbol{k} \tag{1.6}$$

and

$$\boldsymbol{b} = \begin{pmatrix} b_1 \\ b_2 \\ b_3 \end{pmatrix} = b_1\boldsymbol{i} + b_2\boldsymbol{j} + b_3\boldsymbol{k} \tag{1.7}$$

represent those vectors by their components, it holds that $\boldsymbol{a} \cdot \boldsymbol{b} = a_1b_1 + a_2b_2 + a_3b_3$.

The outer product of $\boldsymbol{a}$ and $\boldsymbol{b}$ is the vector denoted by $\boldsymbol{a} \times \boldsymbol{b}$ satisfying the following property.

1 Its length is equal to the area of the parallelogram made by $\boldsymbol{a}$ and $\boldsymbol{b}$.
2 It is perpendicular to $\boldsymbol{a}$ and $\boldsymbol{b}$.
3 $\boldsymbol{a}$, $\boldsymbol{b}$, and $\boldsymbol{a} \times \boldsymbol{b}$ are right-handed.

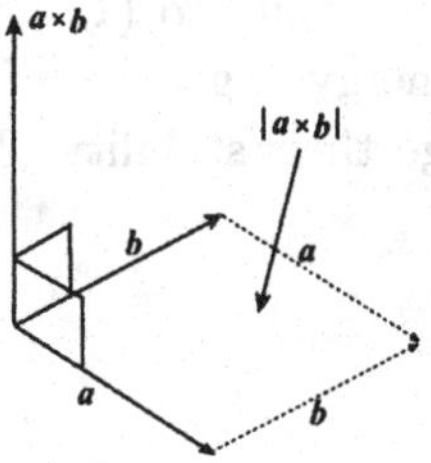

Fig. 1.2

Then, we have the following.

Theorem 1.1 *The operation $(\boldsymbol{a}, \boldsymbol{b}) \mapsto \boldsymbol{a} \times \boldsymbol{b}$ is subject to the following laws.*

1 *(commutative):* $\boldsymbol{b} \times \boldsymbol{a} = -\boldsymbol{a} \times \boldsymbol{b}$.
2 *(associative):* $c(\boldsymbol{a} \times \boldsymbol{b}) = (c\boldsymbol{a}) \times \boldsymbol{b}$.
3 *(distributive):* $\boldsymbol{a} \times (\boldsymbol{b} + \boldsymbol{c}) = \boldsymbol{a} \times \boldsymbol{b} + \boldsymbol{a} \times \boldsymbol{c}$.

Proof. We shall show the distributive law because the other laws are obvious. First, from the associative law we may suppose that $|\boldsymbol{a}| = 1$. We take the plane π containing the origin whose normal vector is $\boldsymbol{a}$. Look down π so that $\boldsymbol{a}$ is upward. Let $\boldsymbol{b}'$, $\boldsymbol{c}'$, and $(\boldsymbol{b} + \boldsymbol{c})'$ be the projections to π of $\boldsymbol{b}$, $\boldsymbol{c}$, and $\boldsymbol{b} + \boldsymbol{c}$, respectively. Then, by the definition we have

$$\boldsymbol{a} \times \boldsymbol{b} = \boldsymbol{a} \times \boldsymbol{b}', \qquad \boldsymbol{a} \times \boldsymbol{c} = \boldsymbol{a} \times \boldsymbol{c}',$$

and

$$\boldsymbol{a} \times (\boldsymbol{b} + \boldsymbol{c}) = \boldsymbol{a} \times (\boldsymbol{b} + \boldsymbol{c})'.$$

Here, we have $(\boldsymbol{b} + \boldsymbol{c})' = \boldsymbol{b}' + \boldsymbol{c}'$, so that the equality

$$\boldsymbol{a} \times (\boldsymbol{b} + \boldsymbol{c}) = \boldsymbol{a} \times \boldsymbol{b} + \boldsymbol{a} \times \boldsymbol{c},$$

to be proven, is equivalent to

$$\boldsymbol{a} \times (\boldsymbol{b}' + \boldsymbol{c}') = \boldsymbol{a} \times \boldsymbol{b}' + \boldsymbol{a} \times \boldsymbol{c}'. \tag{1.8}$$

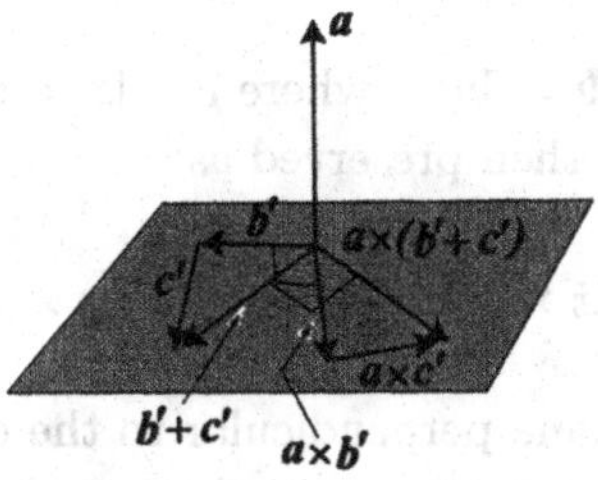

Fig. 1.3

Because $|\boldsymbol{a}| = 1$, the vector $\boldsymbol{a} \times \boldsymbol{b}'$ is nothing but the vector on π with $\boldsymbol{b}'$ rotated counter-clockwise by 90 degrees. The same is true for $\boldsymbol{a} \times \boldsymbol{c}'$ and $\boldsymbol{a} \times (\boldsymbol{b}' + \boldsymbol{c}')$. Therefore, (1.8) is obtained by rotating the parallelogram made by $\boldsymbol{b}'$ and $\boldsymbol{c}'$ on π counter-clockwisely by 90 degrees. □

By the definition, it holds that

$$\boldsymbol{i} \times \boldsymbol{j} = \boldsymbol{k}, \qquad \boldsymbol{j} \times \boldsymbol{k} = \boldsymbol{i}, \qquad \boldsymbol{k} \times \boldsymbol{i} = \boldsymbol{j}.$$

Therefore, three laws in Theorem 1.1 imply

$$\begin{pmatrix} a_1 \\ a_2 \\ a_3 \end{pmatrix} \times \begin{pmatrix} b_1 \\ b_2 \\ b_3 \end{pmatrix} = \begin{pmatrix} a_2 b_3 - a_3 b_2 \\ a_3 b_1 - a_1 b_3 \\ a_1 b_2 - a_2 b_1 \end{pmatrix} \tag{1.9}$$

by (1.6) and (1.7).

Exercise 1.2 Prove (1.9).

Exercise 1.3 Show the *Lagrange identity*

$$(\boldsymbol{a} \times \boldsymbol{b}) \cdot (\boldsymbol{c} \times \boldsymbol{d}) = (\boldsymbol{a} \cdot \boldsymbol{c})(\boldsymbol{b} \cdot \boldsymbol{d}) - (\boldsymbol{a} \cdot \boldsymbol{d})(\boldsymbol{b} \cdot \boldsymbol{c}). \tag{1.10}$$

1.1.3 *Motion of Particles*

If the mass point $x = x(t) \in \mathbf{R}^3$ is subject to the *center force*, then the Newton equation takes the form

$$\ddot{x} = \Phi(r)\omega \tag{1.11}$$

with the scalar function $\Phi = \Phi(r)$, where $r = |x|$ and $x = r\omega$. The *angular momentum* $M = x \times \dot{x}$ is then preserved as

$$\frac{dM}{dt} = \dot{x} \times \dot{x} + x \times \ddot{x} = \frac{\Phi(r)}{r} x \times x = 0$$

and hence x lies in the plane perpendicular to the constant vector M. By rotating the axis, therefore, we may put that

$$x = \begin{pmatrix} x_1 \\ x_2 \\ 0 \end{pmatrix} \quad \text{and} \quad \omega = \begin{pmatrix} \cos\theta \\ \sin\theta \\ 0 \end{pmatrix}.$$

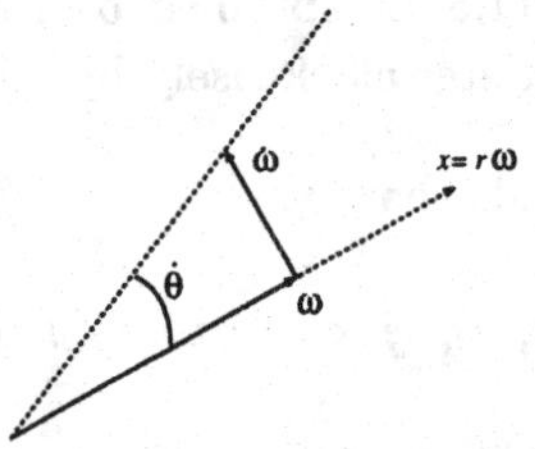

Fig. 1.4

Then it follows that

$$\dot{\omega} = \omega^{\perp}\dot{\theta} \quad \text{with} \quad \omega^{\perp} = \begin{pmatrix} -\sin\theta \\ \cos\theta \\ 0 \end{pmatrix} \tag{1.12}$$

and $x = r\omega$ implies

$$x \times \dot{x} = r\omega \times (\dot{r}\omega + r\omega^{\perp}\dot{\theta}) = r^2\dot{\theta}\omega \times \omega^{\perp} = r^2\dot{\theta}\begin{pmatrix} 0 \\ 0 \\ 1 \end{pmatrix}.$$

In other words, $r^2\dot{\theta}$ is a constant, which is referred to as the constant law of the *area velocity* discovered by Kepler.

If $x = x(t) \in \mathbf{R}^3$ is subject to

$$\ddot{x} = -\nabla U(x) \tag{1.13}$$

with $U = U(r)$, then it holds that

$$\nabla U = \begin{pmatrix} \partial U/\partial x_1 \\ \partial U/\partial x_2 \\ \partial U/\partial x_3 \end{pmatrix} = U'(r)\omega. \tag{1.14}$$

Equation (1.13) with (1.14) takes the form of (1.11) and therefore, $M = r^2\dot{\theta}$ is a constant.

We have from (1.12) that

$$\ddot{\omega} = \dot{\omega}^\perp\dot{\theta} + \omega^\perp\ddot{\theta} \qquad \text{with} \qquad \dot{\omega}^\perp = -\omega\dot{\theta}, \tag{1.15}$$

which implies, from $x = r\omega$ that

$$\begin{aligned} \ddot{x} &= \ddot{r}\omega + 2\dot{r}\dot{\omega} + r\ddot{\omega} = \ddot{r}\omega + 2\dot{r}\dot{\theta}\omega^\perp + r(-\omega\dot{\theta}^2 + \omega^\perp\ddot{\theta}) \\ &= (\ddot{r} - r\dot{\theta}^2)\omega + (2\dot{r}\dot{\theta} + r\ddot{\theta})\omega^\perp \\ &= -U'(r)\omega. \end{aligned}$$

Thus, we obtain

$$\ddot{r} - r\dot{\theta}^2 = -U' \qquad \text{and} \qquad 2\dot{r}\dot{\theta} + r\ddot{\theta} = 0,$$

and from $\dot{\theta} = M/r^2$ it follows that

$$\ddot{r} = -V' \qquad \text{for} \qquad V = U + \frac{M^2}{2r^2}. \tag{1.16}$$

Equation (1.16) takes the form of (1.3) and this $V = V(r)$ is called the *efficient potential energy.*

System of many particles is subject to

$$m_i\ddot{x}_i = \boldsymbol{F}_i \qquad \text{with} \qquad \boldsymbol{F}_i = \sum_{j\neq i} \boldsymbol{F}_{ij} + \boldsymbol{F}'_i,$$

where $\boldsymbol{F}_{ij}$ and $\boldsymbol{F}'_i$ denote the *self-interaction force* and the *outer force*, respectively. Then, the *action-reaction law* of Newton guarantees that

$$\boldsymbol{F}_{ij} = f_{ij}\boldsymbol{e}_{ij}$$

with

$$e_{ij} = \frac{x_i - x_j}{|x_i - x_j|} \qquad \text{and} \qquad f_{ij} = f_{ji} \in \mathbf{R}.$$

Therefore, the *momentum* defined by

$$\boldsymbol{P} = \sum_i m_i \dot{x}_i$$

is invariant in the *closed system* without outer force. In fact, $\boldsymbol{F}'_i = 0$ in this case, and we have

$$\frac{d\boldsymbol{P}}{dt} = \sum_i m_i \ddot{x}_i = \sum_i \boldsymbol{F}_i = \sum_{i,j;i \neq j} \boldsymbol{F}_{ij} = 0$$

by $\boldsymbol{F}_{ij} = -\boldsymbol{F}_{ji}$. The *angular momentum*

$$\boldsymbol{M} = \sum_i x_i \times m_i \dot{x}_i$$

is also preserved in this case, as it holds that

$$\frac{d\boldsymbol{M}}{dt} = \sum_i (\dot{x}_i \times m_i \dot{x}_i + x_i \times m_i \ddot{x}_i) = \sum_{i,j;i \neq j} x_i \times \boldsymbol{F}_{ij}$$

with

$$x_i \times \boldsymbol{F}_{ij} + x_j \times \boldsymbol{F}_{ji} = f_{ij}(x_i \times e_{ij} + x_j \times e_{ji}) = f_{ij}(x_i - x_j) \times e_{ij} = 0.$$

If the self-interaction force is determined by the relative distance of particles, then we have $f_{ij} = f_{ij}(|x_i - x_j|)$. In this case, we have for

$$U_{ij}(r) = -\int^r f_{ij}(\rho) d\rho$$

that

$$-\nabla_{x_i} U_{ij}(|x_i - x_j|) = f_{ij} \nabla_{x_i} |x_i - x_j| = f_{ij} e_{ij}, \tag{1.17}$$

and therefore, it holds that

$$-\nabla_{x_i} U = \sum_{j \neq i} f_{ij} e_{ij}$$

for

$$U(x) = \sum_{k,j;k>j} U_{kj}(|x_k - x_j|) \qquad \text{and} \qquad x = (x_1, x_2, \cdots, x_n) \in \mathbf{R}^{3n}.$$

This implies that

$$m_i \ddot{x}_i = -\nabla_{x_i} U$$

and the *energy conservation law* follows as

$$\frac{dE}{dt} = 0, \tag{1.18}$$

where

$$E = \frac{1}{2} \sum_i m_i |\dot{x}_i|^2 + U(x).$$

Exercise 1.4 Confirm that (1.12) and (1.15) hold true.

Exercise 1.5 Confirm that (1.14) holds.

Exercise 1.6 Confirm that (1.17) and (1.18) hold.

1.1.4 *Gradient*

Chemotaxis is a character that some kind of insects and amoeba are attracted by special chemical substances of their own. If such a life is in $\mathbf{R}^3$, the force that it receives is a vector field denoted by $\boldsymbol{F}$. If $f = f(x_1, x_2, x_3)$ denotes the concentration of the chemical substance at $\boldsymbol{x} = {}^t(x_1, x_2, x_3)$, then the vector $\boldsymbol{F}(x_1, x_2, x_3)$ has the direction where $f(x_1, x_2, x_3)$ increases mostly and $|\boldsymbol{F}(x_1, x_2, x_3)|$ is proportional to its inclination. Letting the rate to be one, we shall give the formula for $\boldsymbol{F}$ to be determined by f.

For this purpose, we take a vector $\boldsymbol{e} = {}^t(e_1, e_2, e_3)$ in $|\boldsymbol{e}| = 1$ arbitrarily. Henceforth, such $\boldsymbol{e}$ is called a *unit vector*. For $|s| \ll 1$, it holds that

$$\begin{aligned} & f(\boldsymbol{x} + s\boldsymbol{e}) - f(\boldsymbol{x}) \\ & \quad = f(x_1 + se_1, x_2 + se_2, x_3 + se_3) - f(x_1, x_2, x_3) \\ & \quad = \{f(x_1 + se_1, x_2 + se_2, x_3 + se_3) - f(x_1, x_2 + se_2, x_3 + se_3)\} \\ & \qquad + \{f(x_1, x_2 + se_2, x_3 + se_3) - f(x_1, x_2, x_3 + se_3)\} \end{aligned}$$

$$
\begin{aligned}
&+\{f(x_1,x_2,x_3+se_3)-f(x_1,x_2,x_3)\}\\
&= se_1\frac{\partial f}{\partial x_1}(x_1+s\xi_1e_1,x_2+se_2,x_3+se_3)\\
&+se_2\frac{\partial f}{\partial x_2}(x_1,x_2+s\xi_2e_2,x_3+se_3)\\
&+se_3\frac{\partial f}{\partial x_3}(x_1,x_2,x_3+s\xi_3e_3)
\end{aligned}
$$

with $0<\xi_1,\xi_2,\xi_3<1$. In the case that $\frac{\partial f}{\partial x_1}$, $\frac{\partial f}{\partial x_2}$, $\frac{\partial f}{\partial x_3}$ are continuous at $\boldsymbol{x}={}^t(x_1,x_2,x_3)$, we have

$$
\lim_{s\to 0}\frac{1}{s}\{f(\boldsymbol{x}+se)-f(\boldsymbol{x})\}=e_1\frac{\partial f}{\partial x_1}(\boldsymbol{x})+e_2\frac{\partial f}{\partial x_2}(\boldsymbol{x})+e_3\frac{\partial f}{\partial x_3}(\boldsymbol{x}).
$$

This relation is written as

$$
f(\boldsymbol{x}+se)=f(\boldsymbol{x})+s\nabla f(\boldsymbol{x})\cdot e+o(s) \tag{1.19}
$$

as $s\to 0$, where

$$
\nabla f=\begin{pmatrix}\partial f/\partial x_1\\ \partial f/\partial x_2\\ \partial f/\partial x_3\end{pmatrix}.
$$

It is a vector field derived from f, called the *gradient* of f. Relation (1.19) means that

$$
\left.\frac{d}{ds}f(\boldsymbol{x}+se)\right|_{s=0}=\nabla f(\boldsymbol{x})\cdot e.
$$

The left-hand side is called the *direction derivative* of f at $\boldsymbol{x}$ toward e.

We are selecting the unit vector e for which the value $f(\boldsymbol{x}+se)-f(\boldsymbol{x})$ increases mostly in $0<s\ll 1$. In fact, from (1.19) this is the case when $\{\nabla f(x)\cdot e \mid |e|=1\}$ attains the maximum. That is,

$$
e=\frac{\nabla f(\boldsymbol{x})}{|\nabla f(\boldsymbol{x})|}
$$

and then the inclination is equal to

$$
\nabla f(\boldsymbol{x})\cdot e=|\nabla f(\boldsymbol{x})|.
$$

Thus, $\boldsymbol{F}(\boldsymbol{x})$ is the vector proportional to e with the length $|\nabla f(\boldsymbol{x})|$, namely, $\nabla f(\boldsymbol{x})$ itself.

Relation (1.19) is written as

$$df = \frac{\partial f}{\partial x_1}dx_1 + \frac{\partial f}{\partial x_2}dx_2 + \frac{\partial f}{\partial x_3}dx_3 \tag{1.20}$$

whose left-hand side is called the total derivative of f. Equality (1.20) shows also that a scalar field with continuous partial derivatives is totally differentiable. Note that it follows from Leibniz' law for the differentiation of composite function,

$$\begin{aligned}\frac{d}{ds}f(x_1(s), x_2(s), x_3(s)) &= f_{x_1}(x_1(s), x_2(s), x_3(s))x_1'(s)\\ &+ f_{x_2}(x_1(s), x_2(s), x_3(s))x_2'(s) + f_{x_3}(x_1(s), x_2(s), x_3(s))x_3'(s),\end{aligned}$$

applied to $\boldsymbol{x}(s) = \boldsymbol{x} + s\boldsymbol{e} = {}^t(x_1(s), x_2(s), x_3(s))$, where f_{x_i} stands for $\frac{\partial f}{\partial x_i}$ for $i = 1, 2, 3$.

If $\nabla f(\boldsymbol{x}_0) = 0$, then the graph of $f(\boldsymbol{x})$ is flat at $\boldsymbol{x} = \boldsymbol{x}_0$. In this case, $\boldsymbol{x}_0$ is called a *critical* (or *stationary*) *point* of f. To examine the behavior of f near by there, we take the matrix

$$\text{Hess } f = \left(\frac{\partial^2 f}{\partial x_i \partial x_j}\right)_{1 \leq i,j \leq 3}.$$

It is called the *Hesse matrix* of f. Actually, if f has continuous second partial derivatives at $\boldsymbol{x} = \boldsymbol{x}_0$, then it holds that

$$f(\boldsymbol{x}_0 + s\boldsymbol{e}) = f(\boldsymbol{x}_0) + s\nabla f(\boldsymbol{x}_0) \cdot \boldsymbol{e} + \frac{s^2}{2}[\text{Hess } f(\boldsymbol{x}_0)]\, \boldsymbol{e} \cdot \boldsymbol{e} + o(s^2) \tag{1.21}$$

as $s \to 0$. Here, we set

$$A\boldsymbol{e} \cdot \boldsymbol{e} = \sum_{i,j=1}^{3} a_{ij}e_i e_j$$

for $A = (a_{ij})_{1 \leq i,j \leq 3}$ and $\boldsymbol{e} = {}^t(e_1, e_2, e_3)$ with ${}^tA = A$. In fact, we have

$$\frac{d}{ds}f(\boldsymbol{x}_0 + s\boldsymbol{e}) = \nabla f(\boldsymbol{x}_0 + s\boldsymbol{e}) \cdot \boldsymbol{e}$$

and hence

$$\left.\frac{d^2}{ds^2}f(\boldsymbol{x}_0 + s\boldsymbol{e})\right|_{s=0} = [\text{Hess } f(\boldsymbol{x}_0)]\, \boldsymbol{e} \cdot \boldsymbol{e}$$

follows.

A critical point x_0 is said to be *non-degenerate* if its Hesse matrix Hess $f(x_0)$ is invertible. In this case, it controls the graph of $f(x)$ near $x = x_0$ in all directions, because real symmetric matrix is diagonalized by orthogonal matrix. In this connection, the number of negative eigenvalues of Hess $f(x_0)$ is called the *Morse index* of f at the critical point x_0. Those notions are extended to any space dimension and the n-dimensional scalar field $f(x_1, x_2, \cdots, x_n)$ is associated with the n-dimensional gradient $\nabla f = {}^t\left(\frac{\partial f}{\partial x_1}, \frac{\partial f}{\partial x_2}, \cdots, \frac{\partial f}{\partial x_n}\right)$.

In the two dimensional case, if its Morse indices are 0, 1, or 2, the non-degenerate critical point in consideration is a local minimum, a saddle, or a local maximum, respectively. At this moment, it is easy to suspect that the Morse index of any critical point cannot be free from those of other critical points. *Morse theory* arises in such a flavor, and for example, if the domain $\Omega \subset \mathbf{R}^2$ is simply connected, if the function $f(x, y)$ defined in Ω has continuous extensions to $\overline{\Omega}$ up to its second partial derivatives, and if any critical point is non-degenerate and is in Ω, then it holds that $m_0 - m_1 + m_2 = 1$, where m_0, m_1, and m_2 denotes the number of critical points with the Morse index 0, 1, and 2, respectively.

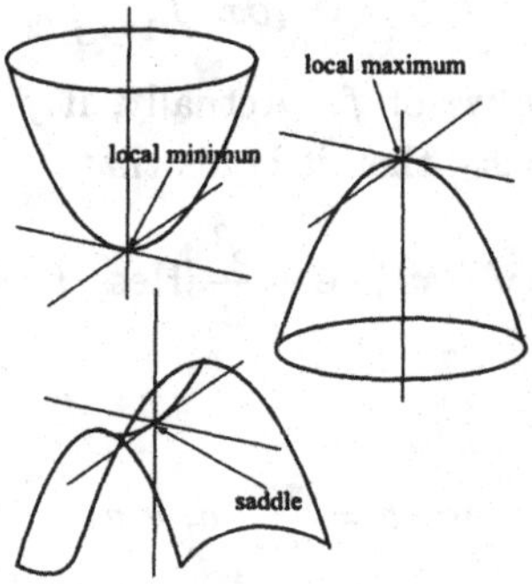

Fig. 1.5

Exercise 1.7 Show that

$$\frac{d^2}{ds^2} f(x_0 + se)\bigg|_{s=0} = [\text{Hess } f(x_0)]e \cdot e$$

if $f = f(x)$ is a C^2 function, and in this way confirm that (1.21) holds true.

Exercise 1.8 Find all critical points of

$$f(x_1, x_2) = \left(1 - x_1^2 - 2x_2^2\right)\left(1 + 3x_1^2\right)\left(1 + 4x_2^2\right)$$

defined on $\left\{(x_1, x_2) \mid x_1^2 + 2x_2^2 \leq 1\right\}$. Classify them into maximum, minimum, and saddle points. Then illustrate the graph

$$\mathcal{G} = \left\{(x_1, x_2, x_3) \mid x_3 = f(x_1, x_2),\ x_1^2 + 2x_2^2 \leq 1\right\}$$

in the three-dimensional space, indicating its level lines.

1.1.5 *Divergence*

The gradient $\nabla f = {}^t\left(\frac{\partial f}{\partial x_1}, \frac{\partial f}{\partial x_2}, \frac{\partial f}{\partial x_3}\right)$ of f is a vector field derived from the scalar field $f = f(x_1, x_2, x_3)$. Then,

$$\nabla = \begin{pmatrix} \dfrac{\partial}{\partial x_1} \\ \\ \dfrac{\partial}{\partial x_2} \\ \\ \dfrac{\partial}{\partial x_3} \end{pmatrix}$$

is regarded as a vector, called the *gradient operator*. Therefore, given a vector field $\boldsymbol{v} = {}^t\left(v^1, v^2, v^3\right)$ with the components $v^1 = v^1(x_1, x_2, x_3)$, $v^2 = v^2(x_1, x_2, x_3)$, $v^3 = v^3(x_1, x_2, x_3)$, we can define

$$\nabla \cdot \boldsymbol{v} = \frac{\partial v^1}{\partial x_1} + \frac{\partial v^2}{\partial x_2} + \frac{\partial v^3}{\partial x_3}$$

and

$$\nabla \times \boldsymbol{v} = \begin{pmatrix} \dfrac{\partial v^3}{\partial x_2} - \dfrac{\partial v^2}{\partial x_3} \\ \\ \dfrac{\partial v^1}{\partial x_3} - \dfrac{\partial v^3}{\partial x_1} \\ \\ \dfrac{\partial v^2}{\partial x_1} - \dfrac{\partial v^1}{\partial x_2} \end{pmatrix}.$$

Those scalar and vector fields are called the *divergence* and the *rotation* of $\boldsymbol{v}$, respectively, which have physical meanings as we are now describing.

To examine $\nabla \cdot \boldsymbol{v}$, first we consider the one-dimensional case. Take a straight uniform pipe, parallel to the x axis. Imagine that a fluid is flowing

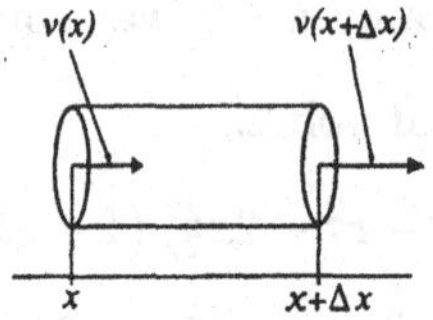

Fig. 1.6

inside and let $v(x)$ be its velocity at $x \in \mathbf{R}$. If no fluid comes in or out from outside of the pipe, then v is uniform so that $\frac{dv}{dx} = 0$ holds. More precisely, the rate of the change of the velocity is proportional to the amount of the fluid flowing into the place in the unit time, and that in $[x, x + \Delta x]$, $v(x + \Delta x) - v(x)$ is approximated by $\frac{dv}{dx}\Delta x$ for $|\Delta x| \ll 1$.

Applying those considerations to the three-dimensional case, we take a small rectangular solid, and suppose that each face of them is perpendicular to one of the x_1, x_2, x_3 axes. If the lengths of its sides parallel to x_1, x_2, x_3 axes are denoted by Δx_1, Δx_2, Δx_3, respectively, then $\left(\frac{\partial v^1}{\partial x_1}\Delta x_1\right) \cdot \Delta x_2 \Delta x_3$ approximates the amount of the fluid flowing into it in the unit time from the direction of the x_1 axis. Similarly, those from the directions of the x_2, x_3 axes are approximated by $\left(\frac{\partial v^2}{\partial x_2}\Delta x_2\right) \cdot \Delta x_3 \Delta x_1$ and $\left(\frac{\partial v^3}{\partial x_3}\Delta x_3\right) \cdot \Delta x_1 \Delta x_2$, respectively. Thus, totally, the amount of the fluid flowing into this rectangular solid in the unit time is approximated by $(\nabla \cdot \boldsymbol{v})\Delta x_1 \Delta x_2 \Delta x_3$.

The above rough descriptions are justified in the following way.

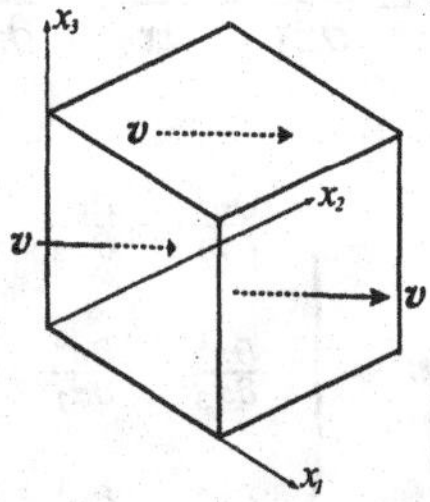

Fig. 1.7

Theorem 1.2 *Let $\boldsymbol{v}$ be a continuously differentiable vector field describ-*

ing the velocity of a fluid, and $\{\omega\}$ be a family of domains shrinking to a point, denoted by $\boldsymbol{x}$, in $\mathbf{R}^3$. Then, it holds that

$$\nabla \cdot \boldsymbol{v}(\boldsymbol{x}) = \lim_{|\omega| \to 0} \frac{Q(\omega)}{|\omega|}, \tag{1.22}$$

where $Q(\omega)$ denotes the amount of the fluid flowing into ω per unit time and $|\omega|$ the volume of ω.

Proof. If the initial value $\boldsymbol{x}_0$ is given, the autonomous ordinary differential equation

$$\frac{d\boldsymbol{x}}{dt} = \boldsymbol{v}(\boldsymbol{x}) \qquad \text{with} \qquad \boldsymbol{x}(0) = \boldsymbol{x}_0 \tag{1.23}$$

has the unique solution $\boldsymbol{x} = \boldsymbol{x}(t)$ locally in time. Then, writing $\boldsymbol{x}(t) = T_t\boldsymbol{x}_0$, we can define a family of operators $\{T_t\}$, called a (local) *dynamical system*. It has the properties that $(\boldsymbol{x}, t) \mapsto T_t\boldsymbol{x}$ is continuous, and $T_t \circ T_s = T_{t+s}$, and $T_0 = Id$, where they are defined. Here and henceforth, $\circ$ and Id denote the composition operation and the identity operator, respectively.

The particles in $\omega \subset \mathbf{R}^3$ at $t = 0$ are carried to $T_t(\omega)$ at $t = t$ by the flow whose velocity is $\boldsymbol{v}$. Then, it holds that

$$Q(\omega) = \frac{d}{dt}|T_t(\omega)|\bigg|_{t=0}.$$

Here, we make use of the transformation $\boldsymbol{\xi} = T_t\boldsymbol{x}$ of variables. Thus, if $J_t(\boldsymbol{x})$ denotes its Jacobian, the volume of $T_t(\omega)$ is given by

$$|T_t(\omega)| = \int_{T_t(\omega)} d\boldsymbol{\xi} = \int_\omega |J_t(\boldsymbol{x})|\, d\boldsymbol{x}.$$

If $\boldsymbol{\xi} = (\xi_1, \xi_2, \xi_3)$ is regarded as a function of t, then it holds that

$$\frac{d\boldsymbol{\xi}}{dt} = \boldsymbol{v}(\boldsymbol{\xi}) \qquad \text{and} \qquad \boldsymbol{\xi}(0) = \boldsymbol{x},$$

or

$$\frac{d\xi_i}{dt} = v^i(\boldsymbol{\xi}) \qquad \text{and} \qquad \xi_i(0) = x_i$$

for $i = 1, 2, 3$. Therefore, writing $y^i_j = \frac{\partial \xi_i}{\partial x_j}$, we have

$$\frac{dy^i_j}{dt} = \sum_{k=1}^{3} \frac{\partial v^i}{\partial \xi_k}(\boldsymbol{\xi}) y^k_j \qquad \text{and} \qquad y^i_j(0) = \delta_{ij},$$

where

$$\delta_{ij} = \begin{cases} 1 & (i = j) \\ 0 & (i \neq j). \end{cases}$$

This implies

$$\left. \frac{dy_j^i}{dt} \right|_{t=0} = \frac{\partial v^i}{\partial x_j}(\boldsymbol{x})$$

and hence

$$\frac{\partial \xi_i}{\partial x_j} = \delta_{ij} + t\frac{\partial v^i}{\partial x_j}(\boldsymbol{x}) + o(t)$$

follows as $t \to 0$. We obtain

$$\begin{aligned} J_t(\boldsymbol{x}) &= \begin{vmatrix} \frac{\partial \xi_1}{\partial x_1} & \frac{\partial \xi_1}{\partial x_2} & \frac{\partial \xi_1}{\partial x_3} \\ \frac{\partial \xi_2}{\partial x_1} & \frac{\partial \xi_2}{\partial x_2} & \frac{\partial \xi_2}{\partial x_3} \\ \frac{\partial \xi_3}{\partial x_1} & \frac{\partial \xi_3}{\partial x_2} & \frac{\partial \xi_3}{\partial x_3} \end{vmatrix} \\ &= \begin{vmatrix} 1 + t\frac{\partial v^1}{\partial x_1} + o(t) & t\frac{\partial v^1}{\partial x_2} + o(t) & t\frac{\partial v^1}{\partial x_3} + o(t) \\ t\frac{\partial v^2}{\partial x_1} + o(t) & 1 + t\frac{\partial v^2}{\partial x_2} + o(t) & t\frac{\partial v^2}{\partial x_3} + o(t) \\ t\frac{\partial v^3}{\partial x_1} + o(t) & t\frac{\partial v^3}{\partial x_2} + o(t) & 1 + t\frac{\partial v^3}{\partial x_3} + o(t) \end{vmatrix} \\ &= 1 + t\nabla \cdot \boldsymbol{v} + o(t). \end{aligned}$$

This implies $J_t(\boldsymbol{x}) > 0$ for $|t| \ll 1$ and also

$$Q(\omega) = \left. \frac{d}{dt} |T_t(\omega)| \right|_{t=0} = \int_\omega \nabla \cdot \boldsymbol{v} d\boldsymbol{x}. \tag{1.24}$$

Then, (1.22) follows. □

The *divergence formula of Gauss* follows from the *conservation of mass.* That is,

$$\int_\omega \nabla \cdot \boldsymbol{v} d\boldsymbol{x} = \int_{\partial\omega} \boldsymbol{\nu} \cdot \boldsymbol{v} dS, \tag{1.25}$$

where ω is a domain in $\mathbf{R}^3$ with C^1 boundary $\partial\omega$, $\boldsymbol{v}$ is a C^1 vector field on $\overline{\omega}$, $\boldsymbol{\nu}$ is the unit outer normal vector on $\partial\omega$, and dS is the area element of $\partial\omega$ described in the following chapter. In fact, both sides represent the mass of the fluid with the velocity $\boldsymbol{v}$ flowing out of ω per unit time. Equality (1.25) is equivalent to the *formula of integration by parts*, or,

$$\int_\omega \frac{\partial}{\partial x_j} v dx = \int_{\partial\omega} \nu_j v dS$$

for $j = 1, 2, 3$, where $v = v(\boldsymbol{x})$ is a C^1 function on $\overline{\omega}$ and $\boldsymbol{\nu} = {}^t(\nu_1, \nu_2, \nu_3)$. Here, we do not provide the mathematical proof of (1.25), but it is based on the *fundamental theorem of analysis* concerning the function of one variable:

$$\int_a^b f'(x)dx = \sum_{x=a,b} (\nu \cdot f)(x),$$

where

$$\nu(x) = \begin{cases} -1 & (x = a) \\ +1 & (x = b). \end{cases}$$

Here and henceforth, $\overline{\omega}$ denotes the *closure* of ω.

We also mention the relation between the dynamical system derived from the vector field and a partial differential equation of the first order. In fact, if (1.23) admits a solution $\boldsymbol{x} = \boldsymbol{x}(t)$ globally in time for any $\boldsymbol{x}_0 \in \mathbf{R}^3$, it generates a (global) dynamical system $\{T_t\}_{t\in\mathbf{R}}$ on $\mathbf{R}^3$. Then, the single linear partial equation of the first order,

$$\frac{\partial u}{\partial t} + \sum_{j=1}^3 v^j(\boldsymbol{x}) \frac{\partial u}{\partial x_j} = 0 \qquad (\boldsymbol{x} \in \mathbf{R}^3,\ t \in \mathbf{R}) \tag{1.26}$$

with

$$u|_{t=0} = f(\boldsymbol{x}) \qquad (\boldsymbol{x} \in \mathbf{R}^3) \tag{1.27}$$

admits a unique solution $u = u(\boldsymbol{x}, t)$, where

$$\boldsymbol{v}(\boldsymbol{x}) = {}^t\left(v^1(x_1, x_2, x_3), v^2(x_1, x_2, x_3), v^3(x_1, x_2, x_3)\right)$$

and f is a continuously differentiable function. Actually, it is given by $u(\boldsymbol{x}, t) = f(T_{-t}\boldsymbol{x})$, explicitly.

In fact, if $u(\boldsymbol{x},t) = f(T_{-t}\boldsymbol{x})$ then

$$w(\boldsymbol{y},t) \equiv u(T_t\boldsymbol{y},t) = f(\boldsymbol{y})$$

satisfies that

$$\begin{aligned} 0 &= \frac{\partial w}{\partial t} = \nabla u(T_t\boldsymbol{y},t)\cdot\frac{\partial}{\partial t}(T_t\boldsymbol{y}) + \frac{\partial u}{\partial t}(T_t\boldsymbol{y},t) \\ &= \frac{\partial u}{\partial t}(T_t\boldsymbol{y},t) + \sum_{j=1}^{3} v^j(T_t\boldsymbol{y})\frac{\partial u}{\partial x_j}(T_t\boldsymbol{y},t) \end{aligned}$$

by $\frac{\partial}{\partial t}T_t\boldsymbol{y} = \boldsymbol{v}(T_t\boldsymbol{y})$. Given $\boldsymbol{x}$, we take $\boldsymbol{y} = T_{-t}\boldsymbol{x}$ and put this into the above relation. Then, (1.26) follows for this $u(\boldsymbol{x},t)$, while (1.27) is obvious.

Conversely, if $u(\boldsymbol{x},t)$ is continuously differentiable and satisfies (1.26) with (1.27), then $w(\boldsymbol{x},t) = u(T_t\boldsymbol{x},t)$ solves

$$\frac{\partial w}{\partial t} = \nabla u(T_t\boldsymbol{x},t)\cdot\frac{\partial}{\partial t}(T_t\boldsymbol{x}) + \frac{\partial u}{\partial t}(T_t\boldsymbol{x},t) = 0$$

and therefore, we obtain

$$w(\boldsymbol{x},t) = w(\boldsymbol{x},0) = u(\boldsymbol{x},0) = f(\boldsymbol{x}).$$

This implies

$$u(\boldsymbol{x},t) = u(T_t(T_{-t}\boldsymbol{x}),t) = f(T_{-t}\boldsymbol{x}).$$

The quantity

$$\begin{aligned} \frac{Du}{Dt} &= \frac{\partial u}{\partial t} + \sum_{j=1}^{3} v^j(\boldsymbol{x})\frac{\partial u}{\partial x_j} \\ &= \frac{\partial u}{\partial t} + \boldsymbol{v}\cdot\nabla u \end{aligned}$$

is called the *material derivative* of u subject to the flow $\{T_t\}$.

Exercise 1.9 Confirm that (1.22) follows from (1.24).

Exercise 1.10 If the fluid is *incompressible*, the velocity $\boldsymbol{v}$ satisfies

$$\nabla\cdot\boldsymbol{v} = 0.$$

Such a vector field is called *solenoidal*. Suppose that $\mathbf{R}^3$ is occupied with the water, that the origin is a unique source, and that the amount of the fluid coming from there is a constant per unit time, denoted by Q. If it spreads radially, then the velocity $\boldsymbol{v}$ at $\boldsymbol{x}$ is given as

$$\boldsymbol{v} = v\omega,$$

where $v = |\boldsymbol{v}|$, $\boldsymbol{x} = r\omega$ with $r = |\boldsymbol{x}|$, and v is a function of r. In the case that the density of the water is equal to one, we have

$$4\pi r^2 v = Q$$

because $4\pi r^2$ indicates the surface area of the ball with the radius r. This means that

$$\boldsymbol{v} = \frac{Q}{4\pi}\frac{\boldsymbol{x}}{|\boldsymbol{x}|^3}.$$

Confirm that this vector field is solenoidal except for the origin.

Exercise 1.11 Henceforth,

$$\Delta = \nabla \cdot \nabla = \sum_{i=1}^{3} \frac{\partial^2}{\partial x_i^2}$$

and

$$\frac{\partial}{\partial \nu} = \nu \cdot \nabla$$

are called the *Laplacian* and the *(outer) normal derivative*, respectively. Remember that ν indicates the outer unit normal vector on the boundary, and $\nu \cdot \nabla$ is nothing but the direction derivative toward ν:

$$(\nu \cdot \nabla) f(\boldsymbol{x}) = \left. \frac{d}{ds} f(\boldsymbol{x} + s\nu) \right|_{s=0}.$$

Now, derive *Green's formula* from that of Gauss:

$$\int_\Omega ((\Delta u)v - u(\Delta v))\, d\boldsymbol{x} = \int_{\partial\Omega} \left(\frac{\partial u}{\partial \nu} v - u \frac{\partial v}{\partial \nu} \right) dS, \tag{1.28}$$

where $\partial\Omega$ is C^2 and u and v are C^2 functions on $\overline{\Omega}$.

1.1.6 *Rotation*

Let $\mathcal{O}$ be a rigid body moving in the three-dimensional space with a point $p \in \mathcal{O}$ fixed. Assume that p coincides with the origin. We take an orthonormal basis fixed on $\mathcal{O}$, and let $\{\boldsymbol{i}(t), \boldsymbol{j}(t), \boldsymbol{k}(t)\}$ be its position at time t. Because $\mathcal{O}$ is rigid, we have

$$\boldsymbol{i}(t) \cdot \boldsymbol{j}(t) = \boldsymbol{j}(t) \cdot \boldsymbol{k}(t) = \boldsymbol{k}(t) \cdot \boldsymbol{i}(t) = 0$$

and

$$\boldsymbol{i}(t) \cdot \boldsymbol{i}(t) = \boldsymbol{j}(t) \cdot \boldsymbol{j}(t) = \boldsymbol{k}(t) \cdot \boldsymbol{k}(t) = 1.$$

This implies

$$\begin{aligned} &\boldsymbol{i}' \cdot \boldsymbol{j} + \boldsymbol{i} \cdot \boldsymbol{j}' = \boldsymbol{j}' \cdot \boldsymbol{k} + \boldsymbol{j} \cdot \boldsymbol{k}' = \boldsymbol{k}' \cdot \boldsymbol{i} + \boldsymbol{k} \cdot \boldsymbol{i}' = 0 \\ &\boldsymbol{i}' \cdot \boldsymbol{i} = \boldsymbol{j}' \cdot \boldsymbol{j} = \boldsymbol{k}' \cdot \boldsymbol{k} = 0 \end{aligned} \tag{1.29}$$

by the differentiation in t.

If we represent $\{\boldsymbol{i}'(t), \boldsymbol{j}'(t), \boldsymbol{k}'(t)\}$ by $\{\boldsymbol{i}(t), \boldsymbol{j}(t), \boldsymbol{k}(t)\}$ as

$$\begin{aligned} \boldsymbol{i}' &= c_{11}\boldsymbol{i} + c_{12}\boldsymbol{j} + c_{13}\boldsymbol{k} \\ \boldsymbol{j}' &= c_{21}\boldsymbol{i} + c_{22}\boldsymbol{j} + c_{23}\boldsymbol{k} \\ \boldsymbol{k}' &= c_{31}\boldsymbol{i} + c_{32}\boldsymbol{j} + c_{33}\boldsymbol{k}, \end{aligned}$$

then we get from (1.29) that

$$\begin{aligned} &c_{23} + c_{32} = c_{31} + c_{13} = c_{12} + c_{21} = 0 \\ &c_{11} = c_{22} = c_{33} = 0. \end{aligned}$$

Therefore, letting $c_1 = c_{23} = -c_{32}$, $c_2 = c_{31} = -c_{13}$, $c_3 = c_{12} = -c_{21}$, we obtain

$$\begin{array}{llll} \boldsymbol{i}' = & & c_3\boldsymbol{j} & -c_2\boldsymbol{k} \\ \boldsymbol{j}' = & -c_3\boldsymbol{i} & & +c_1\boldsymbol{k} \\ \boldsymbol{k}' = & c_2\boldsymbol{i} & -c_1\boldsymbol{j}. & \end{array} \tag{1.30}$$

The vector $\boldsymbol{\omega} = {}^t(c_1, c_2, c_3)$ is called the *angular velocity*, which depends on t and is determined by the movement of $\mathcal{O}$.

We now take a fixed point in $\mathcal{O}$. Let $\boldsymbol{x}(t)$ be its position at time t. Putting $\boldsymbol{x}(0) = {}^t(x_1(0), x_2(0), x_3(0))$, we have

$$\boldsymbol{x}(t) = x_1(0)\boldsymbol{i}(t) + x_2(0)\boldsymbol{j}(t) + x_3(0)\boldsymbol{k}(t)$$

because $\mathcal{O}$ is a rigid body. In the use of the representation of components by $\{\boldsymbol{i}(0), \boldsymbol{j}(0), \boldsymbol{k}(0)\}$, this implies that

$$\begin{aligned}
\boldsymbol{x}'(0) &= x_1(0)\boldsymbol{i}'(0) + x_2(0)\boldsymbol{j}'(0) + x_3(0)\boldsymbol{k}'(0) \\
&= \begin{pmatrix} c_2(0)x_3(0) - c_3(0)x_2(0) \\ c_3(0)x_1(0) - c_1(0)x_3(0) \\ c_1(0)x_2(0) - c_2(0)x_1(0) \end{pmatrix} \\
&= \boldsymbol{\omega} \times \boldsymbol{x}\big|_{t=0}.
\end{aligned}$$

Because the above relation holds at any time t, the velocity $\boldsymbol{v}$ is given by the position $\boldsymbol{x}$ and the angular velocity $\boldsymbol{\omega}$ at that moment in such a way as

$$\boldsymbol{v} = \frac{d\boldsymbol{x}}{dt} = \boldsymbol{\omega} \times \boldsymbol{x}. \tag{1.31}$$

This means that infinitesimally, the rigid body $\mathcal{O}$ is rotating along $\boldsymbol{\omega}$ with the speed $|\boldsymbol{\omega}|$ in the direction where $\boldsymbol{\omega}$, $\boldsymbol{x}$, and $\boldsymbol{v}$ form a right-handed coordinate system. Because $\boldsymbol{\omega}$ is independent of $\boldsymbol{x}$, it follows from (1.31) that

$$\nabla \times \boldsymbol{v} = 2\boldsymbol{\omega}. \tag{1.32}$$

In this way, rotation of the velocity is two times the angular velocity in the rigid body. This is true even when the fixed point $p \in \mathcal{O}$ regarded as the origin is a function of t, if we take the relative coordinate. In the case that $\boldsymbol{v}$ stands for the velocity of fluid, the vector $\nabla \times \boldsymbol{v}$ picks up its rigid movement, and in this sense it is reasonable to be called the *vorticity*.

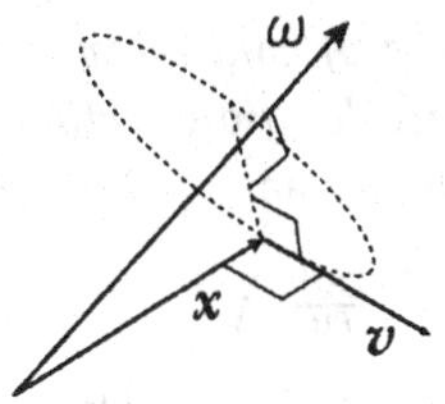

Fig. 1.8

Exercise 1.12 Confirm that the angular velocity of the rigid body is independent of the choice of $\{\boldsymbol{i}(t), \boldsymbol{j}(t), \boldsymbol{k}(t)\}$.

Exercise 1.13 Confirm that (1.32) holds true.

Exercise 1.14 If the rigid body $\mathcal{O}$ is rotating around x_3 axis, then the position at time t of a fixed point on $\mathcal{O}$ is given as

$$\boldsymbol{x}(t) = \begin{pmatrix} r\cos\omega t \\ r\sin\omega t \\ x_3 \end{pmatrix},$$

where $r > 0$ and $x_3 \in \mathbf{R}$ are constants. In using (1.32), we show that the angular velocity is given by $\boldsymbol{\omega} = {}^t(0, 0, \omega)$. Then, we confirm that (1.31) is valid in the case.

1.1.7 *Motion of Fluid*

Motion of fluid is described by the velocity

$$\boldsymbol{v} = \boldsymbol{v}(\boldsymbol{x}, t) = \begin{pmatrix} v^1(x_1, x_2, x_3, t) \\ v^2(x_1, x_2, x_3, t) \\ v^3(x_1, x_2, x_3, t) \end{pmatrix}$$

and the pressure $p = p(\boldsymbol{x}, t)$, regarded as time dependent vector and scalar fields, respectively. If it is *incompressible*, then the density ρ is a constant and the rate of change of the infinitesimal volume is zero, so that it holds that

$$\nabla \cdot \boldsymbol{v} = 0. \tag{1.33}$$

It is referred to as the *equation of continuity*.

On the other hand, the *acceleration vector* of the fluid is given by the material derivative of the velocity as

$$\begin{pmatrix} \frac{Dv^1}{Dt} \\ \frac{Dv^2}{Dt} \\ \frac{Dv^3}{Dt} \end{pmatrix} = \frac{D\boldsymbol{v}}{Dt},$$

where

$$\frac{Dv^i}{Dt} = \frac{\partial v^i}{\partial t} + \boldsymbol{v} \cdot \nabla v^i$$

for $i = 1, 2, 3$, and hence it follows that

$$\frac{D\boldsymbol{v}}{Dt} = \frac{\partial \boldsymbol{v}}{\partial t} + (\boldsymbol{v} \cdot \nabla)\boldsymbol{v}.$$

Therefore, *Newton's equation of motion* is given by

$$\rho \frac{Dv^i}{Dt} = \rho \boldsymbol{F}_i - \frac{\partial p}{\partial x_i} \qquad (i = 1, 2, 3),$$

where $\boldsymbol{F} = {}^t(F_1, F_2, F_3)$ denotes the outer force and the second term of the right-hand side indicates the force acting to the fluid caused by the difference of the pressure. In this way, we get *Euler's equation of motion*,

$$\frac{\partial \boldsymbol{v}}{\partial t} + (\boldsymbol{v} \cdot \nabla)\boldsymbol{v} = \boldsymbol{F} - \nabla \left(\frac{p}{\rho}\right). \tag{1.34}$$

Equation (1.34) is regarded as the *fundamental equation* for the incompressible, non-viscous fluid, which is referred to as the *perfect fluid.*

If $\nabla \times \boldsymbol{v} = 0$, then this fluid is said to be *rotation free.* In this case, we have a scalar function, called the *velocity potential,* $\Phi = \Phi(\boldsymbol{x}, t)$ satisfying

$$\boldsymbol{v} = \nabla \Phi. \tag{1.35}$$

If we take the two-dimensional *steady flow,* then it holds that

$$\boldsymbol{v} = \begin{pmatrix} u(x, y) \\ v(x, y) \end{pmatrix} \in \mathbf{R}^2.$$

Equation (1.33) is now reduced to

$$\frac{\partial u}{\partial x} + \frac{\partial v}{\partial y} = 0, \tag{1.36}$$

from which we obtain the *stream function* $\Psi = \Psi(x, y)$ satisfying

$$\frac{\partial \Psi}{\partial x} = -v, \qquad \frac{\partial \Psi}{\partial y} = u. \tag{1.37}$$

Along the curve that $\Psi =$ constant we have

$$0 = d\Psi = \frac{\partial \Psi}{\partial x} dx + \frac{\partial \Psi}{\partial y} dy = -v dx + u dy$$

and hence (u, v) is parallel to the velocity field (u, v). This means that this curve is the stream line. On the other hand, the rotation free velocity satisfies (1.35), and therefore we obtain the scalar function $\Phi = \Phi(x, y)$ such that

$$u = \frac{\partial \Phi}{\partial x}, \qquad v = \frac{\partial \Phi}{\partial y}. \tag{1.38}$$

Equations (1.37) and (1.38) guarantee the *Cauchy-Riemann's relation*

$$\frac{\partial \Phi}{\partial x} = \frac{\partial \Psi}{\partial y}, \qquad \frac{\partial \Phi}{\partial y} = -\frac{\partial \Psi}{\partial x}$$

and hence $f = \Phi(x, y) + \imath\Psi(x, y)$ is a holomorphic function of $z = x + \imath y$. It also indicates the *equipotential line* Φ = constant is perpendicular to the stream line. If $z = g(\zeta)$ is a *holomorphic function* of ζ, then so is $F(\zeta) = f(g(\zeta))$. Furthermore, equipotential lines are mapped to those and the same is true for the stream lines. In the steady state the boundary itself is a stream line. Thus, if one can find a holomorphic function $F(\zeta)$ with the mapped boundary regarded as a stream line, then $f(z) = F(\zeta)$ gives the status of the physical flow. Thus, this case is reduced to the problem of finding a *conformal mapping* with appropriate conditions.

Exercise 1.15 Show that (1.35) implies $\nabla \times \boldsymbol{v} = 0$. Confirm also that (1.37) implies (1.36).

Exercise 1.16 To find the two-dimensional stationary flow outside the unit circle satisfying $\boldsymbol{v} \to {}^t(1, 0)$ at infinity, let us note that

$$\frac{df}{dz} = \frac{\partial \Phi}{\partial x} + \imath\frac{\partial \Psi}{\partial x} = \frac{\partial \Psi}{\partial y} - \imath\frac{\partial \Psi}{\partial x} = u + \imath v$$

is identified with $\boldsymbol{v}$. Confirm that *Joukowski transformation*

$$\zeta = z + \frac{1}{z}$$

maps outside the unit ball conformally to the whole plane except for the segment $[-2, 2]$ on the real axis and that

$$\frac{df}{dz} = \frac{dF}{d\zeta}\left(1 - \frac{1}{z^2}\right)$$

holds true. This implies $\frac{dF}{d\zeta} \to 1$ as $|\zeta| \to +\infty$, $F(\zeta) = \zeta$, and $f(z) = z + \frac{1}{z}$. Illustrate the stream and the equipotential lines in the physical plane.

1.2 Curvature

1.2.1 *Quadratic Surfaces*

Insect or amoeba is attracted to the place where the local maximum of the concentration $f(\boldsymbol{x})$ of chemical material is attained. This force is proportional to the gradient of f, but meanwhile it feels how that scalar field is twisted. This is the curvature of the level set $\{x \in \mathbf{R}^3 \mid f(\boldsymbol{x}) = s\}$. Now, we provide several examples of surfaces.

Exercise 1.17 Illustrate the following surfaces in x, y, z space, where a, b, c, d, p, q are positive constants.

(1) (sphere) $x^2 + y^2 + z^2 = a^2$

(2) (ellipsoid) $(x^2/a^2) + (y^2/b^2) + (z^2/c^2) = 1$

(3) (elliptic paraboloid) $(x^2/2p) + (y^2/2q) = z$

(4) (hyperboloid with one leaf) $(x^2/a^2) + (y^2/b^2) - (z^2/c^2) = 1$

(5) (hyperboloid with two leaves) $(x^2/a^2) + (y^2/b^2) - (z^2/c^2) = -1$

(6) (hyperbolic paraboloid) $(x^2/2p) - (y^2/2q) = z$

1.2.2 *First Fundamental Form*

Generally, surface is a set of points in $\mathbf{R}^3$ indicated by two parameters, say u and v:

$$\boldsymbol{x}(u,v) = \begin{pmatrix} x_1(u,v) \\ x_2(u,v) \\ x_3(u,v) \end{pmatrix}.$$

Thus, $\boldsymbol{x} = \boldsymbol{x}(u,v)$ is regarded as a mapping $\Omega \to \mathbf{R}^3$, where $\Omega \subset \mathbf{R}^2$ denotes the parameter region and $\boldsymbol{x}(\Omega) = \mathcal{M}$ is the surface in consideration. We get a family of curves on $\mathcal{M}$, putting u or v to be constant.

Let the distance between two points on $\mathcal{M}$, $\boldsymbol{x}(u,v)$ and $\boldsymbol{x}(u+\Delta u, v+\Delta v)$, be Δs, the area of the parallelogram made by

$$\boldsymbol{a} = \boldsymbol{x}(u+\Delta u, v) - \boldsymbol{x}(u,v)$$

and

$$\boldsymbol{b} = \boldsymbol{x}(u, v + \Delta v) - \boldsymbol{x}(u, v)$$

be ΔS, and the normal unit vector on $\mathcal{M}$ at $\boldsymbol{x}(u, v)$ be $\boldsymbol{n}$.

First, we have

$$\begin{aligned}
&\boldsymbol{x}(u + \Delta u, v + \Delta v) - \boldsymbol{x}(u, v) \\
&\quad = \boldsymbol{x}_u(u, v)\Delta u + \boldsymbol{x}_v(u, v)\Delta v + o\left(\sqrt{\Delta u^2 + \Delta v^2}\right)
\end{aligned}$$

similarly to (1.19), and hence it holds that

$$\begin{aligned}
(\Delta s)^2 &= |\boldsymbol{x}(u + \Delta u, v + \Delta v) - \boldsymbol{x}(u, v)|^2 \\
&= \left|\boldsymbol{x}_u \Delta u + \boldsymbol{x}_v \Delta v + o\left(\sqrt{\Delta u^2 + \Delta v^2}\right)\right|^2 \\
&= |\boldsymbol{x}_u \Delta u + \boldsymbol{x}_v \Delta v|^2 + o\left(\Delta u^2 + \Delta v^2\right) \\
&= |\boldsymbol{x}_u|^2 \Delta u^2 + 2\boldsymbol{x}_u \cdot \boldsymbol{x}_v \Delta u \Delta v + |\boldsymbol{x}_v|^2 \Delta v^2 \\
&\quad + o\left(\Delta u^2 + \Delta v^2\right),
\end{aligned} \tag{1.39}$$

where

$$\boldsymbol{x}_u = \boldsymbol{x}_u(u, v) = \frac{\partial \boldsymbol{x}}{\partial u}(u, v)$$

and

$$\boldsymbol{x}_v = \boldsymbol{x}_v(u, v) = \frac{\partial \boldsymbol{x}}{\partial v}(u, v).$$

This relation is expressed in the infinitesimal limit as the *first fundamental form*,

$$ds^2 = E du^2 + 2F du dv + G dv^2,$$

with

$$E = |\boldsymbol{x}_u|^2, \qquad F = \boldsymbol{x}_u \cdot \boldsymbol{x}_v, \qquad G = |\boldsymbol{x}_v|^2,$$

where ds and E, F, G are called the *line element* and the *first fundamental quantities*, respectively. Next, noting

$$\begin{aligned}
\boldsymbol{a} &= \boldsymbol{x}(u + \Delta u, v) - \boldsymbol{x}(u, v) \\
&= \boldsymbol{x}_u(u, v)\Delta u + o(\Delta u)
\end{aligned}$$

and

$$\begin{aligned} \boldsymbol{b} &= \boldsymbol{x}(u, v+\Delta v) - \boldsymbol{x}(u, v) \\ &= \boldsymbol{x}_v(u, v)\Delta v + o(\Delta v), \end{aligned}$$

we have

$$\begin{aligned} \Delta S &= |\boldsymbol{a} \times \boldsymbol{b}| \\ &= |\boldsymbol{x}_u(u, v) \times \boldsymbol{x}_v(u, v)|\, \Delta u \Delta v \\ &\qquad + o\left(\Delta u^2 + \Delta v^2\right). \end{aligned} \tag{1.40}$$

This relation may be written as

$$dS = |\boldsymbol{x}_u \times \boldsymbol{x}_v|\, du dv,$$

call the *area element*. Then, the *vector area element* is defined by

$$d\boldsymbol{S} = (\boldsymbol{x}_u \times \boldsymbol{x}_v) du dv.$$

Finally, $\boldsymbol{x}_u(u, v)$ and $\boldsymbol{x}_v(u, v)$ are tangent to the curves v and u constants on $\mathcal{M}$, so that $\boldsymbol{n}$ is a unit vector perpendicular to those vectors. Thus, we take

$$\boldsymbol{n} = \frac{\boldsymbol{x}_u \times \boldsymbol{x}_v}{|\boldsymbol{x}_u \times \boldsymbol{x}_v|}.$$

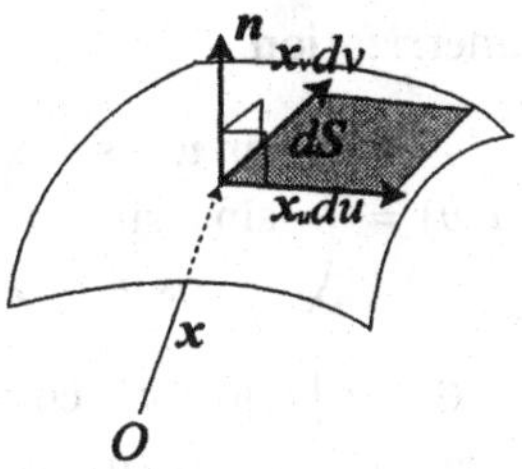

Fig. 1.9

Putting

$$d\boldsymbol{x} = \boldsymbol{x}_u du + \boldsymbol{x}_v dv, \tag{1.41}$$

we can write the first fundamental form as

$$\begin{aligned} ds^2 &= d\boldsymbol{x}\cdot d\boldsymbol{x} \\ &= |\boldsymbol{x}_u|^2\,du^2 + 2\boldsymbol{x}_u\cdot\boldsymbol{x}_v dudv + |\boldsymbol{x}_v|^2\,dv^2. \end{aligned}$$

On the other hand, using (1.10), we obtain

$$\begin{aligned} |\boldsymbol{x}_u\times\boldsymbol{x}_v|^2 &= (\boldsymbol{x}_u\times\boldsymbol{x}_v)\cdot(\boldsymbol{x}_u\times\boldsymbol{x}_v) \\ &= (\boldsymbol{x}_u\cdot\boldsymbol{x}_u)(\boldsymbol{x}_v\cdot\boldsymbol{x}_v) - (\boldsymbol{x}_u\cdot\boldsymbol{x}_v)^2 \\ &= EG - F^2. \end{aligned}$$

This implies

$$dS = \sqrt{EG-F^2}dudv, \qquad \boldsymbol{n} = \frac{\boldsymbol{x}_u\times\boldsymbol{x}_v}{\sqrt{EG-F^2}}, \qquad d\boldsymbol{S} = \boldsymbol{n}dS.$$

We call $d\boldsymbol{x}$ of (1.41) the *infinitesimal vector* of direction $k = dv/du$.

Exercise 1.18 Confirm the third and the second equalities in (1.39) and (1.40), respectively, using $|\mathbf{c}-\mathbf{d}| \geq ||\mathbf{c}| - |\mathbf{d}||$.

Exercise 1.19 Express the first fundamental quantities E, F, G by $p = \partial z/\partial x$ and $q = \partial z/\partial y$ when the surface $\mathcal{M}$ is a graph,

$$\boldsymbol{x} = \begin{pmatrix} x \\ y \\ z(x,y) \end{pmatrix}.$$

Exercise 1.20 In the parametrization

$$\boldsymbol{x}(u,v) = \begin{pmatrix} \sin u\cos v \\ \sin u\sin v \\ \cos u \end{pmatrix}$$

of the unit sphere with $(u,v)\in[0,\pi]\times[0,2\pi)$, compute its fist fundamental form and the unit normal vector. Then, in use of this parametization, compute its total surface area.

1.2.3 *Curves*

We take the plane curve $\mathcal{C}$ on $\mathbf{R}^2$, indicated as $y = f(x)$. Let θ be the inclination of the tangential line at $P(x,y)\in\mathcal{C}$. Then it holds that $\tan\theta =$

$f'(x)$. If $Q(x+\Delta x, y+\Delta y)$ is on C, we have

$$\begin{aligned}\Delta y &= f(x+\Delta x) - f(x) \\ &= f'(x)\Delta x + o(\Delta x)\end{aligned}$$

as $\Delta x \downarrow 0$. Then the length Δs of the segment PQ is given by

$$\begin{aligned}\Delta s &= \left(\Delta x^2 + \Delta y^2\right)^{1/2} \\ &= \left(1 + f'(x)^2\right)^{1/2} \Delta x + o(\Delta x). \end{aligned} \tag{1.42}$$

The second equality of (1.42) justifies that the length of C cut by $x = a$ and $x = b$ is given by

$$\int_a^b \sqrt{1 + f'(x)^2} dx.$$

The first equality of (1.42), on the other hand, justifies the relation

$$nds = \begin{pmatrix} dy \\ -dx \end{pmatrix}$$

on the plane. In particular, the divergence formula of Gauss (1.25) is reduced to the Green's formula,

$$\int_{\partial D} (a_1 dx + a_2 dy) = \int_D \left(-\frac{\partial a_1}{\partial y} + \frac{\partial a_2}{\partial x} \right) dx dy, \tag{1.43}$$

where $D \subset \mathbf{R}^2$ is a domain with C^1 boundary ∂D and a_1, a_2 are C^1 functions on $\overline{D}$.

Let the inclination of the tangent line of C at $Q(x + \Delta x, y + \Delta y)$ be $\theta + \Delta\theta$. Rotating tangent lines by 90 degrees, we get normal lines of C at P and Q. Note that $\Delta\theta$ coincides with the angle made by those normal lines. Their crossing point is called the *center of curvature*. Then, the *curvature* of C at P is defined by

$$\lim_{\Delta s \to 0} \frac{\Delta \theta}{\Delta s} = \frac{1}{\rho}$$

and ρ is called the *curvature radius*. Because Δs is approximated by $\rho\Delta\theta$ if Q is close to P, curve C is approximated by circle, with the radius equal to its curvature radius and the center coinciding with the center of curvature at P.

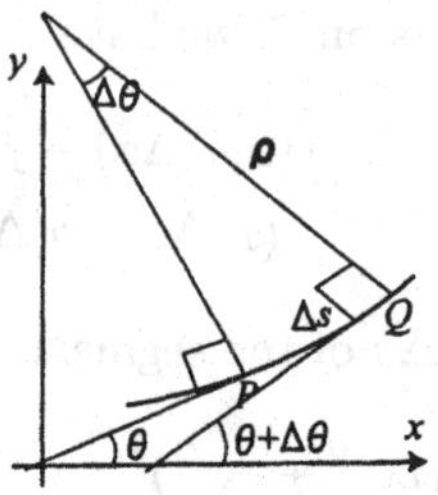

Fig. 1.10

Actual computation of ρ is performed as follows. First, we have

$$\begin{aligned}\tan(\theta+\Delta\theta) &= f'(x+\Delta x)\\ &= f'(x)+f''(x)\Delta x+o(\Delta x)\\ &= \tan\theta+f''(x)\Delta x+o(\Delta x).\end{aligned}$$

On the other hand, we have

$$\tan(\theta+\Delta\theta)=\tan\theta+(1+\tan^2\theta)\Delta\theta+o(\Delta\theta)$$

by $\tan'\theta = 1+\tan^2\theta$. This implies

$$\begin{aligned}\frac{1}{\rho} &= \frac{\lim_{\Delta x\to 0}(\Delta\theta/\Delta x)}{\lim_{\Delta x\to 0}(\Delta s/\Delta x)}\\ &= \frac{f''}{1+f'^2}\cdot\frac{1}{(1+f'^2)^{1/2}}\\ &= \frac{f''}{(1+f'^2)^{3/2}}.\end{aligned}$$

We take a fixed point on C and set s to be the length along C between $\boldsymbol{P}\in C$ and that point. In that way we parametrize C by s, writing the former as $\boldsymbol{x}=\boldsymbol{x}(s)$. Under this parametrization, it holds that

$$\lim_{\Delta s\to 0}\frac{1}{\Delta s}\left|\boldsymbol{x}(s+\Delta s)-\boldsymbol{x}(s)\right|=1,$$

or equivalently,

$$|d\boldsymbol{x}/ds|=1.$$

Hence

$$t = \frac{dx}{ds}$$

coincides with the unit tangent vector of C at P.

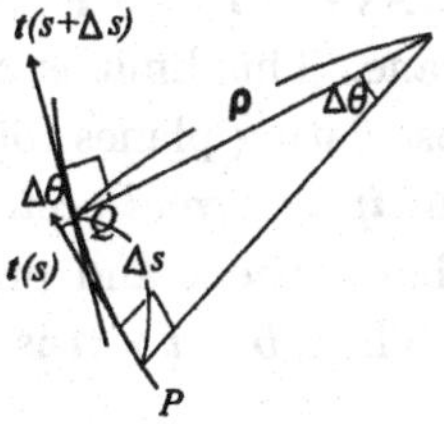

Fig. 1.11

First, $t \cdot t = 1$ gives

$$\frac{dt}{ds} \cdot t = 0.$$

Thus, dt/ds is parallel to the unit normal vector n of C at P. On the other hand, because the angle made by $t(s + \Delta s)$ and $t(s)$ is equal to $\Delta\theta$ and those two vectors have length one, it holds that

$$|t(s + \Delta s) - t(s)| = \Delta\theta + o(\Delta\theta).$$

This implies that

$$\begin{aligned} \left|\frac{dt}{ds}\right| &= \lim_{\Delta s \to 0} \left|\frac{1}{\Delta s}(t(s + \Delta s) - t(s))\right| \\ &= \lim_{\Delta s \to 0} \left|\frac{\Delta\theta}{\Delta s}\right| = \left|\frac{1}{\rho}\right|. \end{aligned}$$

If the direction of the normal vector n is taken to be the center of curvature, then it follows that

$$\frac{dt}{ds} = \frac{1}{\rho}n.$$

Now, we proceed to the space curve. Actually, three-dimensional vectors $\{x(t)\}$, varying as the time t changes, draw a curve C in $\mathbf{R}^3$. We take the length parameter s from a fixed point on C. Thus, C is expressed as $\boldsymbol{x} = \boldsymbol{x}(s)$. Then, $\boldsymbol{t} = d\boldsymbol{x}/ds$ is the unit tangent vector, similarly to the case of plane curves. The normal plane indicates the one containing $\boldsymbol{P} = \boldsymbol{x}(s)$ and perpendicular to $\boldsymbol{t}$.

We take $\boldsymbol{P}_1$, $\boldsymbol{P}_2$ on C near $\boldsymbol{P}$ and set π to be the plane made by those three points, $\boldsymbol{P}_1$, $\boldsymbol{P}_2$, and $\boldsymbol{P}$. As $\boldsymbol{P}_1, \boldsymbol{P}_2 \to \boldsymbol{P}$, the plane π converges to the one orthogonal to normal plane. This limit is called the *osculating plane.* Intersection of normal and osculating planes forms a line. We take a unit vector $\boldsymbol{n}$, called the unit *principal normal vector* on it, of which direction is determined later. Note that tangent and principal normal vectors are perpendicular to each other. Then, $\boldsymbol{b} = \boldsymbol{t} \times \boldsymbol{n}$ is called the unit *bi-principal normal vector.*

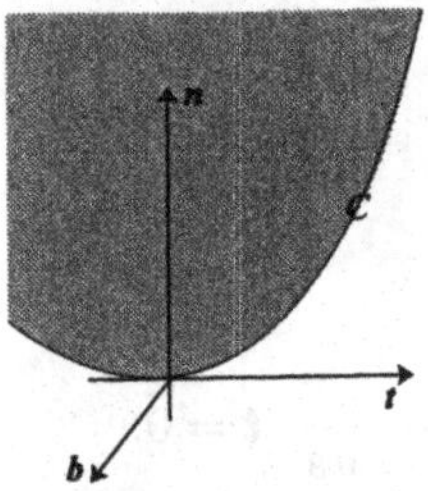

Fig. 1.12

Near $\boldsymbol{P}$, C is approximated by a circle on the osculating plane. If its radius is denoted by ρ and the direction of $\boldsymbol{n}$ is taken toward its center, then it holds that

$$\frac{d\boldsymbol{t}}{ds} = \frac{1}{\rho}\boldsymbol{n}$$

as in the case of plane curves. Thus, $1/\rho$ is called the curvature of C at $\boldsymbol{P}$.

Bi-principle normal vector $\boldsymbol{b}$ changes its direction if C twists. The *torsion* τ indicates how it does. Namely, it is a scalar, positive if $\boldsymbol{b}$ twists clockwisely and satisfies that

$$\left|\frac{d\boldsymbol{b}}{ds}\right| = |\tau|.$$

The role of torsion may be clarified in the following way. In fact, $\boldsymbol{b} = \boldsymbol{t} \times \boldsymbol{n}$

implies

$$\begin{aligned}
\frac{d\boldsymbol{b}}{ds} &= \frac{d\boldsymbol{t}}{ds} \times \boldsymbol{n} + \boldsymbol{t} \times \frac{d\boldsymbol{n}}{ds} \\
&= \frac{1}{\rho}\boldsymbol{n} \times \boldsymbol{n} + \boldsymbol{t} \times \frac{d\boldsymbol{n}}{ds} \\
&= \boldsymbol{t} \times \frac{d\boldsymbol{n}}{ds}.
\end{aligned}$$

On the other hand, $\boldsymbol{b} \cdot \boldsymbol{b} = 1$ implies

$$\boldsymbol{b} \cdot \frac{d\boldsymbol{b}}{ds} = 0.$$

Therefore, $d\boldsymbol{b}/ds$ is perpendicular to $\boldsymbol{t}$ and $\boldsymbol{b}$. Regarding the sign of torsion, we get that

$$\frac{d\boldsymbol{b}}{ds} = -\tau\boldsymbol{n}.$$

Similarly, $\boldsymbol{n} = \boldsymbol{b} \times \boldsymbol{t}$ implies that

$$\begin{aligned}
\frac{d\boldsymbol{n}}{ds} &= \frac{d\boldsymbol{b}}{ds} \times \boldsymbol{t} + \boldsymbol{b} \times \frac{d\boldsymbol{t}}{ds} \\
&= -\tau\boldsymbol{n} \times \boldsymbol{t} + \boldsymbol{b} \times \frac{1}{\rho}\boldsymbol{n} \\
&= \tau\boldsymbol{b} - \frac{1}{\rho}\boldsymbol{t}
\end{aligned}$$

because $\boldsymbol{n} \times \boldsymbol{t} = -\boldsymbol{b}$ and $\boldsymbol{b} \times \boldsymbol{n} = -\boldsymbol{t}$. Those relations are summarized as the *Frenet-Serret formula*,

$$\begin{aligned}
&\frac{d\boldsymbol{t}}{ds} = \frac{1}{\rho}\boldsymbol{n} \\
&\frac{d\boldsymbol{n}}{ds} = -\frac{1}{\rho}\boldsymbol{t} + \tau\boldsymbol{b} \\
&\frac{d\boldsymbol{b}}{ds} = -\tau\boldsymbol{n},
\end{aligned}$$

where ρ, τ, and s are curvature radius, torsion, and length parameter, respectively.

Exercise 1.21 Draw the following curve and compute $ds/dt = |\boldsymbol{x}'(t)|$, where s is the length parameter and $a > b > 0$ are constants:

$$x(t) = \begin{pmatrix} a\cos t \\ a\sin t \\ bt \end{pmatrix}.$$

Then, seek unit tangential vector, unit principal normal vector, unit bi-principal normal vector, curvature, and torsion.

1.2.4 *Second Fundamental Form*

Let $\boldsymbol{x} = \boldsymbol{x}(u, v)$ be a parameter representation of the surface $\mathcal{M}$, and take a curve C on it. It is represented as $u = u(s)$ and $v = v(s)$, or $\boldsymbol{x} = \boldsymbol{x}(u(s), v(s))$, using the length parameter s of C.

The unit tangential vector of C is given as

$$\boldsymbol{t} = \frac{d\boldsymbol{x}}{ds} = \boldsymbol{x}_u \frac{du}{ds} + \boldsymbol{x}_v \frac{dv}{ds} \tag{1.44}$$

and it holds that

$$\frac{d\boldsymbol{t}}{ds} = \frac{1}{\rho_C}\boldsymbol{n}_C,$$

where ρ_C and $\boldsymbol{n}_C$ denote the curvature radius and the principal normal unit vector of C, respectively. If ψ_C denotes the angle between $\boldsymbol{n}$ and $\boldsymbol{n}_C$, then it holds that $\cos\psi_C = \boldsymbol{n}\cdot\boldsymbol{n}_C$, or

$$\frac{\cos\psi_C}{\rho_C} = \boldsymbol{n}\cdot\frac{d\boldsymbol{t}}{ds},$$

where $\boldsymbol{n} = \frac{\boldsymbol{x}_u \times \boldsymbol{x}_v}{|\boldsymbol{x}_u \times \boldsymbol{x}_v|}$ denotes the unit normal vector on $\mathcal{M}$.

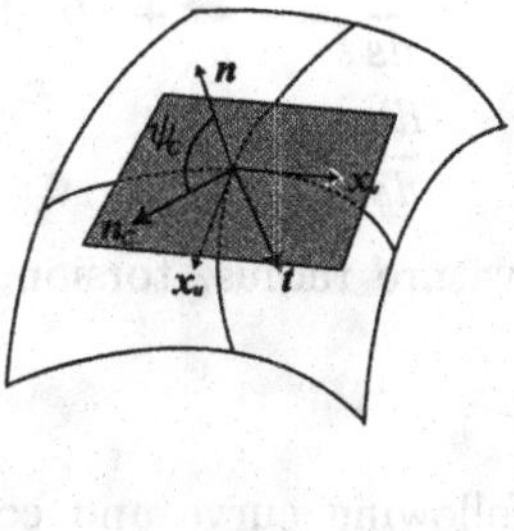

Fig. 1.13

Here, we have

$$\frac{d\boldsymbol{t}}{ds} = \boldsymbol{x}_{uu}\left(\frac{du}{ds}\right)^2 + 2\boldsymbol{x}_{uv}\left(\frac{du}{ds}\right)\left(\frac{dv}{ds}\right) + \boldsymbol{x}_{vv}\left(\frac{dv}{ds}\right)^2 + \boldsymbol{x}_u\frac{d^2u}{ds^2} + \boldsymbol{x}_v\frac{d^2v}{ds^2}$$

by (1.44). This implies that

$$\boldsymbol{n}\cdot\frac{d\boldsymbol{t}}{ds} = \boldsymbol{n}\cdot\left\{\boldsymbol{x}_{uu}\left(\frac{du}{ds}\right)^2 + 2\boldsymbol{x}_{uv}\left(\frac{du}{ds}\right)\left(\frac{dv}{ds}\right) + \boldsymbol{x}_{vv}\left(\frac{dv}{ds}\right)^2\right\}$$

by $\boldsymbol{n}\cdot\boldsymbol{x}_u = \boldsymbol{n}\cdot\boldsymbol{x}_v = 0$.

The *second fundamental quantities* are defined by

$$L = \boldsymbol{x}_{uu}\cdot\boldsymbol{n}, \qquad M = \boldsymbol{x}_{uv}\cdot\boldsymbol{n}, \qquad N = \boldsymbol{x}_{vv}\cdot\boldsymbol{n}.$$

Writing

$$\begin{aligned}\frac{\cos\psi_C}{\rho_C} &= L\left(\frac{du}{ds}\right)^2 + 2M\frac{du}{ds}\frac{dv}{ds} + N\left(\frac{dv}{ds}\right)^2\\ &= \frac{Ldu^2 + 2Mdudv + Ndv^2}{Edu^2 + 2Fdudv + Gdv^2},\end{aligned}$$

we call

$$Ldu^2 + 2Mdudv + Ndv^2$$

the *second fundamental form*.

Recall that $k = dv/du$ is the direction of infinitesimal vector $d\boldsymbol{x}$. Then, we have

$$\frac{\cos\psi_C}{\rho_C} = \frac{L + 2Mk + Nk^2}{E + 2Fk + Gk^2}. \tag{1.45}$$

Here, L, M, N, E, F, G are determined by $\boldsymbol{P}\in\mathcal{M}$. On the other hand,

$$k = \frac{dv}{du} = \frac{dv/ds}{du/ds}$$

is determined by the unit tangential vector $\boldsymbol{t}$ of C as well as by $\boldsymbol{P}$, because of

$$\boldsymbol{t} = \frac{du}{ds}\boldsymbol{x}_u + \frac{dv}{ds}\boldsymbol{x}_v.$$

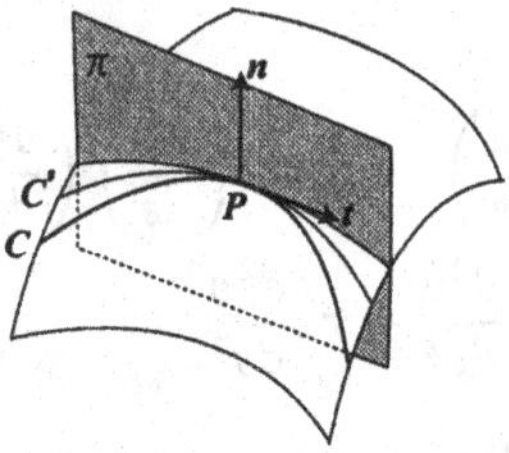

Fig. 1.14

Let C' be the curve on $\mathcal{M}$ cut by the plane π containing $\boldsymbol{P}$ made by the unit tangential vector $\boldsymbol{t}$ of C and the unit normal vector $\boldsymbol{n}$ of $\mathcal{M}$. Let R be the curvature of C' at $\boldsymbol{P} \in \mathcal{M}$. Because the center of curvature of C' at $\boldsymbol{P}$ is on π, it holds that

$$(\psi_{C'}, \rho_{C'}) = (0, R) \qquad \text{or} \qquad (\psi_{C'}, \rho_{C'}) = (\pi, -R).$$

On the other hand the right-hand side of (1.45) is determined by $\boldsymbol{P}$ and $\boldsymbol{t}$ so that we have

$$\frac{\cos \psi_C}{\rho_C} = \frac{\cos \psi_{C'}}{\rho_{C'}} = \frac{1}{R}$$

at $\boldsymbol{P}$ for any curve C on $\mathcal{M}$ passing through $\boldsymbol{P}$ with the unit tangential vector $\boldsymbol{t}$. Henceforth, $1/R$ is called the *normal curvature* of $\mathcal{M}$ at $\boldsymbol{P}$ with the direction $\boldsymbol{t}$.

Fixing $\boldsymbol{P}$, let us seek $\boldsymbol{t}$ such that the normal curvature attains minimum or maximum. In those cases, $\boldsymbol{t}$ and $1/R$ are called the *principal direction* and the *principal curvature*, respectively. Actually, they are obtained by putting

$$\frac{d}{dk}\left(\frac{1}{R}\right) = \frac{d}{dk} \cdot \frac{L + 2Mk + Nk^2}{E + 2Fk + Gk^2} = 0,$$

or equivalently,

$$\frac{1}{R}(F + Gk) = M + Nk. \tag{1.46}$$

This equality gives from

$$\frac{1}{R}(E + 2Fk + Gk^2) = L + 2Mk + Nk^2$$

that

$$\frac{1}{R}(E + Fk) = L + Mk. \tag{1.47}$$

Again by (1.46) we have

$$k = -\frac{F/R - M}{G/R - N}$$

and then

$$\left(\frac{F}{R} - M\right)\left(\frac{F}{R} - M\right) - \left(\frac{E}{R} - L\right)\left(\frac{G}{R} - N\right) = 0 \tag{1.48}$$

follows from (1.47).

In this way, we get two principal curvatures $1/R_1$ and $1/R_2$ by (1.48). Writing (1.48) as

$$(EG - F^2)\frac{1}{R^2} - (GL + EN - 2FM)\frac{1}{R} + LN - M^2 = 0, \tag{1.49}$$

we have

$$2H \equiv \frac{1}{R_1} + \frac{1}{R_2} = \frac{GL + EN - 2FM}{EG - F^2}$$

and

$$K \equiv \frac{1}{R_1 \cdot R_2} = \frac{LN - M^2}{EG - F^2}.$$

The quantities

$$2H = \frac{1}{R_1} + \frac{1}{R_2} \quad \text{and} \quad K = \frac{1}{R_1 R_2}$$

are called the *mean curvature* and the *Gaussian curvature* of $\mathcal{M}$ at $\boldsymbol{P}$, respectively. If the Gaussian curvature is positive, then the surface is convex to one side at that point. If it is negative, then it looks like a saddle there. It is known that the Gaussian curvature is determined by the first fundamental quantities.

Equation (1.48) on $1/R$ has an equivalent form on k. In fact, we have

$$FL - EM + (GL - EN)k + (GM - FN)k^2 = 0 \tag{1.50}$$

by (1.46) and (1.47). Equation (1.50) provides the principal directions k_1 and k_2 as the solution, which give the infinitesimal vectors

$$\begin{aligned} d\boldsymbol{x}_1 &= \boldsymbol{x}_u du_1 + \boldsymbol{x}_v dv_1 \\ d\boldsymbol{x}_2 &= \boldsymbol{x}_u du_2 + \boldsymbol{x}_v dv_2, \end{aligned}$$

where $k_1 = dv_1/du_1$ and $k_2 = dv_2/du_2$.

Here, we have

$$\begin{aligned} d\boldsymbol{x}_1 \cdot d\boldsymbol{x}_2 &= Edu_1du_2 + F(du_2dv_1 + du_1dv_2) + Gdv_1dv_2 \\ &= (E + F(k_1 + k_2) + Gk_1k_2)\, du_1du_2. \end{aligned}$$

Furthermore, it follows from (1.50) that

$$k_1 + k_2 = -\frac{GL - EN}{GM - FN} \quad \text{and} \quad k_1k_2 = \frac{FL - EM}{GM - FN}$$

and hence we obtain

$$\begin{aligned} E + F(k_1 + k_2) + Gk_1k_2 &= \frac{1}{GM - FN} \\ &\cdot \{(GM - FN)E - F(GL - EN) + G(FL - EM)\} = 0. \end{aligned}$$

This means that

$$d\boldsymbol{x}_1 \cdot d\boldsymbol{x}_2 = 0$$

and the principal directions are perpendicular to each other.

Exercise 1.22 The surface is parametrized as $\boldsymbol{x} = \boldsymbol{x}(u, v)$. Show

$$\begin{aligned} L &= -\boldsymbol{n}_u \cdot \boldsymbol{x}_u, \\ M &= -\boldsymbol{n}_u \cdot \boldsymbol{x}_v = -\boldsymbol{n}_v \cdot \boldsymbol{x}_u, \\ N &= -\boldsymbol{n}_v \cdot \boldsymbol{x}_v. \end{aligned}$$

Exercise 1.23 Surface of revolution is parametrized as

$$\begin{pmatrix} f(u)\cos v \\ f(u)\sin v \\ g(u) \end{pmatrix}$$

for $u \in (-\infty, +\infty)$ and $v \in [0, 2\pi)$ with $f(u) > 0$. Express fundamental quantities E, F, G, L, M, N and the principal curvature radii R_1, R_2 in terms of f and g. Next, assume $f'(u)^2 + g'(u)^2 = 1$ in the parametrization, and show that mean and Gaussian curvatures are given by

$$2H = \frac{g'}{f} - \frac{f''}{g'} \quad \text{and} \quad K = -\frac{f''}{f},$$

respectively. Finally, using this expression, construct surfaces with $K = -1/c^2$ and $H = 0$, respectively, where $c > 0$ is a constant.

Exercise 1.24 Fix a point $\boldsymbol{x}_0 = \boldsymbol{x}(u_0, v_0)$ on $\mathcal{M}$, and take the outer unit normal vector $\boldsymbol{a}$ at it. Then, compute Hesse matrix of $f(u,v) = \boldsymbol{a} \cdot \boldsymbol{x}(u,v)$ at $(u,v) = (u_0, v_0)$ in use of the second fundamental quantities, and give the explanation to the geometric meaning of the Gaussian curvature.

Exercise 1.25 If a surface $\mathcal{M}$ is locally expressed by $\psi(\boldsymbol{x}) = 0$, then the unit normal vector is given as

$$\boldsymbol{n} = \frac{\nabla\psi}{|\nabla\psi|}.$$

Take a point on $\mathcal{M}$ and let the principal directions be parallel to x_1 and x_2 coordinates, with the unit normal vector $\boldsymbol{n} = {}^t(n_1, n_2, n_3)$ parallel to x_3 coordinate. Confirming that

$$\frac{\partial n_j}{\partial x_i} = \frac{\delta_{ij}}{R_i}$$

for $i, j = 1, 2$ and

$$\frac{\partial n_3}{\partial x_i} = 0$$

for $i = 1, 2, 3$, show the relation $2H = \nabla \cdot \boldsymbol{n}$, where H indicates the mean curvature.

1.3 Extremals

1.3.1 *Lagrange Multiplier*

Given $y = f(x)$ defined on $a \le x \le b$, determine its maximum and minimum values. For this problem, we may seek all critical points $x_j \in (a, b)$ in $f'(x_j) = 0$ to compare $f(a)$, $f(b)$, and $f(x_j)$'s. Functions with multiple variables are similarly treated. If $z = f(x, y)$ has two variables $a \le x \le b$ and $c \le y \le d$, then we may seek all interior points (x_j, y_j) in $f_x(x_j, y_j) =$

$f_y(x_j, y_j) = 0$. Then, maximum and minimum values are in the boundary values, $f(x,y)$ for $x = a, b$ with $c \leq y \leq d$, and $f(x,y)$ for $y = c, d$ with $a \leq x \leq b$, and the critical values $f(x_j, y_j)$'s. The third problem is to seek maximum and minimum values of $z = g(x,y)$, under the constraint that $f(x,y) = 0$. Then, solving $f(x,y) = 0$ as $y = h(x)$, for example, we may obtain them by $z = g(x, h(x))$ defined on $a \leq x \leq b$.

Those rough answers are justified in the following way. In the first problem, Weierstrass' theorem guarantees that if $y = f(x)$ is continuous, then its maximum and minimum are attained. If $f(x)$ is differentiable at $x = x_0 \in (a,b)$, then it holds that

$$f(x_0 + \Delta x) = f(x_0) + f'(x_0)\Delta x + o(\Delta x)$$

as $\Delta x \to 0$. Therefore, if $f'(x_0) \neq 0$, then $f(x_0)$ cannot be a maximum or minimal value.

The second problem is treated similarly. Any continuous function on a compact set attains its maximum and minimum. If f is *totally differentiable* at the interior point (x_0, y_0), then it has *partial derivatives* there. It holds that

$$\begin{aligned} f(x_0 + \Delta x, y_0 + \Delta y) &= f(x_0, y_0) + f_x(x_0, y_0)\Delta x + f_y(x_0, y_0)\Delta y \\ &\quad + o\left(\sqrt{\Delta x^2 + \Delta y^2}\right) \end{aligned} \tag{1.51}$$

as $\Delta x, \Delta y \to 0$. Therefore, (x_0, y_0) cannot attain maximum or minimum unless $f_x(x_0, y_0) = f_y(x_0, y_0) = 0$.

Here, we may note the following. First, if $f(x,y)$ has continuous partial derivatives $f_x(x,y)$, $f_y(x,y)$ in the domain Ω, then it is totally differentiable there. If f has continuous second derivatives there, then we can make use of (1.21) to examine its local behavior around the critical point, where the Morse index plays a fundamental role. On the other hand, the direction derivatives are useful to investigate boundary values.

For instance, if the boundary $\Gamma = \partial\Omega$ is (piecewisely) C^2 and so is f up to there, we may pick up the boundary point (x_0, y_0) satisfying

$$\left.\frac{d}{ds} f(x(s), y(s))\right|_{s=0} = \boldsymbol{t} \cdot \nabla f(x_0, y_0) = 0,$$

where s is the length parameter with $(x(s), y(s)) \in \Gamma$ and $(x(0), y(0)) = (x_0, y_0)$ so that ${}^t(x'(0), y'(0)) = \boldsymbol{t}$ is equal to the unit tangential vector on

Γ at (x_0, y_0). Then, the local behavior of $f(x,y)$ around (x_0, y_0) can be examined by

$$\frac{d^2}{ds^2} f\left(x(s), y(s)\right)\bigg|_{s=0} = [\text{Hess } f(x_0, y_0)]\, \boldsymbol{t} \cdot \boldsymbol{t} + \frac{1}{\rho} \boldsymbol{n} \cdot \nabla f(x_0, y_0) \qquad (1.52)$$

and

$$\frac{d}{ds} f((x_0, y_0) + s\boldsymbol{n})\bigg|_{s=0} = \boldsymbol{n} \cdot \nabla f(x_0, y_0),$$

where ρ and $\boldsymbol{n}$ denote the curvature radius and the unit normal vector of Γ at (x_0, y_0), respectively, and

$$\text{Hess } f = \begin{pmatrix} \frac{\partial^2 f}{\partial x^2} & \frac{\partial^2 f}{\partial x \partial y} \\ \frac{\partial^2 f}{\partial x \partial y} & \frac{\partial^2 f}{\partial y^2} \end{pmatrix}.$$

Actually, signs of those values determine whether (x_0, y_0) attains local maximum or local minimum of f on the closed region in consideration.

Remember that for the third problem, elimination of one variable by the constraint is proposed. This idea is valid essentially, if the local resolution of $f(x,y) = 0$ is admitted. This is assured by the *implicit function theorem.* The simplest form is stated as follows.

Theorem 1.3 *Let $\Omega \subset \mathbf{R}^2$ be a domain, and $f = f(x,y)$ a continuous function in Ω with $f_y = \partial f / \partial y$ continuous there, and $f(x_0, y_0) = 0$ and $f_y(x_0, y_0) \neq 0$ hold for $(x_0, y_0) \in \Omega$. Then, there exists a unique $y = h(x)$, continuous near x_0 satisfying*

$$y_0 = h(x_0) \qquad \text{and} \qquad f(x, h(x)) = 0. \qquad (1.53)$$

If f_x exists in Ω, then $h(x)$ is differentiable near $x = x_0$ and it holds that

$$h'(x) = -f_x(x,y)/f_y(x,y). \qquad (1.54)$$

The above theorem justifies the following fact, called the *Lagrange multiplier principle.*

Theorem 1.4 *Let $f(x,y)$ and $g(x,y)$ be C^1 functions defined in a domain $\Omega \subset \mathbf{R}^2$, $(x_0, y_0) \in \Omega$ attain maximum or minimum (or just extremal) of*

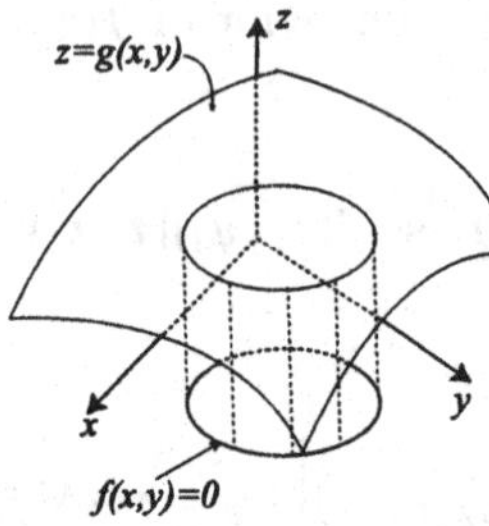

Fig. 1.15

$z = g(x, y)$ under the constraint $f(x, y) = 0$, and $\nabla f(x_0, y_0) \neq 0$. Then it holds that

$$\nabla g(x_0, y_0) = \lambda \nabla f(x_0, y_0) \tag{1.55}$$

with some $\lambda \in \mathbf{R}$.

Proof. Without loss of generality, we suppose that $f_y(x_0, y_0) \neq 0$. Then, by the implicit function theorem 1.3, there exists a unique continuous function $y = h(x)$ near $x = x_0$ satisfying (1.53). Furthermore, $h(x)$ is differentiable at $x = x_0$, and $\varphi(x) = g(x, h(x))$ attains a critical value at $x = x_0$ from the assumption. Therefore, it holds that

$$\varphi'(x_0) = g_x(x_0, h(x_0)) + g_y(x_0, h(x_0))h'(x_0) = 0.$$

In use of (1.54) with $x = x_0$, we have

$$g_x(x_0, y_0) - g_y(x_0, y_0) f_x(x_0, y_0)/f_y(x_0, y_0) = 0.$$

Writing $\lambda = g_y(x_0, y_0)/f_y(x_0, y_0)$, we get that

$$g_x(x_0, y_0) = \lambda f_x(x_0, y_0) \quad \text{and} \quad g_y(x_0, y_0) = \lambda f_y(x_0, y_0).$$

This means (1.55). □

Theorems 1.3 and 1.4 have higher dimensional versions.

Exercise 1.26 Compute maximum and minimum values of $z = xy$ in $\Omega = \mathbf{R}^2$ under the constraint $x^2 + y^2 = 1$ in the following way. First, put $f(x, y) = x^2 + y^2 - 1$ and $g(x, y) = xy$. Confirm that $\nabla f(x, y) \neq 0$ holds if $f(x, y) = 0$. Then show that (1.55) and $f(x_0, y_0) = 0$ imply $\lambda = \pm 1/2$ and

$x_0y_0 = \lambda$. Finally, show that maximum and minimum values are attained from the compactness of $\{(x,y) \in \mathbf{R}^2 \mid f(x,y) = 0\}$ and the continuity of $z = xy$.

1.3.2 *Implicit Function Theorem*

This paragraph is devoted to the proof of Theorem 1.3, valid to the case of higher dimensions. The reader can skip this section if he is familiar with the first course of analysis.

Recall that $f_y(x_0, y_0) \neq 0$ is supposed. Without loss of generality, we assume $f_y(x_0, y_0) > 0$. Henceforth, $B(z_0, r)$ denotes the open ball in $\mathbf{R}^2$ with the center z_0 and the radius r. Because f_y is continuous, there exists $r > 0$ sufficiently small such that $f_y > 0$ in $B(z_0, r)$, where $z_0 = (x_0, y_0)$. The continuously differentiable function $\varphi(t) = f(x_0, y_0 + t)$ satisfies that

$$\varphi'(t) = f_y(x_0, y_0 + t) > 0 \qquad \text{and} \qquad \varphi(0) = f(x_0, y_0) = 0$$

with $|t| < r$. We have

$$\varphi(-r/2) = f(x_0, y_0 - r/2) < 0 < \varphi(r/2) = f(x_0, y_0 + r/2).$$

Because $f(x, y)$ is continuous in Ω, there exists $\delta > 0$ sufficiently small such that

$$f(x, y_0 - r/2) < 0 < f(x, y_0 + r/2) \qquad \text{if} \qquad |x - x_0| < \delta \tag{1.56}$$

and $D \equiv [x_0 - \delta, x_0 + \delta] \times [y_0 - r/2, y_0 + r/2] \subset B(z_0, r)$.

Taking $x \in I \equiv (x_0 - \delta, x_0 + \delta)$, we set $\varphi^x(t) = f(x, y_0 + t)$. Because of $D \subset B(z_0, r)$, it holds that

$$(\varphi^x)'(t) = f_y(x_0, y_0 + t) > 0 \qquad \text{for} \qquad |t| \leq r/2.$$

On the other hand, relation (1.56) implies

$$\varphi^x(-r/2) < 0 < \varphi^x(r/2).$$

Therefore, there is a unique $t_x \in [-r/2, r/2]$ satisfying $\varphi^x(t_x) = 0$. In other words, each $x \in I$ admits a unique $y \in [y_0 - r/2, y_0 + r/2]$ such that $f(x, y) = 0$. Let us write this y as $h(x)$. From the uniqueness of such y, we have $h(x_0) = y_0$.

Now, we prove the continuity of $h(x)$. We shall show that $x_* \in I = (x_0 - \delta, x_0 + \delta)$ and $x_j \to x_*$ imply $h(x_j) \to h(x_*)$. This means that any

$\varepsilon > 0$ admits k such that $j \geq k$ implies

$$|h(x_j) - h(x_*)| < \varepsilon.$$

If this is not the case, there is $\varepsilon_0 > 0$ such that any $k = 1, 2, \cdots$ admits $j(k) \geq k$ such that

$$\left|h(x_{j(k)}) - h(x_*)\right| \geq \varepsilon_0. \tag{1.57}$$

Put $x'_k = x_{j(k)}$ for simplicity. We still have $x'_k \to x_* \in I$, and hence $x'_k \in I$ for k sufficiently large. From the definition of $h(x)$, it holds that

$$h(x'_k) \in [y_0 - r/2, y_0 + r/2].$$

This makes it possible to extract a subsequence $\{x''_k\} \subset \{x'_k\}$ such that

$$h(x''_k) \quad \to \quad y_* \in [y_0 - r/2, y_0 + r/2] \tag{1.58}$$

for some y_*. This, together with $f\left(x''_k, h(x''_k)\right) = 0$, implies

$$f(x_*, y_*) = 0,$$

because f is continuous. Therefore, from the uniqueness of $h(x)$ we get that $y_* = h(x_*)$. Thus, (1.57) with (1.58) is a contradiction.

Finally, we show that if f_x exists in Ω, then $h(x)$ is differentiable at $x = x_0$ and equality (1.54) holds with $x = x_0$. Actually, the other cases of x are proven similarly.

For this purpose, given $|h| \ll 1$, we put

$$\Delta x = h \qquad \text{and} \qquad \Delta y = h(x_0 + \Delta x) - h(x_0).$$

Then, it holds that

$$f(x_0, y_0) = 0 \qquad \text{and} \qquad f(x_0 + \Delta x, y_0 + \Delta y) = 0.$$

Furthermore, $\Delta x \to 0$ implies $\Delta y \to 0$ because $h(x)$ is continuous.

From the mean value theorem, we have

$$f(x_0 + \Delta x, y_0 + \Delta y) = f(x_0 + \Delta x, y_0) + f_y(x_0 + \Delta x, y_0 + \theta \Delta y)\Delta y$$

with $\theta \in (0, 1)$. On the other hand, the relation

$$f(x_0 + \Delta x, y_0) = f(x_0, y_0) + f_x(x_0, y_0)\Delta x + o(\Delta x)$$

is valid from the assumption to f. Those relations imply

$$0 = f_x(x_0, y_0)\Delta x + f_y(x_0 + \Delta x, y_0 + \theta\Delta y)\Delta y + o(\Delta x)$$

and hence

$$\frac{\Delta y}{\Delta x} = -\frac{f_x(x_0, y_0)}{f_y(x_0 + \Delta x, y_0 + \Delta y)} + o(1)$$

follows. Here, f_y is continuous and $f_y(x_0, y_0) \neq 0$, so that we have the existence of $h'(x_0)$ with the relation

$$h'(x_0) = \lim_{\Delta x \to 0} \frac{\Delta y}{\Delta x} = -\frac{f_x(x_0, y_0)}{f_y(x_0, y_0)}$$

and the proof is complete.

1.3.3 *Convex Functions*

We say that a domain $\Omega \subset \mathbf{R}^n$ $(n = 1, 2, \cdots)$ is *convex* if $x, y \in \Omega$ and $\alpha \in [0, 1]$ imply $\alpha x + (1 - \alpha)y \in \Omega$. If Ω is convex, then a function $f = f(x)$ of $x \in \Omega$ is said to be convex if $x, y \in \Omega$ and $\alpha \in [0, 1]$ imply

$$f(\alpha x + (1 - \alpha)y) \leq \alpha f(x) + (1 - \alpha)f(y).$$

This inequality implies for $z = \alpha x + (1 - \alpha)y$ that

$$\alpha \{f(x) - f(z)\} \geq (1 - \alpha)f(z) - (1 - \alpha)f(y).$$

Taking $\alpha \in (0, 1)$, we obtain

$$f(x) - f(z) \geq \frac{1 - \alpha}{\alpha} \{f(z) - f(y)\}.$$

Let us make $\alpha \downarrow 0$. Then, if f is differentiable at y, it holds that

$$f(x) - f(y) \geq \frac{d}{d\alpha} f(\alpha x + (1 - \alpha)y)\Big|_{\alpha=0} = \nabla f(y) \cdot (x - y). \qquad (1.59)$$

This indicates that the graph of a convex function is always over the tangential space.

This observation is the starting point of the *convex analysis.* A convex function $f : \mathbf{R}^n \to (-\infty, +\infty]$ is said to be *proper* if its *effective domain* $D(f) = \{x \in \mathbf{R}^n \mid f(x) \neq +\infty\}$ is non-empty. It is *lower semi-continuous* if $x_j \to x_*$ implies $f(x_*) \leq \liminf f(x_j)$, or equivalently,

$\{\boldsymbol{x} \in \mathbf{R}^n \mid f(\boldsymbol{x}) \le c\}$ is closed for any $c \in \mathbf{R}$. Given a proper, lower semi-continuous, convex function f, its *conjugate* or *Legendre transformation* is given by

$$f^*(\boldsymbol{\xi}) = \sup_{\boldsymbol{x} \in \mathbf{R}^n} \{\boldsymbol{x} \cdot \boldsymbol{\xi} - f(\boldsymbol{x})\}.$$

It is again a proper, lower semi-continuous, convex function, although the first property is not trivial. Then, the *duality theorem of Fenchel-Moreau* guarantees that $f^{**} = f$. The *sub-differential* of f at $\boldsymbol{x} \in D(f)$, denoted by $\partial f(\boldsymbol{x})$ is the set of $\boldsymbol{\xi}$ satisfying

$$f(\boldsymbol{y}) - f(\boldsymbol{x}) \ge \boldsymbol{\xi} \cdot (\boldsymbol{y} - \boldsymbol{x})$$

for any $\boldsymbol{y} \in \mathbf{R}^n$. If f is proper, lower semi-continuous, convex, then $\boldsymbol{\xi} \in \partial f(\boldsymbol{x})$ if and only if $\boldsymbol{x} \in \partial f^*(\boldsymbol{\xi})$, and *Fenchel's identity* guarantees that

$$f(\boldsymbol{x}) + f^*(\boldsymbol{\xi}) = \boldsymbol{x} \cdot \boldsymbol{\xi}.$$

If $\varphi = \varphi(x, y) : \mathbf{R}^n \times \mathbf{R}^m \to (-\infty, +\infty]$ is a proper, convex, and lower semi-continuous function, then problems

$$\inf\{\varphi(x, 0) \mid x \in \mathbf{R}^n\} \qquad \text{and} \qquad \sup\{-\varphi^*(0, q) \mid q \in \mathbf{R}^m\}$$

indicated (P) and (P^*) are called the *principal* and the *dual*, respectively, where

$$\begin{aligned} \varphi^*(p, q) &= \sup_{(x,y) \in \mathbf{R}^n \times \mathbf{R}^m} \{x \cdot p + y \cdot q - \varphi(x, y)\} \\ &= \sup_{x \in \mathbf{R}^n} \left\{ x \cdot p + \sup_{y \in \mathbf{R}^m} (y \cdot q - \varphi(x, y)) \right\} \end{aligned}$$

denotes the Legendre transformation of $\varphi = \varphi(x, y)$. Let $\overline{x}$ and $\overline{q}$ be the solutions to (P) and (P^*), respectively, and

$$\Phi(y) = \inf\{\varphi(x, y) \mid x \in \mathbf{R}^n\}$$

be proper, convex, and lower semi-continuous. Then, we have

$$\varphi^*(0, q) = \sup_{x,y}\{y \cdot q - \varphi(x, y)\} = \sup_{y}\{y \cdot q - \Phi(y)\} = \Phi^*(q),$$

and hence it follows that

$$\sup_{q}\{-\varphi^*(0, q)\} = \sup_{q}\{-\Phi^*(q)\} = \sup_{q}\{0 \cdot q - \Phi^*(q)\}$$

$$= \Phi^{**}(0) = \Phi(0) = \inf_x \varphi(x,0). \tag{1.60}$$

Thus, (P) and (P^*) have the same value.

Fixed $x \in \mathbf{R}^n$, the Legendre transformation of $y \mapsto \varphi(x,y)\ (= \varphi_x(y))$ is given as

$$-L(x,q) = \sup_y \{y \cdot q - \varphi(x,y)\},$$

and $L(x,q)$ is called the *Lagrange function*. Then,

$$\varphi^*(p,q) = \sup_x \{x \cdot p - L(x,q)\}$$

is nothing but the Legendre transformation of $x \mapsto L(x,q)$ with $q \in \mathbf{R}^m$ fixed. We have

$$\varphi^*(0,q) = \sup_x \{-L(x,q)\} = -\inf_x L(x,q)$$

and the dual problem is re-formulated as

$$\sup_q \{-\varphi^*(0,q)\} = \sup_q \inf_x L(x,q).$$

On the other hand, we have $\varphi_x^*(q) = -L(x,q)$ as is noticed, and hence it holds that

$$\varphi(x,y) = \varphi_x^{**}(y) = \sup_q \{y \cdot q + L(x,q)\}$$

and $\varphi(x,0) = \sup_q L(x,q)$. The principal problem is re-formulated as

$$\inf_x \varphi(x,0) = \inf_x \sup_q L(x,q).$$

It is obvious that

$$\sup_q \inf_x L(x,q) \le \inf_x \sup_q L(x,q),$$

but the equality holds here by (1.60). Thus, $\overline{x} \in \mathbf{R}^n$ and $\overline{q} \in \mathbf{R}^m$ are the solutions to (P) and (P^*) if and only if

$$L(\overline{x},q) \le L(\overline{x},\overline{q}) \le L(x,\overline{q}) \tag{1.61}$$

holds for any $(x,q) \in \mathbf{R}^n \times \mathbf{R}^m$. This fact is called Kuhn-Tucker's *saddle point theorem*.

An application of this theorem is minimizing f under the constraint $g_i \leq 0$ $(i = 1, \cdots, m)$. Suppose that $f : \mathbf{R}^n \to \mathbf{R}$ and $g_i : \mathbf{R}^n \to \mathbf{R}$ are convex and lower semi-continuous, and that there exists $\boldsymbol{x}_0 \in \mathbf{R}^n$ such that $g_i(\boldsymbol{x}_0) < 0$ for $i = 1, \cdots, m$. The last condition is called Slater's *constraint qualification*. Then, (1.61) guarantees that if $\overline{\boldsymbol{x}}$ attains the minimum, then the function $L(\boldsymbol{x}, \boldsymbol{\lambda}) : \mathbf{R}^n \times \mathbf{R}^m \to \mathbf{R}$ defined by

$$L(\boldsymbol{x}, \boldsymbol{\lambda}) = f(\boldsymbol{x}) + \boldsymbol{\lambda} \cdot \boldsymbol{G}(\boldsymbol{x})$$

for $G(x) = {}^t\,(g_1(x), \cdots, g_m(x))$ admits $\overline{\boldsymbol{\lambda}} = {}^t(\overline{\lambda}_1, \cdots, \overline{\lambda}_m)$ such that $\overline{\lambda}_i \geq 0$ $(i = 1, \cdots, m)$ and

$$L(\overline{\boldsymbol{x}}, \boldsymbol{\lambda}) \leq L(\overline{\boldsymbol{x}}, \overline{\boldsymbol{\lambda}}) \leq L(\boldsymbol{x}, \overline{\boldsymbol{\lambda}})$$

for any $(\boldsymbol{x}, \boldsymbol{\lambda}) \in \mathbf{R}^n \times \mathbf{R}^m$. Concerning the existence of the saddle point, we can make use of the *mini-max principle* of von Neumann. It says that if X_0, Y_0 are topological vector spaces, $X \subset X_0$, $Y \subset Y_0$ are convex, compact subsets, $x \in X \mapsto f(x, y)$ is convex, lower semi-continuous for any $y \in Y$, and $y \mapsto f(x, y)$ is concave, upper semi-continuous for any $x \in X$, then there exists $(\overline{x}, \overline{y}) \in X \times Y$ satisfying

$$f(\overline{x}, y) \leq f(\overline{x}, \overline{y}) \leq f(x, \overline{y})$$

for any $(x, y) \in X \times Y$.

Instead of examining those general theories, we pick up the following example. That is, minimizing a continuously differentiable convex function $f(\boldsymbol{x})$ of $\boldsymbol{x} \in \mathbf{R}^2$ under the constraint $\boldsymbol{x} \cdot \boldsymbol{e} = c$, where $\boldsymbol{e}$ is a unit vector and $c \in \mathbf{R}$. In fact, first we take $\boldsymbol{a} \in \mathbf{R}^2$ in $\boldsymbol{a} \cdot \boldsymbol{e} = c$ and set $\boldsymbol{x}' = \boldsymbol{x} - \boldsymbol{a}$. Then, it holds that $\boldsymbol{x}' \cdot \boldsymbol{e} = 0$ and $f'(\boldsymbol{x}') = f(\boldsymbol{x}' + \boldsymbol{a})$ is a convex function of $\boldsymbol{x}'$. In other words, we can assume $c = 0$ without loss of generality.

Let $\mathcal{M}$ be the set of $\boldsymbol{x}$ satisfying $\boldsymbol{e} \cdot \boldsymbol{x} = 0$. It is the (one-dimensional) vector space orthogonal to $\boldsymbol{e}$. Let $\mathcal{N}$ be the set of $\boldsymbol{y}$ such that $\nabla f(\boldsymbol{y})$ is parallel to $\boldsymbol{e}$. This means that $\boldsymbol{x} \in \mathcal{M}$ and $\boldsymbol{y} \in \mathcal{N}$ implies $\nabla f(\boldsymbol{y}) \cdot \boldsymbol{x} = 0$ and hence

$$f(\boldsymbol{x}) \geq g(\boldsymbol{y}) \qquad (\boldsymbol{x} \in \mathcal{M},\ \boldsymbol{y} \in \mathcal{N}) \tag{1.62}$$

holds for

$$g(\boldsymbol{y}) = f(\boldsymbol{y}) - \nabla f(\boldsymbol{y}) \cdot \boldsymbol{y}$$

by (1.59).

If $\overline{\boldsymbol{x}} \in \mathcal{M}$ is the solution to the problem, then from the Lagrange multiplier principle there is $\lambda \in \mathcal{R}$ such that

$$\nabla f(\overline{\boldsymbol{x}}) = \lambda \boldsymbol{e}.$$

This means $\overline{\boldsymbol{x}} \in \mathcal{M} \cap \mathcal{N}$, and therefore, $\nabla f(\overline{\boldsymbol{x}}) \cdot \overline{\boldsymbol{x}} = 0$ and $g(\overline{\boldsymbol{x}}) = f(\overline{\boldsymbol{x}})$ follow. It follows from (1.62) that

$$f(\boldsymbol{x}) \geq g(\overline{\boldsymbol{x}}) = f(\overline{\boldsymbol{x}}) \geq g(\boldsymbol{y})$$

for any $\boldsymbol{x} \in \mathcal{M}$ and $\boldsymbol{y} \in \mathcal{N}$. This means that $f(\overline{\boldsymbol{x}}) = g(\overline{\boldsymbol{x}})$ attains the minimum and the maximum of $f(\boldsymbol{x})$ in $\boldsymbol{x} \in \mathcal{M}$ and $g(\boldsymbol{y})$ in $\boldsymbol{y} \in \mathcal{N}$, respectively. This conclusion may be regarded as a saddle point theorem.

Exercise 1.27 Given $\boldsymbol{a}, \boldsymbol{b} \in \mathbf{R}^2$, minimize

$$f(\boldsymbol{u}) = \frac{1}{2}\left|\boldsymbol{u} - \boldsymbol{b}\right|^2 \left(= \frac{1}{2}(\boldsymbol{u} - \boldsymbol{b}) \cdot (\boldsymbol{u} - \boldsymbol{b}) \right)$$

under the constraint that $(\boldsymbol{u} - \boldsymbol{a}) \cdot \boldsymbol{e} = 0$.

Exercise 1.28 Let f be a convex function defined in an interval $I \subset \mathbf{R}$. For a, b, $c \in I$ with $a < b < c$, show

$$\frac{f(b) - f(a)}{b - a} \leq \frac{f(c) - f(a)}{c - a} \leq \frac{f(c) - f(b)}{c - b}.$$

Then, prove that this f is continuous.

Chapter 2

Calculus of Variation

From the analytic point of view, geometric quantities such as length, area, volume, are regarded as the value determined by the function parametrizing the object, and therefore, each of them induces a mapping from the set of functions into $\mathbf{R}$. Sometimes, such a mapping is called the *functional* because it is a function defined on function spaces. In the calculus of variation, a functional is given, and it is required to find its extremal functions. This formulation can describe physical problems if the functional is taken as *energy*, *Lagrangian*, *free energy*, and so forth.

2.1 Isoperimetric Inequality

2.1.1 *Analytic Proof*

The Jordan curve indicates a closed non-self-intersecting curve, and a connected open set is referred to as the domain. We can observe that a Jordan curve Γ on the plane $\mathbf{R}^2$ encloses there a simply connected domain D. The question studied here is referred to as the *isoperimetric problem*. When is the area A of D minimized if the length L of Γ is prescribed ?

The answer is a circle. Analytic proof is as follows. First, we parametrize Γ as $(x(t), y(t))$ in $t \in [a, b]$. This implies that $(x(a), y(a)) = (x(b), y(b))$,

$$(x(t), y(t)) \neq (x(t'), y(t')) \qquad \text{for} \qquad t \neq t' \quad \text{in} \quad t, t' \in [a, b),$$

and

$$L = \int_a^b \sqrt{x'(t)^2 + y'(t)^2} dt.$$

Putting $a_1(x,y) = y$, $a_2(x,y) = 0$ in the Green's formula (1.43), we obtain

$$A = \int_D dxdy = -\int_\Gamma ydx = -\int_a^b y(t)x'(t)dt.$$

Here, we take the parametrization $t = (2\pi s)/L \in [0, 2\pi]$ for the length parameter s. This means that

$$s = \int_0^t \sqrt{x'(t)^2 + y'(t)^2} dt$$

and hence

$$\left(\frac{ds}{dt}\right)^2 = x'(t)^2 + y'(t)^2 = \left(\frac{L}{2\pi}\right)^2$$

holds. We have

$$\int_0^{2\pi} \left(x'(t)^2 + y'(t)^2\right) dt = \int_0^{2\pi} \left(\frac{ds}{dt}\right)^2 dt = \left(\frac{L}{2\pi}\right)^2 \cdot 2\pi = \frac{L^2}{2\pi}$$

and therefore,

$$\begin{aligned} L^2 - 4\pi A &= 2\pi \int_0^{2\pi} \left(x'(t)^2 + y'(t)^2 + 2y(t)x'(t)\right) dt \\ &= 2\pi \int_0^{2\pi} \left(x'(t) + y(t)\right)^2 dt + 2\pi \int_0^{2\pi} \left(y'(t)^2 - y^2(t)\right) dt \\ &\geq 2\pi \int_0^{2\pi} \left(y'(t)^2 - y(t)^2\right) dt \end{aligned}$$

follows.

We may assume that

$$\int_0^{2\pi} y(t)dt = 0$$

by translating Γ parallel to y axis. Then, from the following fact, referred to as *Wirtinger's inequality*, we have $L^2 \geq 4\pi A$ with the equality if and only if Γ is a circle. In this way, we can give a proof for that well-known fact.

Theorem 2.1 *If $y(t)$ is a smooth periodic function with period 2π such that*

$$\int_0^{2\pi} y(t)dt = 0 \qquad \textit{and} \qquad y(t) \neq \alpha \sin t + \beta \cos t,$$

then it holds that

$$\int_0^{2\pi} y'(t)^2 dt > \int_0^{2\pi} y(t)^2 dt,$$

where α, β are constants.

Proof. We apply the theory of Fourier series developed in §2.4.2. Note that a *periodic function* with *period* 2π is the one, denoted by $y(t)$, satisfying

$$y(t + 2\pi) = y(t) \qquad \text{for} \qquad t \in \mathbf{R}.$$

Given such a continuous function $y(t)$, we put

$$a_n = \frac{1}{\pi}\int_0^{2\pi} y(t)\cos nt dt \qquad \text{and} \qquad b_n = \frac{1}{\pi}\int_0^{2\pi} y(t)\sin nt dt$$

for $n = 0, 1, 2, \cdots$. Then, if $y(t)$ is sufficiently smooth, say, C^1 and piecewise C^2 in $t \in \mathbf{R}$, it holds that

$$y(t) = \frac{a_0}{2} + \sum_{n=1}^{\infty} (a_n \cos nt + b_n \sin nt) \tag{2.1}$$

and

$$y'(t) = \sum_{n=1}^{\infty} (-na_n \sin nt + nb_n \cos nt),$$

where the right-hand sides converge absolutely and uniformly in $t \in [0, 2\pi]$. Those relations imply, because of

$$\int_0^{2\pi} 1 \cdot dt = 2\pi,$$

$$\int_0^{2\pi} \sin nt \sin mt dt = \int_0^{2\pi} \cos nt \cos mt dt = 0 \qquad (n \neq m)$$

$$\int_0^{2\pi} \cos nt \sin mt dt = 0,$$

$$\int_0^{2\pi} \sin^2 nt dt = \int_0^{2\pi} \cos^2 nt dt = \pi \ (n \geq 1)$$

that

$$\int_0^{2\pi} y(t)^2 dt = \int_0^{2\pi} \left\{ \left(\frac{a_0}{2}\right)^2 + \sum_{n=1}^{\infty} (a_n^2 \cos^2 nt + b_n^2 \sin^2 nt) \right\} dt$$

$$= \pi\left\{\frac{a_0^2}{2}+\sum_{n=1}^{\infty}\left(a_n^2+b_n^2\right)\right\}$$

and

$$\begin{aligned}\int_0^{2\pi} y'(t)^2 dt &= \sum_{n=1}^{\infty}\int_0^{2\pi}\left(n^2a_n^2\sin^2 nt+n^2b_n^2\cos^2 nt\right)dt \\ &= \pi\sum_{n=1}^{\infty}\left(n^2a_n^2+n^2b_n^2\right).\end{aligned}$$

Here, from the assumption it follows that

$$a_0=\frac{1}{\pi}\int_0^{2\pi} y(t)dt=0.$$

Thus, we obtain

$$\begin{aligned}\int_0^{2\pi} y(t)^2 dt &= \pi\sum_{n=1}^{\infty}\left(a_n^2+b_n^2\right) \\ &\leq \pi\sum_{n=1}^{\infty}\left(n^2a_n^2+n^2b_n^2\right)=\int_0^{2\pi} y'(t)^2 dt.\end{aligned}$$

Here, the equality holds if and only if $a_n = b_n = 0$ for $n \geq 2$, or equivalently, $y(t) = a_1 \cos t + b_1 \sin t$. The proof is complete. □

We have proven

$$L^2 \geq 4\pi A \tag{2.2}$$

with the equality if and only if Γ is a circle, where Γ is a Jordan curve on $\mathbf{R}^2$, L is its length, and A denotes the area of the domain D enclosed by Γ. It is called the *isoperimetric inequality.* Examining the proof, we see that it is valid if Γ is C^1 and piecewise C^2, which, however, is a technical assumption. Actually, the geometric proof guarantees (2.2) for any continuous Jordan curve Γ.

From the analytic point of view, it may be worth mentioning that Theorem 2.1 is extended to the case that $y(t)$ and $y'(t)$ are *quadratic summable* because of *Parseval's equality.* Here, the integration and the differentiation are taken in the sense of *Lebesgue* and that of *distributions*, respectively. The set of such functions, generally referred to as a *function space*, forms the *Sobolev space*, of which details are described in later sections.

Those geometric objects are thus extended by the generalization of integration and differentiation of functions, but more direct ways are possible. First, the area A of the set $D \subset \mathbf{R}^2$ is definite if it is Lebesgue measurable. Correspondingly, the length L of its boundary $\Gamma = \partial D$ is definite if D has the *finite perimeter*. In this case, we still have (2.2) with the equality if and only if Γ is a circle. However, generally, it may be difficult to solve variational problems through an a priori insight as (2.2).

Exercise 2.1 Confirm that the equality $L^2 = 4\pi A$ implies that Γ is a circle in use of Theorem 2.1.

2.1.2 *Geometric Proof*

Several geometric proofs to (2.2) are known. Here, we describe the method of symmetrization. Such a technique is important in the study of partial differential equations arising in mathematical physics, although the reader can skip this paragraph first.

Taking the convex hull of D, we see that the problem is reduced to the case that D is convex. Note that the convex domain in the plane always admits right and left tangential lines. In fact, given $P \in \Gamma$, we take $P' \in \Gamma \setminus \{P\}$ and the line $\ell_{P'}$ connecting P' and P. As P' approaches P from one side, the limiting line ℓ of $\ell_{P'}$ exists, because the inclination of the latter is monotone from the convexity of D. Remember that if $y = f(x)$ is a convex function, its right derivative at $x = x_0$ is given by

$$m(x_0) = \lim_{\Delta x \downarrow 0} \frac{f(x_0 + \Delta x) - f(x_0)}{\Delta x}.$$

If $P(x_0, f(x_0))$, $P'(x_0 + \Delta x, f(x_0 + \Delta x))$, and R' denotes the crossing point of the right tangential line and $x = x_0 + \Delta x$, then it holds that

$$PP' \leq (PP')_C \leq PR' + R'P',$$

where $(PP')_C$ denotes the arc length of Γ between P and P'. From those relations we have

$$(PP')_C = PP' \{1 + o(1)\} \tag{2.3}$$

as $P' \to P$.

We deform the convex body D in the following way, called the *Steiner symmetrization*. Namely, we fix a line ℓ in the plane, and take another one ℓ' perpendicular to ℓ. If $\ell' \cap \overline{D} \neq \emptyset$, it forms a segment $[P, Q]$. Then, we translate it as $[P^*, Q^*]$ on ℓ' so that the middle point the latter is on ℓ and $PQ = P^*Q^*$. Then, sliding ℓ', we see that those $\{[P, Q]\}$ form a domain denoted by D^*, which is said to be the Steiner symmetrization of D. It holds that D^* is convex and symmetric with respect to ℓ. Also, if a family of convex bodies $\{D_k\}$ converges to D_∞ as $k \to \infty$, then so does $\{D_k^*\}$ to D_∞^*.

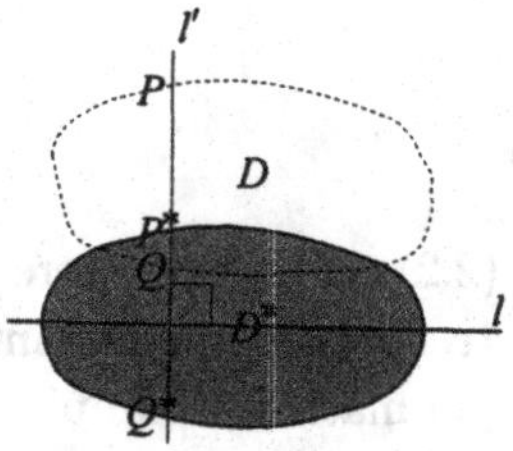

Fig. 2.1

Theorem 2.2 *We have the following.*

1. *The area of D^* is equal to that of D.*

2. *We have $|\partial D| \geq |\partial D^*|$, where $|\partial D|$ and $|\partial D^*|$ denote the lengths of ∂D and ∂D^*, respectively.*

3. *The equality $|\partial D| = |\partial D^*|$ is valid if and only if D^* is a translation of D.*

Proof. To show the first item, we may assume that ℓ coincides with x axis. Let the projection of $\overline{D}$ to ℓ be $[a, b]$, and

$$\Delta : a = x_0 < x_1 < \cdots < x_n = b$$

be its division with the mesh size $\|\Delta\| = \max_{1 \leq i \leq n}(x_i - x_{i-1})$. The line parallel to y axis and passing through x_i, denoted by ℓ_i, cuts a segment from D, denoted by $[P_i, Q_i]$. Then, the area of D is the limit of

$$S_\Delta = \sum_{i=1}^{n} \Delta S_i$$

as $\|\Delta\| \to 0$, where ΔS_i is the area of the trapezoid made by P_i, P_{i-1}, Q_{i-1}, Q_i. Let P_i^*, P_{i-1}^*, Q_{i-1}^*, Q_i^* be the corresponding points on ∂D^* of P_i, P_{i-1}, Q_i, Q_{i-1}, respectively. Then, ΔS_i is equal to the area of the trapezoid made by P_i^*, P_{i-1}^*, Q_i^*, Q_{i-1}^* by the definition, and therefore, the area of D,

$$\lim_{\|\Delta\| \to 0} S_\Delta,$$

is equal to that of D^*.

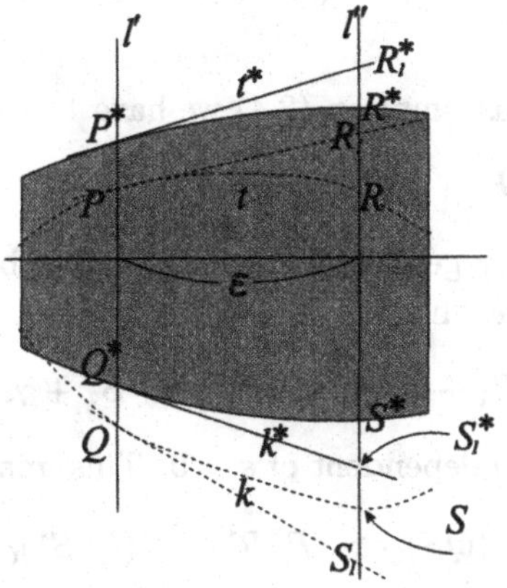

Fig. 2.2

For the second item to prove, we note that $|\partial D|$ is given by the limit of

$$s_\Delta = \sum_{i=1}^{n} \Delta s_i$$

as $\|\Delta\| \to 0$, where $\Delta s_i = P_i P_{i-1} + Q_i Q_{i-1}$. Here, it follows from the elementary geometry that

$$\Delta s_i \geq \Delta s_i^* = P_i^* P_{i-1}^* + Q_i^* Q_{i-1}^*$$

and hence $|\partial D| \geq |\partial D^*|$ follows.

To show the third item, we assume that D and D^* is not congruent. In this case, there is ℓ' such that

$$m_+(P) \neq m_+(P^*), \tag{2.4}$$

where $m_+(P)$, $m_+(P^*)$ denote the inclinations of the right tangential lines of D, D^* at P, P^*, respectively. Let Q^* be the corresponding point of Q on ∂D^*, where $[PQ] = \ell' \cap \overline{D}$. Let the right tangential lines at P, Q of ∂D be t, k, and those at P^*, Q^* of ∂D^* be t^*, k^*, respectively. We take a parallel line ℓ'' to ℓ' in the right, with the distance $\varepsilon > 0$. Its crossing points between t, k, t^*, k^* are denoted by R_1, S_1, R_1^*, S_1^*. Furthermore, let $[R, S] = \ell'' \cap \overline{D}$ and $[R^*, S^*] = \ell'' \cap \overline{D^*}$. Then, it holds that

$$\begin{aligned} (PR)_C &= PR_1 + o(\varepsilon) \qquad & (QS)_C &= QS_1 + o(\varepsilon) \\ (P^*R^*)_C &= P^*R_1^* + o(\varepsilon) \qquad & (Q^*S^*)_C &= Q^*S_1^* + o(\varepsilon) \end{aligned}$$

as $\varepsilon \downarrow 0$ by (2.3).

However, from the assumption (2.4) we have

$$PR_1 + QS_1 > P^*R_1^* + Q^*S_1^*.$$

Again by the elementary geometry the difference of both sides is homogeneous in ε of degree one, that is,

$$PR_1 + QS_1 = P^*R_1^* + Q^*S_1^* + \gamma\varepsilon$$

with a constant $\gamma > 0$ independent of $\varepsilon > 0$. This implies that

$$(PR)_C + (QS)_C > (P^*R^*)_C + (Q^*S^*)_C + \frac{\gamma\varepsilon}{2}$$

for $\varepsilon > 0$ sufficiently small. Then, $|\partial D| > |\partial D^*|$ follows. □

For the moment $|C|$ denotes the length of the curve C. In the case that the convex body D is not a disc, we take outscribing disc B, line ℓ passing through its center O, and largest concentric disc E contained in D. Then, it holds that $|\partial B| > |\partial D| > |\partial E|$. The Steiner symmetrization D^* of D with respect to ℓ is convex, and it holds that $E \subset D^* \subset B$. Therefore, $|\partial B| > |\partial D^*| > |\partial E|$ follows.

We shall show that any convex domain D admits a sequence of convex domains $\{D_k\}_{k=0}^{\infty}$ such that $D_0 = D$, D_{k+1} is a Steiner symmetrization of D_k, and D_k converges to a disc. Then by Theorem 2.2, inequality (2.2) follows for D with the equality if and only if D is a disc.

Henceforth, $\mathbf{Q}$ denotes the set of rational numbers. We note that D is a disc if it is symmetric with respect to two lines passing through the origin, denoted by O, with the angle made by them not in $\mathbf{Q}/(2\pi)$. Letting ℓ_1, ℓ_2 be such lines, we take $\{D_k\}$ by the Steiner symmetrization with respect to ℓ_1 and ℓ_2, successively. Then, we have $E \subset D_k \subset B$ for any k. There is a countable dense set C_0 on ∂B. Each $q \in C_0$ determines $p_k(q) \in [O, q] \cap \partial D_k$ and $\gamma_k = \{p_k(q) \mid q \in C_0\} \subset \partial D_k$ is a countable set. In use of the *diagonal argument* of Cantor, we get a subsequence of $\{D_k\}$, still denoted by the same symbol, such that any $p \in \gamma_k$ converges. Furthermore, it can be required that both directions of symmetrization, with respect to ℓ_1, ℓ_2, are contained infinitely many times in this sequence. Let the closure in $\mathbf{R}^2$ of the set of those limiting points be γ.

Again by the diagonal argument, there is a subsequence of $\{D_k\}$, still denoted by the same symbol, such that any $\varepsilon = 1/n$ $(n = 1, 2, \cdots)$ admits k_0 such that γ_k with $k \geq k_0$ lies in the $\varepsilon/2$ neighborhood of γ. Similarly, this sequence can contain both directions of symmetrization infinitely many times. This implies that ∂D_k is in the ε neighborhood of γ, and it follows that ∂D_k converges to γ. Because the former is a closed convex closed curve, so is γ. It encloses a domain in $\mathbf{R}^2$.

Let * indicates the symmetrization with respect to ℓ_1. From the above description, there is $\left\{\tilde{D}_k\right\} \subset \{D_k\}$ such that

$$\lim \left|\partial \tilde{D}_k^*\right| \geq |\partial D_0|$$

from the monotonicity and also

$$\lim \left|\partial \tilde{D}_k^*\right| = |\partial D_0^*|$$

from the continuity of symmetrization. This implies $|\partial D_0^*| \geq |\partial D_0| \geq |\partial D_0^*|$ and D_0 is symmetric with respect to ℓ_1. Similarly, it is symmetric with respect to ℓ_2, and the proof is complete.

Exercise 2.2 Confirm that if the convex domain D admits a sequence of convex domains $\{D_k\}_{k=0}^{\infty}$ such that $D_0 = D$, D_{k+1} is a Steiner symmetrization of D_k, and D_k converges to a disc, then inequality (2.2) follows for D with the equality if and only if D is a disc.

2.2 Indirect Method

2.2.1 *Euler Equation*

Structure of the isoperimetric problem is formed as follows. First, quantities A and L determined by the function $(x(t), y(t))$ are given. Regarding them as functionals, we are asked to maximize A under the constraint that L is a constant. In such a situation, it seems to be natural to take "critical" functions first, where the "derivative" of the functional vanishes. Even in this case with the constraint, the "Lagrange multiplier principle" will be applicable.

Generally, the functions in consideration are called *admissible*, and such a problem of finding extremal functions for the given functional is called the *variational problem.* Furthermore, such a critical function is to satisfy the *Euler equation.* Thus, in the variational problem, a functional is given and its extremal functions are required among the admissible functions.

We illustrate the story for the problem to minimize the functional

$$I(\varphi) = \int_{x_1}^{x_2} f\left(x, \varphi(x), \varphi'(x)\right) dx$$

defined for the function $y = \varphi(x)$ passing the fixed points $P_1(x_1, y_1)$, $P_2(x_2, y_2)$ in $\mathbf{R}^2$, where $f(x, y, y')$ is a given function.

To fix the idea, we suppose that f is continuous in (x, y, y'), and $y = \varphi(x)$ is admissible if it is C^1 on $[x_1, x_2]$ and satisfies

$$\varphi(x_1) = y_1 \qquad \text{and} \qquad \varphi(x_2) = y_2.$$

Under such a situation, let us suppose that the minimum in attained by an admissible function, denoted by $y = \varphi_0(x)$. Then $\varphi(x) = \varphi_0(x) + s\eta(x)$ is also admissible for any C^1 function $y = \eta(x)$ with $\eta(x_0) = \eta(x_1) = 0$ and $s \in \mathbf{R}$. Namely, this φ is C^1 on $[x_1, x_2]$ and satisfies $\varphi(x_1) = y_1$ and $\varphi(x_2) = y_2$. This implies

$$I(\varphi_0 + s\eta) \geq I(\varphi_0)$$

and therefore, $s \in \mathbf{R} \mapsto I(\varphi_0 + s\eta)$ attains the minimum at $s = 0$. In particular,

$$\left.\frac{d}{dt} I(\varphi_0 + s\eta)\right|_{s=0} = 0$$

follows if the left-hand side exists. Because of

$$I(\varphi_0 + s\eta) = \int_{x_1}^{x_2} f(x, \varphi_0(x) + s\eta(x), \varphi_0'(x) + s\eta'(x))dx,$$

we have, formally that

$$\begin{aligned}
&\left.\frac{d}{ds} I(\varphi_0 + s\eta)\right|_{s=0} \\
&= \int_{x_1}^{x_2} \left.\frac{\partial}{\partial s} f(x, \varphi_0(x) + s\eta(x), \varphi_0'(x) + s\eta'(x))\right|_{s=0} dx \\
&= \int_{x_1}^{x_2} \{f_y(x, \varphi_0(x), \varphi_0'(x))\eta(x) + f_{y'}(x, \varphi_0(x), \varphi_0'(x))\eta'(x)\} \, dx \\
&= \int_{x_1}^{x_2} \left\{ f_y(x, \varphi_0(x), \varphi_0'(x)) - \frac{d}{dx} f_{y'}(x, \varphi_0(x), \varphi_0'(x)) \right\} \\
&\quad \cdot \eta(x)dx
\end{aligned} \tag{2.5}$$

by $\eta(x_1) = \eta(x_2) = 0$. Because such $\eta(x)$ is arbitrary, it holds that

$$\frac{d}{dx}\left(\frac{\partial f}{\partial y'}\right) - \frac{\partial f}{\partial y} = 0 \tag{2.6}$$

for $y = \varphi_0(x)$ and $y' = \varphi_0'(x)$. Based on those considerations, we say that equation (2.6) is the Euler equation for this problem. Generally, each variational problem is associated with the Euler equation of its own, and *indirect method* in calculus of variation is to solve it.

The above derivation of (2.6) is justified if f_y is continuous, $f_{y'}$ is C^1 in (x, y, y'), and the extremal function $\varphi(x)$ is C^2 in x. The first two conditions are concerned on the variational problem itself, and it is possible to examine them in advance. On the other hand, the last condition is on the extremal function which we are seeking. We may be able to impose it as an admissibility. However, in this case the admissibility looks too restrictive to define the functional $I(\varphi)$. Because the variational problem is to find the extremal solution within admissible functions, that discrepancy leads us to the question that the extremal function can exist actually, or, what is the appropriate admissibility for the existence of the extremal function.

Actually, even the assumption $\varphi \in C^1[x_1, x_2]$ is restrictive for the admissibility to assure the existence of the extremal function φ_0 of the functional I. Eventually, this problem of existence is overcome by replacing the notions of integration and differentiation from those of Riemann to those of

Lebesgue. On the other hand, such an admissibility is too rough to justify the Euler equation. The second difficulty is overcome by introducing the notions of *weak solution* and its *regularity*. Those stories are realized in later sections.

Exercise 2.3 Derive the Euler equation for $f(x, y, y') = \sqrt{1 + y'^2}$ and confirm that the shortest curve connecting two fixed points in $\mathbf{R}^2$ is a segment.

2.2.2 *Lagrange Mechanics*

As is described in §1.1.3, if mass particles $x_i = x_i(t) \in \mathbf{R}^3$ $(i = 1, 2, \cdots, f)$ are subjected to the *potential energy* $U = U(x_1, x_2, \cdots, x_f)$, then Newton's equation of motion takes the form

$$\frac{dp_i}{dt} = -\frac{\partial U}{\partial x_i} \qquad (i = 1, 2, \cdots, f), \tag{2.7}$$

where $p_i = m_i \dot{x}_i$ denotes the momentum. Then, the *kinetic energy* is given by

$$K(\dot{x}_1, \dot{x}_2, \cdots, \dot{x}_f) = \frac{1}{2} \sum_{i=1}^{f} m_i \dot{x}_i^2$$

and it holds that

$$p_i = \frac{\partial K}{\partial \dot{x}_i}.$$

If $(q_1, \cdots, q_f)$ denotes the *generalized coordinate*, then it holds that $x_1 = x_1(q_1, \cdots, q_f; t), \cdots, x_f = x_f(q_1, \cdots, q_f; t)$ and we obtain

$$\dot{x}_i = \sum_j \frac{\partial x_i}{\partial q_j} \dot{q}_j + \frac{\partial x_i}{\partial t}. \tag{2.8}$$

Regarding $(q, \dot{q}, t)$ as independent variables, we differentiate (2.8) with respect to $\dot{q}_j$, and get that

$$\frac{\partial \dot{x}_i}{\partial \dot{q}_j} = \frac{\partial x_i}{\partial q_j}.$$

Hence we have

$$\frac{\partial K}{\partial \dot{q}_j} = \sum_k \frac{\partial K}{\partial \dot{x}_k}\frac{\partial \dot{x}_k}{\partial \dot{q}_j} = \sum_k p_k \frac{\partial x_k}{\partial q_j},$$

and then it follows that

$$\begin{aligned} \frac{d}{dt}\frac{\partial K}{\partial \dot{q}_j} &= \sum_k \left(\frac{dp_k}{dt}\frac{\partial x_k}{\partial q_j} + p_k \frac{d}{dt}\frac{\partial x_k}{\partial q_j} \right) \\ &= \sum_k \left(-\frac{\partial U}{\partial x_k}\frac{\partial x_k}{\partial q_j} + \frac{\partial K}{\partial \dot{x}_k}\frac{\partial \dot{x}_k}{\partial q_j} \right) = -\frac{\partial U}{\partial q_k} + \frac{\partial K}{\partial q_k} \end{aligned}$$

by

$$\frac{d}{dt}\frac{\partial x_k}{\partial q_j} = \frac{\partial \dot{x}_k}{\partial q_j}.$$

Therefore, if we regard K as a function of $(q, \dot{q}, t)$ by (2.8) and take the *Lagrange function*

$$L(q, \dot{q}, t) = K(q, \dot{q}, t) - U(q, t)$$

then it follows that

$$\frac{d}{dt}\frac{\partial L}{\partial \dot{q}_j} = \frac{\partial L}{\partial q_j} \qquad (j = 1, \cdots, f). \tag{2.9}$$

Equation (2.9) is called *Lagrange's equation of motion*, and now we know that it is the Euler equation for the variational problem $\delta S = 0$ under the constraint that $\delta q(t_1) = \delta q(t_2) = 0$, where

$$S = \int_{t_1}^{t_2} L\left(q(t), \dot{q}(t), t\right) dt$$

denotes the *action integral* defined for $q = q(t) \in \mathbf{R}^{3f}$ with $t \in [t_1, t_2]$. This fact is called *Hamilton's principle of least action.*

The *Legendre transformation* is generally taken to $L = L(\dot{q})$ by

$$L^*(p) = p\dot{q} - L \qquad \text{with} \qquad p = \frac{\partial L}{\partial \dot{q}}.$$

The Legendre transformation $\dot{q} \mapsto p$ to the Lagrangian $L = L(q, \dot{q}, t)$ is called the *Hamiltonian* with the *general momentum* p:

$$H = H(p, q, t) = \sum_j p_j \dot{q}_j - L \qquad \text{with} \qquad p_j = \frac{\partial L}{\partial \dot{q}_j}.$$

From (2.9) we have

$$\begin{aligned} dL &= \frac{\partial L}{\partial q}dq + \frac{\partial L}{\partial \dot{q}}d\dot{q} + \frac{\partial L}{\partial t}dt \\ &= \frac{d}{dt}\left(\frac{\partial L}{\partial \dot{q}}\right)dq + \frac{\partial L}{\partial \dot{q}}d\dot{q} + \frac{\partial L}{\partial t}dt = \dot{p}dq + pd\dot{q} + \frac{\partial L}{\partial t}dt, \end{aligned}$$

which implies that

$$\begin{aligned} dH &= \dot{q}dp + pd\dot{q} - dL = \dot{q}dp - \dot{p}dq - \frac{\partial L}{\partial t}dt \\ &= \frac{\partial H}{\partial p}dp + \frac{\partial H}{\partial q}dq + \frac{\partial H}{\partial L}dt, \end{aligned}$$

or *Hamilton's cannonical equation*

$$\dot{q} = \frac{\partial H}{\partial p} \quad \text{and} \quad \dot{p} = -\frac{\partial H}{\partial q}.$$

Exercise 2.4 Confirm that (2.9) is the Euler equation for $\delta S = 0$ under the constraint that $\delta q(t_1) = \delta q(t_2) = 0$.

2.2.3 *Minimal Surfaces*

The reader can skip this paragraph first. Given a surface $\mathcal{M}$, let $\boldsymbol{x} = \boldsymbol{x}(u,v) : \Omega \subset \mathbf{R}^2 \to \mathcal{M} \subset \mathbf{R}^3$ be its parametrization. The first fundamental form

$$I = Edu^2 + 2Fdudv + Gdv^2$$

induces the inner product in $\mathbf{R}^2$,

$$(U,V)_I = \sum_{i,j=1}^{2} I_{ij}u_iv_j,$$

where $I_{11} = E$, $I_{12} = I_{21} = F$, $I_{22} = G$, $U = {}^t(u_1,u_2) \in \mathbf{R}^2$, and $V = {}^t(v_1,v_2) \in \mathbf{R}^2$. So does the second fundamental form

$$II = Ldu^2 + 2Mdudv + Ndv^2$$

and those inner products are so related as

$$(AU,V)_I = (U,V)_{II}$$

for

$$A = \begin{pmatrix} E & F \\ F & G \end{pmatrix}^{-1} \begin{pmatrix} L & M \\ M & N \end{pmatrix}, \tag{2.10}$$

because

$$(U, V)_I = \begin{pmatrix} E & F \\ F & G \end{pmatrix} U \cdot V$$

and

$$(U, V)_{II} = \begin{pmatrix} L & M \\ M & N \end{pmatrix} U \cdot V$$

hold. Then, we can confirm that mean and Gaussian curvatures H, K are equal to $\frac{1}{2}\mathrm{tr}\ A$ and $\det A$, where tr and det indicate trace and determinant of matrices, respectively. Therefore, (1.49) is regarded as the eigenequation of A, and at $\boldsymbol{P} \in \mathcal{M}$ if the parametrization is so taken as

$$|\boldsymbol{x}_u| = |\boldsymbol{x}_v| = 1, \qquad \boldsymbol{x}_u \cdot \boldsymbol{x}_v = 0, \tag{2.11}$$

and $\boldsymbol{x}_u$, $\boldsymbol{x}_v$ to be principal directions, then it holds that

$$A\boldsymbol{x}_u = k_1 \boldsymbol{x}_u \qquad \text{and} \qquad A\boldsymbol{x}_v = k_2 \boldsymbol{x}_v,$$

where k_1, k_2 are the principal curvatures.

On the other hand, we have

$$L = \boldsymbol{x}_{uu} \cdot \boldsymbol{n} = -\boldsymbol{x}_u \cdot \boldsymbol{n}_u$$

by $\boldsymbol{n} \cdot \boldsymbol{x}_u = 0$, and similarly,

$$M = \boldsymbol{x}_{uv} \cdot \boldsymbol{n} = -\boldsymbol{x}_u \cdot \boldsymbol{n}_v = -\boldsymbol{x}_v \cdot \boldsymbol{n}_u$$
$$N = \boldsymbol{x}_{vv} \cdot \boldsymbol{n} = -\boldsymbol{x}_v \cdot \boldsymbol{n}_v.$$

From those relations we get that

$$\boldsymbol{n}_u = -k_1 \boldsymbol{x}_u, \qquad \boldsymbol{n}_v = -k_2 \boldsymbol{x}_v$$

at $\boldsymbol{P} \in \mathcal{M}$, because A is diagonalized by $\boldsymbol{x}_u$, $\boldsymbol{x}_v$ and $E = 1$, $F = 0$, $G = 1$, $L = k_1$, $M = 0$, and $N = k_2$ there.

Now, we take its deformation, a family of surfaces $\mathcal{M}_\varepsilon$ parametrized by $\boldsymbol{x}^\varepsilon = \boldsymbol{x} + \varepsilon f \boldsymbol{n}$, where f is a smooth function defined on $\mathcal{M}$ (or Ω). We wish to seek $\mathcal{M}$ such that

$$\frac{d}{d\varepsilon}A(\varepsilon)\bigg|_{\varepsilon=0} = 0 \tag{2.12}$$

for any f, where $A(\varepsilon)$ denotes the area of $\mathcal{M}_\varepsilon$. Such $\mathcal{M}$ is called the *minimal surface.*

In fact, we have

$$A(\varepsilon) = \int_{\mathcal{M}_\varepsilon} dS_\varepsilon \qquad \text{with} \qquad dS_\varepsilon = |\boldsymbol{x}_u^\varepsilon \times \boldsymbol{x}_v^\varepsilon|\, dudv.$$

However, at $\boldsymbol{P} \in \mathcal{M}$ it holds that

$$\begin{aligned} \boldsymbol{x}_u^\varepsilon &= \boldsymbol{x}_u + \varepsilon f_u \boldsymbol{n} + \varepsilon f \boldsymbol{n}_u \\ &= (1 - \varepsilon f k_2)\boldsymbol{x}_u + \varepsilon f_u \boldsymbol{n} \end{aligned}$$

and similarly,

$$\boldsymbol{x}_v^\varepsilon = (1 - \varepsilon f k_2)\boldsymbol{x}_v + \varepsilon f_v \boldsymbol{n}.$$

Because (2.11) implies

$$\boldsymbol{x}_u \times \boldsymbol{x}_v = \boldsymbol{n}, \qquad \boldsymbol{x}_v \times \boldsymbol{n} = \boldsymbol{x}_u, \qquad \boldsymbol{n} \times \boldsymbol{x}_u = \boldsymbol{x}_v,$$

we have

$$\begin{aligned} \boldsymbol{x}_u^\varepsilon \times \boldsymbol{x}_v^\varepsilon &= (1 - \varepsilon f k_1)(1 - \varepsilon f k_2)\boldsymbol{n} \\ &\quad -\varepsilon(1 - \varepsilon f k_1) f_v \boldsymbol{x}_v - \varepsilon(1 - \varepsilon f k_2) f_u \boldsymbol{x}_u \\ &= (1 - 2\varepsilon f H)\boldsymbol{n} - \varepsilon(f_v \boldsymbol{x}_v + f_u \boldsymbol{x}_u) + \varepsilon^2 f^2 K \boldsymbol{n} \\ &\quad +\varepsilon^2 f(k_1 f_v \boldsymbol{x}_v + k_2 f_u \boldsymbol{x}_u) \end{aligned}$$

by $2H = k_1 + k_2$ and $K = k_1 k_2$. This implies

$$|\boldsymbol{x}_u^\varepsilon \times \boldsymbol{x}_v^\varepsilon|^2 = 1 - 4\varepsilon f H + \varepsilon^2(4f^2H^2 + 2f^2K + f_u^2 + f_v^2) + O(\varepsilon^3)$$

and hence

$$|\boldsymbol{x}_u^\varepsilon \times \boldsymbol{x}_v^\varepsilon| = 1 - 2\varepsilon f H + \varepsilon^2\left(f^2K + \frac{f_u^2 + f_v^2}{2}\right) + O(\varepsilon^3)$$

follows.

In use of (2.11), we have

$$f_u^2 + f_v^2 = |df|^2$$

at $\boldsymbol{P} \in \mathcal{M}$ with the *one form* $df = f_u \boldsymbol{x}_u + f_v \boldsymbol{x}_v$. The value f^2, $|df|^2$ are free from parametrization and is regarded as a function on $\mathcal{M}$. Thus, we obtain

$$A(\varepsilon) = A(0) - 2\varepsilon \int_{\mathcal{M}} fHdS + \varepsilon^2 \int_{\mathcal{M}} \left(f^2 K + \frac{1}{2} |df|^2 \right) dS + O(\varepsilon^3).$$

Condition (2.12) holds for any f if and only if $H = 0$. Therefore, the minimal surface is characterized by the vanishing of mean curvature. Physically, it is realized as a *soap film*. It is stable in the direction of f if

$$2f^2 K + |df|^2 \geq 0$$

holds.

Soap bubble is realized as a critical closed surface of the area functional under the constraint that the volume of the enclosed body is a constant. In this case, the condition $H =$ constant arises.

Exercise 2.5 Show that tr $A = 2H$ and det $A = K$ for A given by (2.10).

2.3 Direct Method

2.3.1 *Vibrating String*

Let a string indicated as $[0, \pi]$ be given with the endpoints $x = 0, \pi$ fixed, and let $f(x)$ be the outer force acting on it. Suppose that the displacement $u = u(x)$ is so small as $|u'(x)| \ll |u(x)|$ holds and that the tension T is a constant. Then, what is the law to determine $u(x)$?

To answer this question, let θ and $\theta + \Delta\theta$ be the inclinations of the string at x and $x + \Delta x$, respectively. Then, the deformation induces the inner force comparable with the outer force, so that we have

$$T \sin(\theta + \Delta\theta) - T \sin\theta \approx -f(x)\Delta x.$$

On the other hand, from $u'(x) = \tan\theta$ it follows that

$$T\sin\theta = T\frac{u'(x)}{\sqrt{1+u'(x)^2}} \approx Tu'(x). \tag{2.13}$$

Similarly, we obtain

$$T\sin(\theta+\Delta\theta) \approx Tu'(x+\Delta x),$$

which implies the relation

$$T\cdot\frac{u'(x+\Delta x)-u'(x)}{\Delta x} \approx -f(x).$$

Putting $T=1$ for simplicity, we get

$$-\frac{d^2u}{dx^2} = f(x) \quad (0 \le x \le \pi) \qquad \text{with} \qquad u|_{x=0,\pi} = 0, \tag{2.14}$$

because end points of the string are fixed. Thus, the *boundary value problem* arises, when seeking $u(x)$ satisfying (2.14) for given $f(x)$.

While this derivation follows the *Newton mechanics*, the *Lagrange mechanics* asserts that the actual motion is realized as a critical state of *Lagrangian*. It is defined by *potential energy* minus *kinetic energy*. In the equilibrium state of this case, physical parameters are independent of the time variable. Thus, it is realized as a (local) minimum of the potential energy, denoted by E.

Length of the string is equal to

$$\int_0^\pi \sqrt{1+u'(x)^2}dx$$

so that the inner energy cause by the tension T is given by

$$T\int_0^\pi \left(\sqrt{1+u'(x)^2}-1\right)dx \approx \frac{T}{2}\int_0^\pi u_x^2 dx.$$

Because the outer force $f(x)dx$ works by $u(x)$ at $x\in[0,\pi]$, it induces the energy $\int_0^\pi uf dx$. Putting $T=1$, we obtain

$$E = \frac{1}{2}\int_0^\pi u_x^2 dx - \int_0^\pi uf dx. \tag{2.15}$$

Thus, E is a functional because it is determined by the function $u(x)$. Putting $E=E[u]$, we get the variational problem to minimize $E[u]$ between the function $u(x)$ in $u(0)=u(\pi)=0$. We can expect the conclusion that this formulation of Lagrange is equivalent to that of Newton described above.

Exercise 2.6 Confirm that the Euler equation (2.6) for the functional E of (2.15) is equivalent to (2.14). Confirm also that if we adopt the exact formula in (2.13), then the equation corresponding to (2.14) is equivalent to the Euler equation for the functional

$$\tilde{E} = \int_0^\pi \left(\sqrt{1 + u'(x)^2} - 1 \right) dx - \int_0^\pi u f dx.$$

2.3.2 *Minimizing Sequence*

The *Weierstrass principle* says that the continuous function defined on a compact set attains the minimum. Namely, if K is a compact topological space and $f : K \to \mathbf{R}$ is continuous, then there is $\{x_k\} \subset K$ satisfying $\lim_{k\to\infty} f(x_k) = j$, where $j = \inf_K f$ with $j = -\infty$ permitted at this stage. Then, from the compactness of K we can subtract a subsequence, still denoted by $\{x_k\}$, which converges to some $x_* \in K$. Then from the continuity we have $j = f(x_*) > -\infty$ and the minimum of f on K is attained at $x = x_*$.

Is this argument applicable to the variational problem given above ? To examine it, we have to formulate the energy

$$E[u] = \frac{1}{2} \int_0^\pi u_x^2 dx - \int_0^\pi u f dx \tag{2.16}$$

as a functional. Henceforth, $C[0, \pi]$ denotes the set of continuous functions defined on $[0, \pi]$. Let $f \in C[0, \pi]$. If the integral in the right-hand side of (2.16) is taken in the sense of Riemann, then it will be appropriate to assume $u \in C^1[0, \pi]$, where

$$C^1[0, \pi] = \{u \in C[0, \pi] \mid u' \in C[0, \pi]\}.$$

Actually, the set

$$V_1 = \left\{ u \in C^1[0, \pi] \mid u(0) = u(\pi) = 0 \right\}$$

is regarded as a vector space provided with the zero element 0 given by $u(x) \equiv 0$ and with the additive and scalar multiplication operations

$$(u + v)(x) = u(x) + v(x) \qquad \text{and} \qquad (cu)(x) = cu(x)$$

for $u, v \in V_1$ and $c \in \mathbf{R}$, Furthermore,

$$\|u\| = \left(\int_0^\pi u'(x)^2 dx \right)^{1/2} \tag{2.17}$$

provides a *norm* there. This means that $\|u\| \geq 0$ with the equality if and only if $u = 0$, $\|cu\| = |c| \, \|u\|$, and $\|u + v\| \leq \|u\| + \|v\|$, where $c \in \mathbf{R}$ and $u, v \in V_1$. Namely, V_1 forms a *normed space.* This induces the metric to V_1 by

$$\text{dist}(u, v) = \|u - v\|$$

and in particular, V_1 is a topological space. That is, $U \subset V_1$ is open if and only if any $u \in U$ takes $r > 0$ such that $B(u, r) \subset U$, where

$$B(u, r) = \{v \in V_1 \mid \|v - u\| < r\}.$$

This means that $\{v_j\} \subset V_1$ converges to $v \in V_1$ if and only if

$$\lim_{j \to \infty} \|v_j - v\| = 0$$

holds. Under this topology given to V_1, it is not difficult to see that E is a continuous mapping from V_1 to $\mathbf{R}$. Then, is V_1 compact ? The answer is no !

We are seeking the minimum of E on V_1. As the first step we need to know that this functional is bounded from below. In fact, we have from the Schwarz inequality that

$$|u(x)| = \left| \int_0^x u'(y) dy \right| \leq \pi^{1/2} \cdot \left(\int_0^\pi u'(y)^2 dy \right)^{1/2} = \pi^{1/2} \|u\| \tag{2.18}$$

for any $x \in [0, \pi]$. This implies

$$E[u] \geq \frac{1}{2} \|u\|^2 - C \|u\| \tag{2.19}$$

for $C = \pi^{1/2} \int_0^\pi |f(x)| \, dx$. Because the right-hand side is estimated from below by $-C^2$, we obtain

$$\inf_{u \in V_1} E[u] \geq -C^2.$$

This situation allows us to take the *minimizing sequence* $\{u_j\} \subset V_1$ satisfying

$$\lim_{j\to\infty} E[u_j] = \inf_{u\in V_1} E[u] > -\infty.$$

Again by (2.19), it holds that

$$\sup_j \|u_j\| < +\infty,$$

in which case we say that the minimizing sequence $\{u_j\}$ is *bounded* in V_1. Thus, we get the key question. Does any bounded sequence have a converging subsequence ? This is actually the property of compactness. It is true in the finite dimensional Euclidean space. However, here we have two obstructions. That is, the space V_1 is neither *complete*, nor of finite dimension.

Exercise 2.7 Show that $\|\cdot\|$ in (2.17) provides a norm to V_1.

Exercise 2.8 Confirm that a normed space L with the norm $\|\cdot\|$ is a metric space and show that

$$\big|\|u\| - \|v\|\big| \leq \|u - v\|$$

holds for $u, v \in L$. Then, observe that the proof of the continuity of $E : V_1 \to \mathbf{R}$ is reduced to that of

$$v \in V_1 \quad \mapsto \quad \int_0^\pi v f dx.$$

Confirm, finally, that it follows from (2.18).

2.3.3 *Sobolev Spaces*

Remember that a sequence $\{u_j\}$ in a metric space V with the distance dist(,) is said to be a *Cauchy sequence* if it satisfies $\text{dist}(u_j, u_k) \to 0$ as $j, k \to 0$. Then, the metric space (V, dist) is said to be *complete* if any Cauchy sequence converges.

In this sense, $V_1 = \{u \in C^1[0,\pi] \mid u(0) = u(\pi) = 0\}$ provided with the norm

$$\|u\| = \left(\int_0^\pi u'(x)^2 dx\right)^{1/2}$$

is not complete. We have to change the notions of integration and differentiation.

First, integration must be changed from the sense of Riemann to that of Lebesgue. Thus we introduce the space $L^2(0,\pi)$. Namely, $g \in L^2(0,\pi)$ means that $g = g(x)$ is a *measurable function* on $(0,\pi)$ satisfying

$$\|g\|_2 = \left(\int_0^\pi g(x)^2 dx\right)^{1/2} < +\infty,$$

where the integration is taken in the sense of Lebesgue. Such g is said to be *square integrable.* Two measurable functions f, g are identified in $L^2(0,\pi)$ if they are equal to each other *almost everywhere.* Then, $L^2(0,\pi)$ becomes a Banach space under the norm $\|\cdot\|_2$.

On the other hand, differentiation of $u \in L^2(0,\pi)$ is taken in the sense of *distribution.* We shall write u_x for this notion of derivative of u. It has a vast background, but in this case the following notations are enough. Namely, given $v \in L^2(0,\pi)$, we say $v_x \in L^2(0,\pi)$ if there is $w \in L^2(0,\pi)$ such that

$$\int_0^\pi w\varphi dx = -\int_0^\pi v\varphi' dx$$

for any $\varphi \in C_c^1(0,\pi)$, Here, $C_c^1(0,\pi)$ is the set of $\varphi \in C^1(0,\pi)$ satisfying

$$\text{supp } \varphi = \overline{\{x \in (0,\pi) \mid \varphi(x) \neq 0\}} \subset (0,\pi).$$

Henceforth, supp φ is called the *support* of φ. Such w is (if it exists) unique as an element in $L^2(0,1)$, and we write as $w = v_x$. It is called the *distributional derivative* of v. If $v \in C^1[0,\pi]$, and v' denotes the usual derivative of v, then v_x is equal to v' as an element in $L^2(0,\pi)$. We set

$$H^1(0,\pi) = \left\{v \in L^2(0,\pi) \mid v_x \in L^2(0,\pi)\right\}.$$

Two functions equal to each other almost everywhere are identified in $H^1(0,\pi)$. Under this agreement it is shown that

$$H^1(0,\pi) \subset C[0,\pi]. \tag{2.20}$$

Namely, any element $u \in H^1(0,\pi)$ has a representation $\tilde{u} \in C[0,\pi]$, so that $u = \tilde{u}$ almost everywhere. Relation (2.20) is actually the most primitive case of *Sobolev's imbedding theorem,* but because of this, the condition $v(0) = v(\pi) = 0$ has a meaning for $v \in H^1(0,\pi)$. Then, we take $V = H_0^1(0,\pi)$, where

$$H_0^1(0,\pi) = \left\{ v \in H^1(0,\pi) \mid v(0) = v(\pi) = 0 \right\}.$$

Relation (2.20) is a consequence of

$$\varphi(x) - \varphi(y) = \int_x^y \varphi'(s)ds,$$

valid for $\varphi \in C^1[0,\pi]$. From this equality we have

$$|\varphi(x) - \varphi(y)| \leq \left| \int_x^y \varphi'(s)ds \right| \leq |x-y|^{1/2} \|\varphi'\|_2 .$$

It is extended for $u \in H^1(0,\pi)$:

$$|u(x) - u(y)| \leq |x-y|^{1/2} \|u_x\|_2 , \tag{2.21}$$

because $C^1[0,\pi]$ is shown to be dense in $H^1(0,\pi)$. Namely, any $u \in H^1(0,\pi)$ admits $\{\varphi_k\} \subset C^1[0,\pi]$ such that

$$\lim_{k\to\infty} \{\|\varphi_k - u\|_2 + \|\varphi_k' - u_x\|_2\} = 0.$$

Once inequality (2.21) is justified for $u \in H^1(0,\pi)$, then it implies (2.20).

The norm $\|\cdot\|_2$ provides to $L^2(0,\pi)$ with the complete metric. This fact is proven by the convergence theorems on Lebesgue integrals.

A complete normed space is called the *Banach space*.

Exercise 2.9 Confirm that any converging sequence is a Cauchy sequence in the metric space.

Exercise 2.10 Show that the normed space $(L, \|\cdot\|)$ is a Banach space if and only if any *absolutely converging series* converges. This means that if $\{u_k\} \subset L$ satisfies

$$\sum_{k=1}^{\infty} \|u_k\| < +\infty,$$

then there exists $u \in L$ such that

$$\lim_{n\to\infty} \left\| \sum_{k=1}^{n} u_k - u \right\| = 0.$$

Note that any Cauchy sequence in a metric space converges if it has a converging subsequence.

Exercise 2.11 Prove that $H^1(0,\pi)$ is a Banach space under the norm

$$\|u\| = \sqrt{\|u\|_2^2 + \|u_x\|_2^2},$$

making use of the fact that $L^2(0,\pi)$ is so. Show also that $H_0^1(0,\pi)$ is a Banach space.

Exercise 2.12 Prove that the classical derivative v' is identified with the distributional derivative v_x if $v \in C^1[0,\pi]$.

Exercise 2.13 Justify (2.21) for $v \in H^1(0,1)$ in use of the fact that $C^1[0,1]$ is dense in $H^1(0,1)$.

2.3.4 *Lower Semi-Continuity*

The space $V = H_0^1(0,\pi)$ denotes the set of square integrable functions on $(0,\pi)$ with their distributional derivatives. If $u, v \in V$, then their *inner product*

$$(u,v) = \int_0^\pi u_x(x)v_x(x)dx \tag{2.22}$$

is well-defined by the Schwarz inequality and satisfies the axioms that $(u,v) = (v,u)$, that $(\alpha u + \beta v, w) = \alpha(u,w) + \beta(v,w)$, that $(u,u) = \|u\|^2 \geq 0$ with the equality if and only if $u = 0$ in V. Importantly, V is complete with respect to the metric induced by the norm $\|u\| = (u,u)^{1/2}$ so that $\{u_j\} \subset X$ and $\|u_j - u_k\| \to 0$ as $j,k \to +\infty$ imply the existence of $u \in V$ such that $\|u_j - u\| \to 0$ as $j \to +\infty$. Those properties are summarized that $V = H_0^1(0,\pi)$ forms a *Hilbert space* with respect to the inner product $(\ ,\ )$ defined by (2.22).

Let us come back to the problem in §2.3.1. Now, we formulate it as to minimize E on $V = H_0^1(0,\pi)$, where

$$E[u] = \frac{1}{2}\int_0^\pi u_x^2 dx - \int_0^\pi uf dx \tag{2.23}$$

for given $f \in C[0,\pi]$. In use of (2.20), we can extend (2.18) for $u \in H_0^1(0,\pi)$:

$$\max_{x \in [0,1]} |u(x)| \leq \pi^{1/2} \|u\| \qquad (u \in H_0^1(0,1)).$$

This implies (2.19) for $u \in H_0^1(0,\pi)$ and therefore, E is bounded from below and the minimizing sequence $\{u_j\} \subset V$ is bounded in V. The latter means the boundedness of $\{\|u_j\|\}$ so that we have

$$\lim_{j\to\infty} E[u_j] = \inf_{u\in V} E[u] > -\infty, \qquad \sup_j \|u_j\| < +\infty.$$

Here, we apply a theorem of abstract analysis that any bounded sequence $\{u_j\}$ in a Hilbert space V admits a subsequence, converging *weakly*. Here, this subsequence is denoted by the same symbol for simplicity. This means the existence of $u \in V$ such that $(u_j, v) \to (u, v)$ for any $v \in V$, where $(\ ,\ )$ denotes the inner product in V. In this case, from the abstract Schwarz inequality

$$|(u_j, v)| \leq \|u_j\| \cdot \|v\|$$

proven in 3.1.2, we obtain

$$|(u, v)| \leq \liminf_{j\to\infty} \|u_j\| \cdot \|v\|.$$

Then, putting $v = u$, we get that

$$\|u\| \leq \liminf_{j\to\infty} \|u_j\|.$$

This property indicates the *lower semi-continuity* of the norm in the Hilbert space with respect to the weak convergence.

Here, we make use of the following.

Theorem 2.3 *The embedding (2.20) is compact.*

Proof. The statement means that if $\{u_j\} \subset H_0^1(0,\pi)$ converges weakly, then it converges uniformly on $[0,\pi]$. On the other hand, the uniformly bounded principle assures that any weakly converging sequence is bounded in the Hilbert space, we may show that any bounded $\{u_j\} \subset H_0^1(0,\pi)$ admits a subsequence, denoted by the same symbol, converging uniformly on $[0,\pi]$.

To show this, we note that inequality (2.21) implies that if $\sup_j \|u_j\| < +\infty$, then $\{u_j\} \subset C[0,1]$ is *uniformly bounded* and *equi-continuous.* Then, *Ascoli-Arzela's theorem* assures the conclusion. □

Coming back to the problem, we have the minimizing sequence $\{u_j\}$ of E on $V = H_0^1(0,\pi)$ that converges weakly in V and uniformly on $[0,\pi]$ to some $u \in H_0^1(0,\pi) \subset C[0,\pi]$. Those facts imply

$$\liminf_{j\to\infty} \int_0^\pi u_{jx}^2 dx \geq \int_0^\pi u_x^2 dx$$

and

$$\lim_{j\to\infty} \int_0^\pi u_j f dx = \int_0^\pi u f dx,$$

respectively, and hence

$$\lim_{j\to\infty} E[u_j] = \liminf_{j\to\infty} E[u_j] \geq E[u]$$

follows. However, $\{u_j\}$ is a minimizing sequence and we have $E[u_j] \to \inf_V E$. Thus, $E[u] = \inf_V E$ follows from $u \in V$. This means that it attains the minimum of E on V. We have proven the existence of the solution to the variational problem of minimizing E on V.

In other words, $E[v] \geq E[u]$ holds for any $v \in V$ and therefore, we obtain the Euler equation for this u to solve. This is derived from

$$\left. \frac{d}{ds} E[u+sv] \right|_{s=0} = 0,$$

where $v \in H_0^1(0,\pi)$ is arbitrary. This means that

$$\int_0^\pi u_x v_x dx = \int_0^\pi f v dx \tag{2.24}$$

by (2.23), which implies

$$-\frac{d}{dx}(u_x) = f$$

in the sense of distribution. We get from $f \in C[0,\pi]$ that this derivative can be taken in the classical sense and therefore, u_x is continuously. Then, it holds that $u \in C^2[0,\pi]$, and (2.14) follows in the classical sense. This final stage to derive $u \in C^2[0,\pi]$ from $u \in V$ and the Euler equation (2.23) is called to establish *regularity* of *weak solution*. It is discussed in the later chapters more systematically.

In contrast to the weak convergence, usual convergence in norm is referred to as the *strong convergence*. In §3.1.2, we shall show that $\{u_j\}$ actually converges strongly in V and also the uniqueness of the minimizer of E on V.

Euclidean space $\mathbf{R}^n$ is provided with the standard Hilbert space structure and there, two notions of convergence, strong and weak, are equivalent. Therefore, in $\mathbf{R}^n$, the fact that any bounded sequence in Hilbert space admits a subsequence converging weakly indicates the *theorem of Bolzano - Weierstrass.*

For the higher dimensional case with $[0,1]$ replaced by a bounded domain Ω in $\mathbf{R}^n$, the embedding $H_0^1(\Omega) \subset C(\overline{\Omega})$ does not hold any more, and the boundary value of $u \in H_0^1(\Omega)$ must be taken in the sense of *trace*. Even so, some properties on $\partial\Omega$ are necessary to carry out the task, although details are not described here.

Exercise 2.14 Confirm that any bounded sequence in $H^1(0,\pi)$ is uniformly bounded and equi-continuous on $[0,\pi]$.

Exercise 2.15 Show that if $g \in L^2(0,\pi)$ satisfies $f = g_x \in C[0,\pi]$ in the distributional sense, then it is identified as a continuously differentiable function with the derivative f.

Exercise 2.16 As will be shown in §3.1.2, the abstract Schwarz inequality

$$|(u,v)| \leq \|u\| \cdot \|v\|$$

holds in Hilbert space. In use of this fact, show that any strongly convergent sequence converges weakly there. Show, more precisely, that $\{v_j\}$ converges strongly to v if and only if this convergence is weak and also $\|v_j\| \to \|v\|$ holds.

Exercise 2.17 First, show that $\mathbf{R}^n$ becomes a Hilbert space under the standard inner product

$$(x,y) = \sum_{i=1}^{n} x_i y_i$$

for $x = (x_1, x_2, \cdots, x_n)$ and $y = (y_1, y_2, \cdots, y_n)$. Then, in use of the uniform bounded principle and the theorem of Bolzano-Weierstrass, show that weak convergence is equivalent to strong convergence in $\mathbf{R}^n$.

Exercise 2.18 The Hilbert space H is said to be *separable* if it has a countable subset H_0 satisfying $\overline{H_0} = H$. Show that if H is a separable Hilbert space, any bounded subsequence has a weakly converging sequence in use of the diagonal argument, completeness of $\mathbf{R}$, and *Hahn-Banach's theorem*, which says that any bounded linear operator $T_0 : H_0 \to \mathbf{R}$ admits a bounded linear extension $T : H \to \mathbf{R}$ with the operator norm preserved.

2.4 Numerical Schemes

2.4.1 *Finite Difference Method*

Numerical analysis is the mathematical study of numerical schemes. One aspect is the theory and the other the practice. Those are combined and form interfaces between mathematics, applied and theoretical sciences, and technology. Thus, it is concerned with the following items.

1. Derivation of approximate problem (*scheme*) and establishing its unique solvability.

2. Proposal to actual computation (*algorithm*) and examining its practicality.

3. Mathematical study of the scheme, such as *stability*, *convergence*, and *error analysis* to the approximate solution.

Remember the two ways of derivation of (2.14) describing the balance of string, that is, the Newton mechanics and the Lagrange one. Finite difference method is based on the former. We take the integer N sufficiently large, and put $h = \pi/N$ as the *mesh size parameter*. Then, we take the approximation that

$$u'(x) \sim D_h u(x) = \frac{1}{h}\{u(x+h) - u(x)\}$$

or

$$u'(x) \sim \overline{D}_h u(x) = \frac{1}{h}\{u(x) - u(x-h)\}.$$

Then, we take $\overline{D}_h D_h u(x)$ as an approximation of $u''(x)$. Letting $v(x) = \frac{1}{h}\{u(x+h) - u(x)\}$, we have

$$\begin{aligned} u''(x) \sim \overline{D}_h D_h u(x) &= \overline{D}_h v(x) \\ &= \frac{1}{h}\{v(x) - v(x-h)\} \\ &= \frac{1}{h^2}\{u(x+h) + u(x-h) - 2u(x)\}. \end{aligned}$$

Thus, problem (2.14) is replaced by

$$-\frac{1}{h^2}\{u(x+h)+u(x-h)-2u(x)\}=f(x) \tag{2.25}$$

for $x = nh$ with $n = 1, \cdots, N-1$ and $u(0) = u(\pi) = 0$. We regard it as the approximate problem to (2.14). Replacing $u''(x)$ by

$$\frac{1}{h^2}\{u(x+h)+u(x-h)-2u(x)\}$$

is called the *three-point difference*, because the values of u at x, $x-h$, and $x+h$ are used to approximate $u''(x)$.

Putting $u_n = u(nh)$ and $f_n = f(nh)$, we can write (2.25) as

$$-\frac{1}{h^2}(u_{n+1}+u_{n-1}-2u_n)=f_n \qquad (n=1,\cdots,N-1)$$

with $u_0 = u_N = 0$. Known and unknown values are $\{f_n\}_{n=1}^{N-1}$ and $\{u_n\}_{n=1}^{N-1}$, respectively. Thus we get a numerical scheme, of which detailed study is not treated in this monograph.

Exercise 2.19 Prove that $\overline{D}_h D_h = D_h \overline{D}_h$. Then, evaluate it, D_h^2, and $\overline{D}_h^2$ as the approximation of $\frac{d^2}{dx^2}$.

2.4.2 *Finite Element Method*

In the Lagrange mechanics, problem (2.14) is reduced to find

$$u \in V \qquad \text{such that} \qquad \int_0^\pi u_x v_x dx = \int_0^\pi v f dx \quad \text{for any} \quad v \in V, \tag{2.26}$$

where $V = H_0^1(0,\pi)$. *Finite element method* is a discretization of this formulation. More concretely, we take a large integer N and the uniform division of $[0,\pi]$, denoted by

$$\Delta : x_0 = 0 < x_1 < \cdots < x_{N-1} < x_N = \pi$$

with the mesh size parameter $h = \pi/N$, so that $x_n = nh$ for $n = 0, 1, \cdots, N$. Then, the underlying space V in (2.26) is replaced by the finite dimensional vector space,

$$V_h = \{v \in C[0,\pi] \mid v \text{ is piecewise linear and } v(0) = v(\pi) = 0\}.$$

Then, it holds that $V_h \subset V$ and dim $V_h = N-1$. Actually, a basis of V_h is provided with $\{e_n\}_{n=1}^{N-1}$, where $e_n = e_n(x) \in V_h$ and

$$e_n(x) = \begin{cases} 1 & (x = nh) \\ 0 & (x = ih, i \neq n), \end{cases}$$

because each element $v \in V_h$ is determined by its values on the nodal points, $x = x_n$ $(n = 1, 2, \cdots, N-1)$.

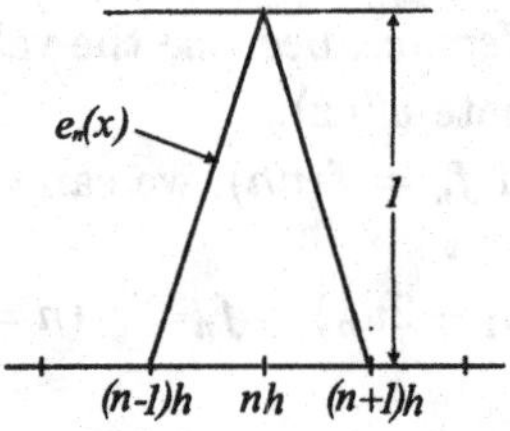

Fig. 2.3

We can reproduce the argument in §2.3.4 with V replacing V_h and get $u_h \in V_h$ that minimizes E on V_h. Then, this u_h satisfies

$$u_h \in V \qquad \text{such that} \qquad \int_0^\pi u_{hx} v_{hx} dx = \int_0^\pi v_h f dx \quad \text{for any} \quad v_h \in V_h. \tag{2.27}$$

Writing $u_h(x) = \sum_{i=1}^{N-1} \xi_n e_n(x)$ in use of the unknown constants $\xi_n \in \mathbf{R}$ for $n = 1, \cdots, N-1$, we see that (2.27) is equivalent to

$$\sum_{n=1}^{N-1} \xi_n \int_0^\pi e_{nx} e_{mx} dx = \int_0^\pi e_m f dx$$

for $m = 1, \cdots, N-1$. This means that

$$A\xi = F$$

for $A = (a_{nm})$ with

$$a_{nm} = \int_0^\pi e_{nx}(x) e_{mx}(x) dx, \tag{2.28}$$

$$\xi = \begin{pmatrix} \xi_1 \\ \cdot \\ \cdot \\ \cdot \\ \xi_{N-1} \end{pmatrix}, \quad \text{and} \quad F = \begin{pmatrix} \int_0^\pi e_1 f dx \\ \cdot \\ \cdot \\ \cdot \\ \int_0^\pi e_{N-1} f dx \end{pmatrix}.$$

This A is a triple diagonal matrix and its inverse is not hard to compute numerically.

Exercise 2.20 Prove that $V_h \subset V$.

Exercise 2.21 Confirm that the matrix A given by (2.28) is a triple diagonal.

Chapter 3

Infinite Dimensional Analysis

In the previous chapter, it was suggested that the infinite dimensional analysis is necessary to make the calculus of variation in a rigorous way. The key word is the completeness and this chapter is devoted to it. Thus, we shall describe the theory of Hilbert spaces, Fourier series, and eigenvalue problems.

3.1 Hilbert Space

3.1.1 *Bounded Linear Operators*

Remember that norm induces metric in the vector space, which is said to be a Banach space if it is complete with respect to that metric. Let $(L, \|\cdot\|)$ be a Banach space and $T : L \to \mathbf{R}$ be a linear mapping. Sometimes T is referred to as an *operator*. It is said to be *bounded* if there is a constant $M > 0$ satisfying

$$|T(f)| \le M \|f\| \tag{3.1}$$

for any $f \in L$.

This is equivalent to saying that T is continuous at any or some element in L because of its linearity. In fact, if (3.1) holds and $f_n \to f$ in L, then it follows that

$$|T(f_n) - T(f)| = |T(f_n - f)| \le M \|f_n - f\| \to 0.$$

Therefore, $T(f_n) \to T(f)$ follows. Conversely, if T is continuous at $f = 0$, then it is bounded. In fact, if this is not the case, there is a sequence

$\{f_n\} \subset L$ such that

$$|T(f_n)| > n \|f_n\|$$

for $n = 1, 2, \cdots$. Because T is linear, it holds that $T(0) = 0$ and hence $f_n \neq 0$. Therefore, $g_n = f_n / (n \|f_n\|) \in L$ is well-defined. However, then we get

$$\|g_n\| = \frac{1}{n} \qquad \text{and} \qquad |T(g_n)| > 1$$

and hence $g_n \to 0$ and $T(g_n) \not\to 0$ hold as $n \to \infty$, a contradiction.

If $T : L \to \mathbf{R}$ is a bounded linear operator, the infimum of $M > 0$ satisfying (3.1) is called the *operator norm* of T and is written as $\|T\|$. Because (3.1) is equivalent to

$$\frac{|T(f)|}{\|f\|} \leq M \qquad (f \in L \setminus \{0\}), \tag{3.2}$$

it holds that

$$\|T\| = \sup\left\{\frac{|T(f)|}{\|f\|} \mid f \in L \setminus \{0\}\right\}.$$

In particular, we have

$$|T(f)| \leq \|T\| \cdot \|f\| \qquad (f \in L)$$

and

$$\|T\| = \sup\left\{\left|T\left(\frac{f}{\|f\|}\right)\right| \mid f \in L \setminus \{0\}\right\} = \sup\left\{|T(g)| \mid g \in L, \|g\| = 1\right\}.$$

In those notions, the target space of the linear operator T may not be $\mathbf{R}$, and the case $T : L \to K$ is admitted, where $(K, |\cdot|)$ stands for a Banach space.

Exercise 3.1 Confirm for the linear operator $T : L \to \mathbf{R}$ that it is continuous at some $f = f_0 \in L$ then it is continuous at $f = 0$. Confirm also that (3.1) for any $f \in L$ is equivalent to (3.2).

Exercise 3.2 Given a Banach space $(L, \|\cdot\|)$, and introduce its *dual space* by

$$L' = \{T : L \to \mathbf{R} \mid \text{bounded linear operators }\}.$$

It is a vector space under the operations

$$(T+S)(f) = T(f) + S(f) \quad \text{and} \quad (cT)(f) = cT(f)$$

for $T, S \in L'$, $f \in L$, and $c \in \mathbf{R}$. Confirm that L' becomes a Banach space under the operator norm.

3.1.2 *Representation Theorem of Riesz*

Remember that inner product of the vector space L is the mapping $(\ ,\ ) : L \times L \to \mathbf{R}$ satisfying the axioms that $(u, v) = (v, u)$, $(\alpha u + \beta v, w) = \alpha(u, w) + \beta(v, w)$, and $(v, v) \geq 0$ with the equality if and only if $v = 0$, where $u, v, w \in L$ and $\alpha, \beta \in \mathbf{R}$. A vector space provided with the inner product $(\ ,\ )$ is called the *pre-Hilbert space*. First, let us confirm the following.

Theorem 3.1 *A pre-Hilbert space is a normed space by* $\|f\| = \sqrt{(f,f)}$. *The abstract Schwarz inequality*

$$|(f,g)| \leq \|f\| \cdot \|g\| \qquad (f, g \in L) \tag{3.3}$$

also holds.

Proof. It is obvious that $\|f\| \geq 0$ with the equality if and only if $f = 0$ and $\|cf\| = |c| \, \|f\|$, where $f \in L$ and $c \in \mathbf{R}$. The Schwarz inequality implies that

$$\|f+g\| \leq \|f\| + \|g\| \qquad (f, g \in L)$$

as

$$\begin{aligned} \|f+g\|^2 &= (f+g, f+g) = \|f\|^2 + 2(f,g) + \|g\|^2 \\ &\leq \|f\| + 2\|f\| \cdot \|g\| + \|g\|^2 = (\|f\| + \|g\|)^2 . \end{aligned}$$

To prove (3.3), we may assume that $f \neq 0$. In this case, we can put $\alpha = (f,g)/\|f\|^2$ in

$$0 \leq (\alpha f - g, \alpha f - g) = |\alpha|^2 \|f\|^2 - 2\alpha(f,g) + \|g\|^2 .$$

This implies

$$|(f,g)|^2 / \|f\|^2 - 2\,|(f,g)|^2 / \|f\|^2 + \|g\|^2 \geq 0,$$

or

$$|(f,g)|^2 \leq \|f\|^2 \cdot \|g\|^2 ,$$

and hence (3.3) follows. □

A pre-Hilbert space is called a *Hilbert space* if it is complete with respect to the metric induced from the inner product. Henceforth, L denotes a Hilbert space with the inner product $(\ ,\ )$. If $g \in L$ is fixed, then

$$T(f) = (f, g) \qquad (f \in L)$$

defines a linear mapping from L to $\mathbf{R}$. Because of (3.3), we have

$$|T(f)| \leq \|g\| \cdot \|f\| \qquad (f \in L),$$

which means that $T : L \to \mathbf{R}$ is a bounded operator satisfying $\|T\| \leq \|g\|$. On the other hand we have

$$T\left(\frac{g}{\|g\|}\right) = \|g\|$$

if $g \neq 0$. This implies that $\|T\| = \|g\|$.

The *representation theorem of Riesz* says that any bounded linear operator is expressed like this. In use of the notion of dual space, it is written as $L' \cong L$. This is done by solving an abstract variational problem. The following theorem provides an abstract version of the argument of §2.3.4. Here, the proof is given without using weak convergence.

Theorem 3.2 *Let L be a Hilbert space with the inner product $(\ ,\ )$ and the norm $\|\cdot\|$, and $T : L \to \mathbf{R}$ be a bounded linear operator. Then, the functional*

$$J(v) = \frac{1}{2}\|v\|^2 - T(v) \qquad (v \in L)$$

attains the minimum with a unique minimizer.

Proof. In use of $|T(v)| \leq \|T\| \cdot \|v\|$ we have

$$J(v) \geq \frac{1}{2}\|v\|^2 - \|T\| \cdot \|v\| \geq -\frac{1}{2}\|T\|^2$$

for any $v \in L$. This implies

$$j = \inf_L J > -\infty$$

and there exists $\{v_n\} \subset L$ satisfying $J(v_n) \to j$. In use of the *parallelogram law* that

$$\left\|\frac{u+v}{2}\right\|^2 + \left\|\frac{u-v}{2}\right\|^2 = \frac{\|u\|^2 + \|v\|^2}{2} \qquad (u, v \in L), \tag{3.4}$$

we have

$$\begin{aligned} &2J\left(\frac{v_n + v_m}{2}\right) + \left\|\frac{v_n - v_m}{2}\right\|^2 \\ &= \left\|\frac{v_n + v_m}{2}\right\|^2 + \left\|\frac{v_n - v_m}{2}\right\|^2 - 2T\left(\frac{v_n + v_m}{2}\right) \\ &= \frac{\|v_n\|^2 + \|v_m\|^2}{2} - T(v_n) - T(v_m) = J(v_n) + J(v_m) \to 2j \end{aligned}$$

as $n, m \to \infty$. On the other hand, the first term of the left-hand side is always greater than or equal to $2j$, and hence $\|v_n - v_m\| \to 0$ follows as $n, m \to \infty$.

Because L is complete with respect to $\|\cdot\|$, there is $u \in L$ such that $\|v_n - u\| \to 0$ as $n \to \infty$. This implies

$$\|v_n\| \to \|u\| \qquad \text{and} \qquad J(v_n) \to J(u) = j$$

in turn. Thus, $j = \inf_L J$ is attained by $u \in L$. The uniqueness of such a u is obtained by the proof of the following theorem. □

We are now able to prove the representation theorem of Riesz indicated as follows.

Theorem 3.3 *If L is a Hilbert space provided with the inner product $(\ ,\ )$ and the norm $\|\ \|$, then any bounded linear operator $T : L \to \mathbf{R}$ admits unique $g \in L$ such that $T(f) = (f, g)$ for any $f \in L$. Furthermore, it holds that $\|T\| = \|g\|$.*

Proof. We have shown that $\|T\| = \|g\|$ holds if such g exists. If $g_1, g_2 \in L$ satisfies the condition, then $(f, g_1 - g_2) = 0$ for any $f \in L$. Putting $f = g_1 - g_2$, we get $g_1 = g_2$. Thus, we only have to show the existence of such g.

Given $T \in L'$, we have a minimizer $g \in L$ of the functional

$$J(v) = \frac{1}{2}\|v\|^2 - T(v) \qquad (v \in L)$$

from the previous theorem. This means $J(g) \leq J(v)$ for any $v \in L$ so that

$$s \mapsto J(g+sf)$$

attains the minimum at $s = 0$, where $f \in L$ is an arbitrary element. However, we have

$$\begin{aligned} J(g+sf) &= \frac{1}{2}\|g+sf\|^2 - T(g+sf) \\ &= \frac{1}{2}s^2\|f\|^2 + s(f,g) + \frac{1}{2}\|g\|^2 - T(g) - sT(f), \end{aligned}$$

and hence

$$\left.\frac{d}{ds}J(g+sf)\right|_{s=0} = (f,g) - T(f)$$

follows. This means $T(f) = (f,g)$ and the proof is complete. □

So far, the vector space which we treat is over **R**, the real numbers. Hilbert space over **C**, the complex numbers, is defined similarly when the underlying vector space is over **C**. In this case the axiom of the symmetry of inner product is changed from $(f,g) = (g,f)$ to $(f,g) = \overline{(g,f)}$, where $\overline{z} = x - \imath y$ denotes the complex conjugate of $z = x + \imath y$ for $x, y \in \mathbf{R}$. Hilbert space over **C** arises in *quantum mechanics*, but we are mostly concentrated on the Hilbert space over **R** in the following.

Exercise 3.3 Prove that the parallelogram law (3.4) holds in Hilbert space.

Exercise 3.4 Give an alternative proof of Theorem 3.2 based on the weak convergence.

3.1.3 *Complete Ortho-Normal Systems*

Let L be a Hilbert space with the inner product $(\ ,\ )$ and the norm $\|\cdot\|$. A family $\{\varphi_i\} \subset L$ is said to be *ortho-normal* if

$$(\varphi_i, \varphi_j) = \delta_{ij} = \begin{cases} 1 & (i = j) \\ 0 & (i \neq j) \end{cases}$$

holds. Because they are linearly independent, if that family is composed of infinitely many elements, then the dimension of L is infinite.

Given a family of ortho-normal system $\{\varphi_i\}_{i=1}^{\infty}$, let L_n be the linear subspace of L spanned by $\{\varphi_i\}_{i=1}^{n}$, where $n = 1, 2, \cdots$. Then, we consider the problem of *least square approximation*, that is, to seek a minimizer of

$$\inf_{g \in L_n} \|g - f\|^2, \tag{3.5}$$

where $f \in L$ is a given element.

For this problem to solve, we put

$$g = \sum_{i=1}^{n} \alpha_i \varphi_i \in L_n$$

with the undetermined coefficients $\{\alpha_i\} \subset \mathbf{R}$. Letting $\beta_i = (f, \varphi_i)$, we set

$$\hat{f} = f - \sum_{i=1}^{n} \beta_i \varphi_i.$$

Then, it holds that

$$(\hat{f}, \varphi_j) = (f, \varphi_j) - \sum_{i=1}^{n} \beta_i (\varphi_i, \varphi_j) = (f, \varphi_j) - \beta_j = 0$$

because $\{\varphi_i\}$ is ortho-normal, and hence

$$\begin{aligned}
\|f - g\|^2 &= \left\| \hat{f} - \sum_{i=1}^{n} (\alpha_i - \beta_i) \varphi_i \right\|^2 \\
&= \left\| \hat{f} \right\|^2 - 2 \sum_{i=1}^{n} (\alpha_i - \beta_i)(\hat{f}, \varphi_i) + \sum_{i,j=1}^{n} (\alpha_i - \beta_i)(\alpha_j - \beta_j)(\varphi_i, \varphi_j) \\
&= \left\| \hat{f} \right\|^2 + \sum_{i=1}^{n} (\alpha_i - \beta_i)^2
\end{aligned}$$

follows. Therefore, $J(g) = \|g - f\|^2$ defined for $g \in L_n$ attains the minimum

$$j_n(f) = \left\| \hat{f} \right\|^2 = \left\| f - \sum_{i=1}^{n} (f, \varphi_i) \varphi_i \right\|^2 \geq 0$$

if and only if $\alpha_i = \beta_i$ for $1 \leq i \leq n$, or equivalently,

$$g = \sum_{i=1}^{n} (f, \varphi_i) \varphi_i.$$

Relation $\lim_{n\to\infty} j_n(f) = 0$ means that

$$\lim_{n\to\infty} \left\| f - \sum_{i=1}^{n} (f, \varphi_i)\varphi_i \right\| = 0. \tag{3.6}$$

If this is the case, f is in the closure of the *linear hull* of $\{\varphi_i\}_{i=1}^{\infty}$, the set of linear combinations of $\{\varphi_i\}_{i=1}^{\infty}$. Henceforth, the linear hull of $\{\varphi_i\}_{i=1}^{\infty}$ is denoted by L_0:

$$L_0 = \left\{ \sum_{i=1}^{\infty} \alpha_i \varphi_i \mid \alpha_i \in \mathbf{R},\ \alpha_i = 0 \text{ except for a finite } i \right\}.$$

Thus, if (3.6) holds for any $f \in L$, then it holds that $\overline{L_0} = L$. In this case we say that $\{\varphi_i\}_{i=1}^{\infty}$ forms a *complete ortho-normal system* of L. This is actually the extension of the notion of *ortho-normal basis* in finite-dimensional vector spaces.

Let $\{\varphi_i\}_{i=1}^{\infty}$ be a complete ortho-normal system of L, and L_{00} be the set of linear combinations of $\{\varphi_i\}_{i=1}^{\infty}$ with coefficients in $\mathbf{Q}$, the set of rational numbers:

$$L_{00} = \left\{ \sum_{i=1}^{\infty} \alpha_i \varphi_i \mid \alpha_i \in \mathbf{Q},\ \alpha_i = 0 \text{ except for a finite } i \right\}.$$

Then, it holds that $L_{00} \subset \overline{L_0}$ and $L_0 \subset \overline{L_{00}}$. Thus, we obtain $\overline{L_{00}} = L$.

A Hilbert space L is said to be *separable* if it has a countable dense subset. Above description guarantees that if L is provided with a complete ortho-normal system (of countable members), then it is separable. However, the converse is also true, and it can be shown that if L is separable then it has a complete orthonormal system.

We have the following.

Theorem 3.4 *Let $\{\varphi_i\}_{i=1}^{\infty}$ be an ortho-normal system in a Hilbert space L and $f \in L$. Then it holds that*

$$\sum_{i=1}^{\infty} |(f, \varphi_i)|^2 \le \|f\|^2. \tag{3.7}$$

Furthermore, the equality in (3.7) is equivalent for (3.6) to hold.

Proof. Inequality (3.7) follows from

$$\left\| f - \sum_{i=1}^{n} (f, \varphi_i)\varphi_i \right\|^2 = \|f\|^2 - \sum_{i=1}^{n} |(f, \varphi_i)|^2 \geq 0,$$

where $n = 1, 2, \cdots$. The latter part is also a direct consequence of this equality. □

Inequality (3.7) is called *Bessel's inequality*. If equality holds, then it is called *Parseval's relation*. If an ortho-normal system is given in a Hilbert space, it is complete if and only if Parseval's relation always holds.

Exercise 3.5 Confirm that $L_0 \subset \overline{L_{00}}$ holds.

Exercise 3.6 Confirm the last statement that an ortho-normal system $\{\varphi_i\}_{i=1}^{\infty}$ in the Hilbert L is complete if and only if Parseval's relation

$$\sum_{i=1}^{\infty} |(f, \varphi_i)|^2 = \|f\|^2$$

holds for any $f \in L$.

3.2 Fourier Series

3.2.1 *Historical Note*

If we take non-stationary state of the string described in §2.3.1, then kinetic energy is taken into account in the Lagrangian. It is given by

$$J[u] = \frac{1}{2}\int_0^T \int_0^\pi u_x^2 dx dt - \int_0^T \int_0^\pi u f dx dt - \frac{1}{2}\int_0^T \int_0^\pi u_t^2 dx dt \quad (3.8)$$

under the agreement that any physical constant is one. If $f = 0$ for simplicity, then the Euler equation is given as

$$u_{tt} = u_{xx}. \quad (3.9)$$

It describes the vibrating motion of the string, and generally is called the *wave equation*. We take the boundary condition $u(0) = u(\pi) = 0$ and the initial condition

$$u|_{t=0} = u_1(x) \quad \text{and} \quad u_t|_{t=0} = u_0(x) \quad (x \in [0, \pi]).$$

d'Alembert observed that (3.9) is satisfied for

$$u(x,t) = \varphi(x-t) + \psi(x+t), \tag{3.10}$$

where φ and ψ are arbitrary C^2 function. We take the odd extension to u with respect to x: $u(-x,t) = -u(x,t)$, and then 2π periodic extension: $u(x+2\pi,t) = u(x,t)$, and in such a way we get a function defined on $\mathbf{R} \times [0,\infty)$ denoted by the same symbol $u = u(x,t)$. Then the boundary condition $u(0,t) = u(\pi,t) = 0$ is satisfied and the extended $u = u(x,t)$ is continuous in $x \in \mathbf{R}$. Putting (3.10), we have

$$u_1(x) = \varphi(x) + \psi(x) \qquad \text{and} \qquad -\varphi'(x) + \psi'(x) = u_0(x).$$

The latter equality gives

$$-\varphi(x) + \psi(x) = \int_0^x u_0(s)ds + c$$

with a constant c, and hence

$$\psi(x) = \frac{1}{2}u_1(x) + \frac{1}{2}\int_0^x u_0(s)ds + \frac{c}{2}$$

and

$$\varphi(x) = \frac{1}{2}u_1(x) - \frac{1}{2}\int_0^x u_0(s)ds - \frac{c}{2}$$

follow. Again by (3.10), we have

$$u(x,t) = \frac{1}{2}\left(u_1(x-t) + u_1(x+t)\right) + \frac{1}{2}\int_{x-t}^{x+t} u_0(s)ds. \tag{3.11}$$

Observe that $u(-x,t) = -u(x,t)$ and $u(x+2\pi,t) = u(x,t)$ hold if $u_1(x)$ and $u_0(x)$ satisfy the same conditions.

D. Bernoulli introduced the method of *super-position.* First, special solution to (3.9) is taken in the form of *separation of variables,* that is, $u(x,t) = \varphi(x)\psi(t)$. This implies that

$$\varphi''(x)/\varphi(x) = \psi''(t)/\psi(t).$$

Because left-hand and right-hand sides are independent of t and x, respectively, this quantity must be a constant, denoted by $-\lambda$. Then, we get the *eigenvalue problem*

$$-\varphi''(x) = \lambda\varphi(x) \quad (0 \le x \le \pi) \qquad \text{with} \qquad \varphi(0) = \varphi(\pi) = 0. \tag{3.12}$$

We see that problem (3.12) has the *trivial solution* $u \equiv 0$ for any λ. In the case that non-trivial solution to this problem exists, then λ and $\varphi(x)$ are called *eigenvalue* and *eigenfunction*, respectively. If $\varphi(x)$ is an eigenfunction, then $c\varphi(x)$ is again so, where $c \in \mathbf{R} \setminus \{0\}$. Therefore, we take *normalization*. Eigenvalues to (3.12) are n^2 and $\varphi_n(x) = \sin nx$ act as eigenfunctions for $n = 1, 2, \cdots$. Then, it follows from $-\psi''(t) = n^2\psi(t)$ that

$$\psi(t) = A_n \cos nt + B_n \sin nt,$$

where $\psi(0) = A_n$ and $\psi'(0)/n = B_n$, and this method leads to

$$u(x,t) = \sum_{n=1}^{\infty} \sin nx \left(A_n \cos nt + B_n \sin nt\right), \tag{3.13}$$

where A_n, B_n ($n = 1, 2, \cdots$) are constants. However, if it is true, then (3.11) implies

$$u_1(x) = \sum_{n=1}^{\infty} A_n \sin nx. \tag{3.14}$$

Thus, two questions arise. Does the right-hand side of (3.14) converge? Is it possible to express any odd 2π periodic function $u_1(x)$ in use of the trigonometric functions like (3.14)? Fourier showed that

$$f(x) = \begin{cases} -1 & (-\pi < x < 0) \\ 1 & (0 < x < \pi) \end{cases}$$

is expressed as the *Fourier series*

$$\frac{4}{\pi}\left[\sin x + \frac{1}{3}\sin 3x + \frac{1}{5}\sin 5x + \cdots + \frac{1}{2n+1}\sin(2n+1)x + \cdots\right].$$

In connection with this, the right-hand side of (3.14) is called generally *formal Fourier series.*

Exercise 3.7 Show that Euler equation to $J[u]$ defined by (3.8) is (3.9).

Exercise 3.8 Show that eigenvalues and eigenfunctions of (3.12) are given by n^2 and $\sin nx$ for $n = 1, 2, \cdots$, and nothing else can be so.

Exercise 3.9 Derive Euler equation for

$$J[u] = \int_0^T \int_0^\pi \sqrt{1 + u_x^2}\, dxdt - \int_0^T \int_0^\pi uf\, dxdt - \frac{1}{2}\int_0^T \int_0^\pi u_t^2\, dxdt.$$

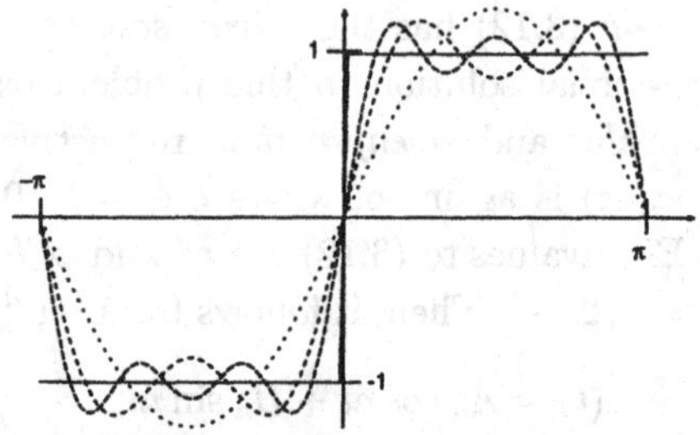

Fig. 3.1

3.2.2 *Completeness*

A measurable function is said to be *locally square integrable* on **R** if it is square integrable on any compact set in **R**. Set of locally square integrable functions on **R** is denoted by $L^2_{loc}(\mathbf{R})$. Then,

$$X = \left\{ v \in L^2_{loc}(\mathbf{R}) \mid v(x + 2\pi) = v(x) \text{ for a.e. } x \in \mathbf{R} \right\}$$

is identified with $L^2(0, 2\pi)$. We shall work on this space, which forms a Hilbert space with the inner product

$$(u, v) = \int_0^{2\pi} u(x)v(x)dx$$

and the norm $\|v\|_2 = \sqrt{(v, v)}$.

We have seen that

$$\left\{ \frac{1}{\sqrt{2\pi}}, \frac{1}{\sqrt{\pi}} \cos mx, \frac{1}{\sqrt{\pi}} \sin nx \mid m, n = 1, 2, \cdots \right\} \tag{3.15}$$

forms an ortho-normal system in $L^2(0, 2\pi)$. Here, we note that odd functions of them are $\frac{1}{\sqrt{\pi}} \sin nx$ with $n = 1, 2, \cdots$ in connection with the argument in the previous paragraph. Here, we show that it is complete in the sense of §3.1.3, namely (3.6), or

$$\lim_{n \to \infty} \|s_n - f\|_2 = 0 \tag{3.16}$$

in this context, where

$$s_n(x) = \frac{a_0}{2} + \sum_{k=1}^{n} (a_k \cos kx + b_k \sin kx) \tag{3.17}$$

with

$$a_k = \frac{1}{\pi}\int_0^{2\pi} f(x)\cos kx dx, \qquad b_k = \frac{1}{\pi}\int_0^{2\pi} f(x)\sin kx dx. \tag{3.18}$$

To prove (3.16) for $f \in X$, we can assume that $f \in X_0$ because X_0 is dense in X, where

$$X_0 = \{f \in C(\mathbf{R}) \mid f(x+2\pi) = f(x) \text{ for any } x \in \mathbf{R}\}.$$

Now, we shall show (3.16) for $f \in X_0$.

For this purpose, first we note that Bessel's inequality holds as

$$\frac{a_0^2}{4} + \sum_{n=1}^{\infty} (a_n^2 + b_n^2) \le \|f\|_2^2,$$

and hence we have

$$\lim_{n\to\infty} a_n = \lim_{n\to\infty} b_n = 0$$

for any $f \in X$, which is referred to as *Riemann-Lebesgue's theorem.* Next, we have

$$\begin{aligned} s_n(x) &= \frac{1}{\pi}\int_0^{2\pi} f(y)\left[\frac{1}{2} + \sum_{k=1}^{n} (\cos ky \cos kx + \sin ky \sin kx)\right] dy \\ &= \frac{1}{\pi}\int_0^{2\pi} f(y)\left[\frac{1}{2} + \sum_{k=1}^{n} \cos k(y-x)\right] dy \\ &= \frac{1}{\pi}\int_{-x}^{2\pi - x} f(x+y)\left[\frac{1}{2} + \sum_{k=1}^{n} \cos ky\right] dy, \end{aligned}$$

where $\int_{-x}^{0}$ is replaced by $\int_{2\pi-x}^{2\pi}$ from the periodicity. We obtain

$$s_n(x) = \frac{1}{\pi}\int_0^{2\pi} f(x+y) D_n(y) dy \tag{3.19}$$

with

$$D_n(y) = \frac{1}{2} + \sum_{k=1}^{n} \cos ky = \frac{1}{2}\cdot\frac{\sin(n+1/2)y}{\sin(y/2)} = \frac{1}{2}\cdot\frac{\cos ny - \cos(n+1)y}{1-\cos y}. \tag{3.20}$$

Here, we take the *arithmetic mean* of $s_n(x)$ as

$$\begin{aligned} S_n(x) &= \frac{1}{n+1}\left(s_0(x) + s_1(x) + \cdots + s_n(x)\right) \\ &= \frac{1}{2\pi(n+1)}\int_0^{2\pi} f(x+y)\frac{1-\cos(n+1)y}{1-\cos y}dy \\ &= \frac{1}{2\pi(n+1)}\int_0^{2\pi} f(x+y)\frac{\sin^2\left[(n+1)y/2\right]}{\sin^2(y/2)}dy. \end{aligned} \tag{3.21}$$

Sometimes, it is referred to as the *Cesáro mean* of the original series (3.17). Letting $f(x) = 1$ in (3.21), we have

$$\frac{1}{2\pi n}\int_0^{2\pi} \frac{\sin^2(ny/2)}{\sin^2(y/2)}dy = 1 \tag{3.22}$$

because $a_0 = 2$ and $a_k = b_k = 0$ for $k \geq 1$ in this case. Equalities (3.21) and (3.22) imply

$$S_{n-1}(x) - f(x) = \frac{1}{2\pi n}\int_0^{2\pi} \left[f(x+y) - f(x)\right]\frac{\sin^2(ny/2)}{\sin^2(y/2)}dy. \tag{3.23}$$

The following theorem is due to Fejér.

Theorem 3.5 *If $f(x)$ is a continuous function with the period 2π, then the Cesáro mean of the formal Fourier series converges to $f(x)$. That is, $S_n(x)$ converges to $f(x)$ uniformly in $x \in \mathbf{R}$.*

Proof. The continuous periodic function f is regarded as a function on $\mathbf{R}/(2\pi\mathbf{Z})$, which is identified with the circle S^1. Because the latter is compact, we see that $f : \mathbf{R} \to \mathbf{R}$ is uniformly continuous. Thus, any $\varepsilon > 0$ admits $\delta \in (0, 2\pi)$ such that

$$|y| < \delta \qquad \Rightarrow \qquad |f(x+y) - f(x)| < \varepsilon \qquad \text{for any} \qquad x \in \mathbf{R}. \tag{3.24}$$

We divide the integral in the right-hand side of (3.23) as

$$\int_0^{\delta} + \int_{\delta}^{2\pi-\delta} + \int_{2\pi-\delta}^{2\pi}.$$

From the periodicity, the third integral is reduced to $\int_{-\delta}^{0}$. Then, this term and the first one are combined as $\int_{-\delta}^{\delta}$. In use of (3.24), we have

$$\begin{aligned}
&\frac{1}{2\pi n}\left|\int_{-\delta}^{\delta}[f(x+y)-f(x)]\frac{\sin^2(ny/2)}{\sin^2(y/2)}dy\right| \\
&\quad\le \frac{\varepsilon}{2\pi n}\int_{-\delta}^{\delta}\frac{\sin^2(ny/2)}{\sin^2(y/2)}dy \\
&\quad< \frac{\varepsilon}{2\pi n}\int_{0}^{2\pi}\frac{\sin^2(ny/2)}{\sin^2(y/2)}dy = \varepsilon.
\end{aligned}$$

On the other hand, we have $\|f\|_\infty = \sup_{x\in[0,2\pi]}|f(x)| < +\infty$ and hence

$$\begin{aligned}
&\left|\frac{1}{2\pi n}\int_{\delta}^{2\pi-\delta}[f(x+y)-f(x)]\frac{\sin^2(ny/2)}{\sin^2(y/2)}dy\right| \\
&\quad\le \frac{2\|f\|_\infty}{2\pi n}\int_{\delta}^{2\pi-\delta}\frac{\sin^2(ny/2)}{\sin^2(y/2)}dy \\
&\quad< \frac{2\|f\|_\infty\cdot 2\pi}{2\pi n\sin^2(\delta/2)} = \frac{2\|f\|_\infty}{n\sin^2(\delta/2)}
\end{aligned}$$

follows from $\sin(y/2) \ge \sin(\delta/2)$ and $|\sin(ny/2)| \le 1$ on $[\delta, 2\pi-\delta]$. Those relations are summarized as

$$|S_{n-1}(x) - f(x)| < \varepsilon + \frac{2\|f\|_\infty}{n\sin^2(\delta/2)}.$$

This implies

$$\sup_{x\in[0,2\pi]}|S_n(x) - f(x)| < \varepsilon + \frac{2\|f\|_\infty}{(n+1)\sin^2(\delta/2)}$$

and

$$\limsup_{n\to\infty}\sup_{x\in[0,2\pi]}|S_n(x) - f(x)| \le \varepsilon.$$

Because $\varepsilon > 0$ is arbitrary, we get $\lim_{n\to\infty}\sup_{x\in[0,2\pi]}|S_n(x) - f(x)| = 0$, or equivalently, $S_n(x)$ converges to $f(x)$ uniformly in x. □

We are ready to give the following.

Theorem 3.6 *Ortho-normal system (3.15) is complete in $L^2(0, 2\pi)$.*

Proof. We only have to show that (3.6) holds for any $f \in X_0$. However, in this case, Theorem 3.5 guarantees that $\lim_{n\to\infty} \|S_n - f\|_\infty = 0$. If L_n denotes the linear subspace of $L^2(0, 2\pi)$ spanned by

$$\left\{ \frac{1}{\sqrt{2\pi}}, \frac{1}{\sqrt{\pi}} \cos kx, \frac{1}{\sqrt{\pi}} \sin kx \mid k = 1, 2, \cdots, n \right\},$$

we have $S_n \in L_n$. This implies that

$$\|s_n - f\|_2 \le \|S_n - f\|_2$$

because $g = s_n$ attains $\inf_{g \in L_n} \|g - f\|_2$. We obtain $\lim_{n\to\infty} \|s_n - f\|_2 = 0$ by

$$\|S_n - f\|_2 \le (2\pi)^{1/2} \|S_n - f\|_\infty \to 0$$

and the proof is complete. □

Exercise 3.10 Confirm that the completeness of (3.15) in X is reduced to (3.6) for $f \in X_0$.

Exercise 3.11 Confirm that (3.20) holds.

Exercise 3.12 Expand $f(x) = \pi - x$ in Fourier series in use of the complete ortho-normal system

$$\left\{ \frac{1}{\sqrt{2\pi}}, \frac{1}{\sqrt{\pi}} \cos mx, \frac{1}{\sqrt{\pi}} \sin nx \mid m, n = 1, 2, \cdots \right\}$$

in $L^2(0, 2\pi)$. Then, show that

$$\sum_{n=1}^{\infty} \frac{1}{n^2} = \frac{\pi^2}{6}$$

follows from Parseval's relation.

3.2.3 *Uniform Convergence*

Even if $f \in X_0$, the formal Fourier series does not necessarily converge uniformly. In this context, first, we note the following.

Theorem 3.7 *If $f \in X$ and a_n, b_n defined by (3.18) satisfies*

$$\sum_{n=1}^{\infty} (|a_n| + |b_n|) < +\infty, \tag{3.25}$$

then f is regarded as an element in X_0 and $s_n(x)$ defined by (3.17) converges uniformly to $f(x)$.

Proof. In fact, in this case the series $s_n(x)$ has the majorant (3.25) and therefore converges uniformly to some function, denoted by $g(x)$. However, this $g(x)$ coincides with $f(x)$ for almost every $x \in (0, 2\pi)$, because s_n converges to f in $L^2(0, 2\pi)$ by Theorem 3.6, and $g(x)$ and $f(x)$ are identified in X. Thus, the proof is complete. □

The criterion given in the previous theorem is assured by the following.

Theorem 3.8 *If $f \in X_0$ is piecewise C^1, then $s_n(x)$ defined by (3.17) converges uniformly to $f(x)$.*

Proof. From the assumption we have

$$\begin{aligned} a_n &= \frac{1}{\pi} \int_0^{2\pi} f(x) \cos nx dx = -\frac{1}{n\pi} \int_0^{2\pi} f'(x) \sin nx dx \\ b_n &= \frac{1}{\pi} \int_0^{2\pi} f(x) \sin nx dx = \frac{1}{n\pi} \int_0^{2\pi} f'(x) \cos nx dx. \end{aligned} \tag{3.26}$$

Because $f' \in L^2(0, 2\pi)$, it holds that

$$\sum_{n=1}^{\infty} \left[(-na_n)^2 + (nb_n)^2 \right] < +\infty$$

by Bessel's inequality. In use of

$$2|a_n| \leq (na_n)^2 + \frac{1}{n^2} \qquad \text{and} \qquad 2|b_n| \leq (nb_n)^2 + \frac{1}{n^2}$$

we obtain

$$\sum_{n=1}^{\infty} (|a_n| + |b_n|) < +\infty$$

and the proof is complete. □

The following theorem is refined in the next paragraph.

Theorem 3.9 *Let $f(x)$ be 2π periodic, bounded, and continuously differentiable except for finite points in $[0, 2\pi]$. Then the formal Fourier series $s_n(x_0)$ converges to $f(x_0)$ as $n \to \infty$, if x_0 is not the exceptional point.*

Proof. Similarly to (3.22), we have

$$\frac{1}{\pi}\int_0^{2\pi} D_n(y)dy = \frac{1}{\pi}\int_{-\pi}^{\pi} D_n(y)dy = 1 \tag{3.27}$$

by (3.19). Therefore, it holds that

$$s_n(x_0) - f(x_0) = \frac{1}{\pi}\int_0^{2\pi} [f(x_0+y) - f(x_0)]\, D_n(y)dy. \tag{3.28}$$

We have

$$\begin{aligned} D_n(y) &= \frac{1}{2} + \sum_{k=1}^{n} \cos ky = \frac{\sin(n+1/2)y}{2\sin(y/2)} \\ &= \frac{1}{2}\left[\frac{\cos(y/2)}{\sin(y/2)}\sin ny + \cos ny\right] \end{aligned}$$

and hence the right-hand side of (3.28) is divided as

$$\begin{aligned} &\frac{1}{2\pi}\int_0^{2\pi} [f(x_0+y) - f(x_0)] \cos nydy \\ &\quad + \frac{1}{2\pi}\int_0^{2\pi} [f(x_0+y) - f(x_0)] \frac{\cos(y/2)}{\sin(y/2)} \sin nydy. \end{aligned}$$

The first term converges to 0 as $n \to 0$ by Riemann-Lebesgue's theorem. The second term is treated similarly, because

$$[f(x_0+y) - f(x_0)]\frac{\cos(y/2)}{\sin(y/2)} = \frac{f(x_0+y) - f(x_0)}{y} \cdot \frac{y/2}{\sin(y/2)} \cdot \cos\frac{y}{2} \tag{3.29}$$

is a bounded function of $y \in [0, 2\pi]$ from the assumption. □

Exercise 3.13 Confirm that (3.26) holds for $f \in X_0$ in piecewise C^1.

Exercise 3.14 Confirm that the quantity indicated by (3.29) is a bounded function of y.

Exercise 3.15 Regard $f(x) = \frac{2\sin(x/2)}{x}$ defined for $x \in [-\pi, \pi)$ as a 2π periodic function and apply Theorem 3.9 with $x_0 = 0$. Confirm that

$$\begin{aligned} s_n(0) &= \frac{2}{\pi}\int_0^{\pi} f(y)D_n(y)dy = \frac{2}{\pi}\int_0^{\pi} \frac{\sin(n+1/2)y}{y}dy \\ &= \frac{2}{\pi}\int_0^{(n+1/2)\pi} \frac{\sin x}{x}dx \end{aligned}$$

and prove

$$\int_0^{\infty} \frac{\sin y}{y}dy = \lim_{R\to+\infty}\int_0^{R} \frac{\sin y}{y}dy = \frac{\pi}{2} \tag{3.30}$$

in this way.

3.2.4 *Pointwise Convergence*

To handle with more rough functions, we make use of real analysis. A function f defined on the compact interval $[a, b]$ is said to be of *bounded variation* if there is $M > 0$ such that

$$\sum_{i=1}^{n} |f(x_i) - f(x_{i-1})| \leq M$$

for any division $a = x_0 < x_1 < \cdots < x_n = b$ of $[a, b]$. Monotone function is of bounded variation. Then, *Jordan's theorem* assures that any function of bounded variation is a difference of two monotone non-decreasing functions. Therefore, a function of bounded variation has $f(x + 0) = \lim_{y\downarrow x} f(y)$ and $f(x - 0) = \lim_{y\uparrow x} f(y)$ for any $x \in [a, b]$, and its discontinuous points are at most countable. Hence it is Riemann integrable.

We shall show that if $f(x)$ is a 2π periodic function of bounded variation, then $s_n(x)$ converges to $(f(x + 0) + f(x - 0))/2$. For this purpose, we make use of the following lemma by Dirichlet.

Lemma 3.1 *If $f(x)$ is of bounded variation in $[a, b]$ and $f(a + 0) = 0$, then it holds that*

$$\lim_{s\to+\infty}\int_a^b f(x)\frac{\sin sx}{x}dx = 0.$$

Proof. We may assume that $f(x)$ is non-decreasing, $a = 0$, and $f(a) = f(a+0) = 0$. Then, any $\varepsilon > 0$ admits $\delta > 0$ such that $x \in [0, \delta]$ implies $0 \le f(x) < \varepsilon$. In use of the *second mean value theorem of Riemann integral*, we have $\xi \in [0, \delta]$ such that

$$\int_0^\delta f(x)\frac{\sin sx}{x}dx = f(\delta)\int_\xi^\delta \frac{\sin sx}{x}dx.$$

The right-hand side is equal to

$$f(\delta)\int_{s\xi}^{s\delta} \frac{\sin y}{y}dy = f(\delta)\left[\int_0^{s\delta} \frac{\sin y}{y}dy - \int_0^{s\xi} \frac{\sin y}{y}dy\right].$$

Because of (3.30), there is $M > 0$ such that

$$\left|\int_0^R \frac{\sin y}{y}dy\right| \le M$$

for any $R > 0$. Thus, we obtain

$$\left|\int_0^\delta f(x)\frac{\sin sx}{x}dx\right| \le 2M\varepsilon$$

by $f(\delta) \in [0, \varepsilon)$.

Similarly, we have $\eta \in [\delta, b]$ satisfying

$$\int_\delta^b f(x)\frac{\sin sx}{x}dx = f(\delta)\int_\delta^\eta \frac{\sin sx}{x}dx + f(b)\int_\eta^b \frac{\sin sx}{x}dx.$$

The first term of the right-hand side is treated similarly, and we have

$$f(\delta)\left|\int_\delta^\eta \frac{\sin sx}{x}dx\right| \le 2M\varepsilon.$$

For the second term we note that

$$\int_\eta^b \frac{\sin sx}{x}dx = \int_{s\eta}^{sb} \frac{\sin y}{y}dy.$$

Because $\eta \ge \delta$, we have

$$\lim_{s\to+\infty} \int_{s\eta}^{sb} \frac{\sin y}{y}dy = 0$$

again by the convergence of (3.30). We obtain

$$\limsup_{s\to+\infty}\left|\int_0^b f(x)\frac{\sin sx}{x}dx\right| \le 4M\varepsilon$$

and get the conclusion because $\varepsilon > 0$ is arbitrary. □

Now, we give the following.

Theorem 3.10 *If $f(x)$ is a 2π periodic function of bounded variation, then the formal Fourier series $s_n(x)$ converges to $\left(f(x-0)+f(x+0)\right)/2$ for any $x \in \mathbf{R}$.*

Proof. In fact, from (3.19) and the periodicity we have

$$\begin{aligned} s_n(x) &= \frac{1}{\pi}\int_{-\pi}^{\pi} f(x+y)D_n(y)dy \\ &= \frac{1}{2\pi}\left(\int_{-\pi}^{0}+\int_0^{\pi}\right) f(x+y)\frac{\sin(n+1/2)y}{\sin(y/2)}dy \\ &= \frac{1}{2\pi}\left[\int_0^{\pi} f(x+y)\frac{\sin(n+1/2)y}{\sin(y/2)}dy + \int_0^{\pi} f(x-y)\frac{\sin(n+1/2)y}{\sin(y/2)}dy\right]. \end{aligned}$$

In use of (3.27) it follows that

$$\begin{aligned} s_n(x) - \frac{f(x+0)+f(x-0)}{2} &= \frac{1}{2\pi}\int_0^{\pi} g(y)\frac{\sin(n+1/2)y}{\sin(y/2)}dy \\ &= \frac{1}{2\pi}\int_0^{\pi} g(y)\frac{y}{\sin(y/2)}\cdot\frac{\sin(n+1/2)y}{y}dy \qquad (3.31) \end{aligned}$$

with $g(y) = f(x+y)+f(x-y)-f(x+0)-f(x-0)$. Because $y/\sin(y/2)$ is smooth in $\mathbf{R}$, $G(y) = g(y)\cdot\frac{y}{\sin(y/2)}$ is of bounded variation on $[0,\pi]$. It satisfies $G(+0) = 0$ and therefore, the right-hand side of (3.31) converges to 0 as $n \to \infty$. This means

$$\lim_{n\to\infty} s_n(x) = \frac{f(x+0)+f(x-0)}{2}$$

and the proof is complete. □

Exercise 3.16 Confirm that monotone functions are of bounded variation.

Exercise 3.17 Show that discontinuous points of a function of bounded variation are at most countable.

Exercise 3.18 Confirm that Lemma 3.1 is reduced to the case that $f(x)$ is non-decreasing, $a = 0$, and $f(a) = 0$.

Exercise 3.19 Confirm that $g(y) \cdot \varphi(y)$ is of bounded variation on $[a, b]$, if g is so and $\varphi(y)$ is C^1 there.

3.3 Eigenvalue Problems

3.3.1 *Vibrating Membrane*

Let $\Omega \subset \mathbf{R}^n$ be a bounded domain with smooth boundary $\partial\Omega$, and take the problem

$$-\Delta\psi = \lambda\psi \quad \text{in } \Omega, \qquad \psi = 0 \quad \text{on } \partial\Omega, \tag{3.32}$$

where λ is a constant. This problem always admits the *trivial solution* $\psi \equiv 0$ and if a non-trivial solution $\psi \not\equiv 0$ exists then such λ and ψ are called the *eigenvalue* and the *eigenfunction*, respectively. One-dimensional problem is studied in the previous chapter, and a general case in this dimension is referred to as the *Sturm-Liouville problem.* If it is symmetric, then eigenvalues are real, countably many, and simple. *Expansion theorem of Mercer* holds so that the eigenfuction expansion is valid with the uniform convergence for C^2 functions satisfying the boundary condition. However, *multiple eigenvalues* can exist if $n \geq 2$, which means that each of them shares plurally linearly independent eigenfunctions. Problem (3.32) arises in the process of separation of variables to the *heat equation*

$$u_t = \Delta u \quad \text{in} \quad \Omega \times (0, T), \qquad u|_{\partial\Omega} = 0$$

or the *wave equation*

$$u_{tt} = \Delta u \quad \text{in} \quad \Omega \times (0, T), \qquad u|_{\partial\Omega} = 0 \tag{3.33}$$

and hence the *expansion theorem* is necessary to justify the method of superposition by eigenfunctions $\psi_k(x)$ for $k = 1, 2, \cdots$.

In the case of (3.33), the solution u is given by

$$u(x,t) = \sum_{k=1}^{\infty} \psi_k(x) \left\{ (u_1, \psi_k) \cos \omega_k t + (u_0, \psi_k) \sin \omega_k t / \omega_k \right\}$$

with the right-hand side converging in $C^1(\mathbf{R}; L^2(\Omega)) \cap C(\mathbf{R}; H_0^1(\Omega))$ if the initial values $u_1 = u|_{t=0}$ and $u_0 = u_t|_{t=0}$ are in $(u_1, u_0) \in H_0^1(\Omega) \times L^2(\Omega)$, where (,) denotes the L^2 inner product and

$$\omega_k = \sqrt{\lambda_k}.$$

In this context, this $\{\omega_k\}_{k=1}^{\infty}$ indicates the *characteristic vibration numbers*, or *tone* if Ω is regarded as a vibrating object. The deformation of the material on the boundary is set to be zero in (3.32), while the Neumann problem arises if it is open:

$$-\Delta\psi = \lambda\psi \quad \text{in} \quad \Omega, \qquad \frac{\partial\psi}{\partial\nu} = 0 \quad \text{on} \quad \partial\Omega, \tag{3.34}$$

where ν denotes the outer unit normal vector.

If $\Omega = \left\{(x,y) \mid x^2+y^2<1\right\} \subset \mathbf{R}^2$ is the unit disc, then in use of the polar coordinate $x = r\cos\theta,\ y = r\sin\theta$ we have from (3.32) that

$$\psi_{rr} + \frac{1}{r}\psi_r + \frac{1}{r^2}\psi_{\theta\theta} + \lambda\psi = 0 \qquad \text{with} \qquad \psi(1,\theta) = 0 \tag{3.35}$$

by

$$\Delta = \frac{\partial^2}{\partial r^2} + \frac{1}{r}\frac{\partial}{\partial r} + \frac{1}{r^2}\frac{\partial^2}{\partial\theta^2}. \tag{3.36}$$

Then, we apply the separation of variables $\psi(r,\theta) = f(r)h(\theta)$ to (3.35) and get that

$$\frac{r^2\left(f'' + \frac{f'}{r} + \lambda f\right)}{f} = -\frac{h''}{h},$$

where the left- and the right-hand sides are independent of θ and r, respectively. Hence this quantity is a constant, and must be k^2 with some integer $k \geq 0$, because $h = h(\theta)$ is 2π periodic. Thus, we obtain

$$h(\theta) = a\cos k\theta + b\sin k\theta$$

and

$$r^2 f'' + rf' + (r^2\lambda - k^2)f = 0 \quad (0 < r < 1) \qquad \text{with} \qquad f(1) = 0, \tag{3.37}$$

where a, b are constants. Confirming that $\lambda > 0$ if it is an eigenvalue of (3.32), we put that $\rho = \lambda^{1/2}r$ and $f(r) = J(\rho)$ and get *Bessel's equation*

$$\frac{d^2J}{d\rho^2} + \frac{1}{\rho}\frac{dJ}{d\rho} + \left(1 - \frac{k^2}{\rho^2}\right)J = 0. \tag{3.38}$$

This equation admits a solution in the form of

$$J(\rho) = \sum_{m=0}^{\infty} a_m \rho^m = \frac{\rho^k}{2^k k!}\left(1 - \frac{\rho^2}{2(2k+2)} + \frac{\rho^4}{2\cdot 4(2k+2)(2k+4)} - \cdots\right),$$

denoted by $J = J_k(\rho)$. It is called the *Bessel function* of k-th order. Writing $\lambda = \omega^2$, we have $f_k(r) = J_k(\omega r)$ and hence (3.37) is equivalent to $J_k(\omega) = 0$.

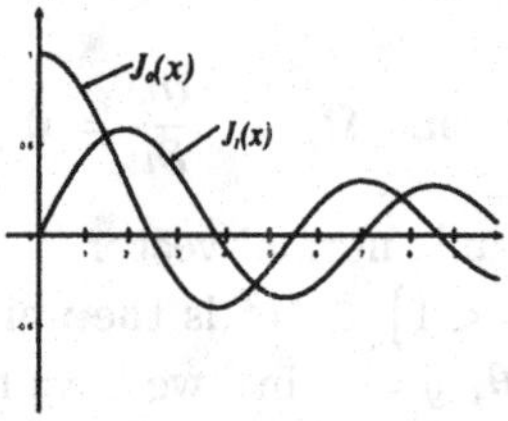

Fig. 3.2

More precisely, if $\{\omega_{k\ell}\}_{\ell=1,2,\cdots}$ denotes the set of zeros of k-th Bessel's function $J_k = J_k(r)$ in $r > 0$, then the eigenvalues of (3.32) on the unit disc are $\{\omega_{k\ell}^2\}_{\ell=1,2,\cdots,k=0,1,\cdots}$, and the eigenfunctions are given by $J_k(\omega_{k\ell}r)\cos k\theta$ and $J_k(\omega_{k\ell}r)\sin k\theta$. Thus, they are double except for the first eigenvalue.

If $\Omega = \{(x,y,z) \mid x^2+y^2+z^2 < 1\}$ is the three-dimensional unit ball, then the polar coordinate is given by $x = r\sin\theta\cos\varphi$, $y = r\sin\theta\sin\varphi$, $z = r\cos\theta$ with $(r,\theta,\varphi) \in (0,1)\times[0,\pi]\times[0,2\pi)$. Noting

$$\Delta\psi = \frac{1}{r^2\sin\theta}\left[\frac{\partial}{\partial r}\left(r^2\psi_r\sin\theta\right) + \frac{\partial}{\partial\varphi}\left(\frac{\psi_\varphi}{\sin\theta}\right) + \frac{\partial}{\partial\theta}\left(\psi_\theta\sin\theta\right)\right] \tag{3.39}$$

we take the separation of variables $\psi = Y(\theta,\varphi)f(r)$, and get that

$$\frac{(r^2f')' + \lambda r^2 f}{f} = -\frac{1}{Y\sin\theta}\left[\frac{\partial}{\partial\varphi}\left(\frac{Y_\varphi}{\sin\theta}\right) + \frac{\partial}{\partial\theta}\left(Y_\theta\sin\theta\right)\right].$$

This quantity is again a constant denoted by μ, and in this way we get

$$\Delta^*Y + \mu Y = 0 \qquad (0 \le \theta \le \pi,\ 0 \le \varphi < 2\pi), \tag{3.40}$$

where Δ^* denotes the *Laplace-Beltrami* operator on the unit sphere:

$$\Delta^*Y = \frac{1}{\sin\theta}\left[\frac{\partial}{\partial\varphi}\left(\frac{Y_\varphi}{\sin\theta}\right) + \frac{\partial}{\partial\theta}\left(Y_\theta\sin\theta\right)\right].$$

Because $Y = Y(\theta, \varphi)$ is regular at $\theta = 0, \pi$ and 2π periodic with respect to φ, it can be shown that $\mu = \ell(\ell+1)$ with $\ell = 0, 1, 2, \cdots$. The solution to (3.40) in this case is referred to as the ℓ-th *spherically harmonic function* and is denoted by $Y_\ell(\theta, \varphi)$.

On the other hand, we have from

$$(r^2 f')' - \ell(\ell+1)f + \lambda r^2 f = 0$$

that $f(r) = S_\ell(\lambda^{1/2} r) \equiv J_{\ell+\frac{1}{2}}(\lambda^{1/2} r)/r^{1/2}$ and eigenvalues are determined by the boundary condition,

$$J_{\ell+\frac{1}{2}}(\lambda^{1/2}) = 0,$$

where $J_k(\rho)$ is the solution to (3.38). Furthermore, if $\{\lambda_{\ell m}\}_{m=1,2,\cdots}$ denote its solutions in $\lambda > 0$, then the eigenfunctions are given by

$$\left\{ Y_\ell(\theta, \varphi) S_\ell(\lambda_{\ell m}^{1/2} r) \mid \ell = 0, 1, \cdots, m = 1, 2, \cdots \right\}.$$

Take the separation of variables $Y(\theta, \varphi) = p(\varphi) q(\theta)$ furthermore to (3.40) with $\mu = \ell(\ell+1)$, then we obtain

$$-\frac{p''}{p} = \frac{(q' \sin\theta)' \sin\theta}{q} + \ell(\ell+1)\sin^2\theta.$$

This quantity is again a constant, and from the 2π periodicity in φ of p it is the form k^2 with $k = 0, 1, 2, \cdots$. On the other hand, by $z = \cos\theta$ the equation to q is transformed into

$$\left((1-z^2)q'\right)' + \left(\ell(\ell+1) - \frac{k^2}{1-z^2}\right) q = 0 \qquad (-1 < z < 1). \qquad (3.41)$$

It is called the *associated Legendre equation.* If $\ell = 0$, then it has a solution $q_{k,m}(z) = (1-z^2)^{m/2} \left(\frac{d}{dz}\right)^m P_m(z)$, regular at $z = \pm 1$ for $m = 0, 1, \cdots$, where $P_m(z)$ is the *Legendre polynomial* defined by

$$P_m(z) = \frac{1}{2^m m!} \left(\frac{d}{dz}\right)^m (z^2 - 1)^m.$$

Even in the general domain, eigenvalues are real, countably many, and bounded from below labeled as $\lambda_1 < \lambda_2 \leq \cdots$ according to their multiplicities in (3.32) or (3.34). Thus, the fist eigenvalue λ_1 is always simple and is positive and zero according to (3.32) and (3.34), respectively. The L^2-normalized eigenfunctions form a complete ortho-normal system in $L^2(\Omega)$. The number of the nodal domains of the k-th eigenfunction is less than or equal to k. Here, *nodal domain* denotes the sub-domain of Ω where ψ_k has a definite sign. *Weyl's formula* says that if $A(\lambda)$ denotes the number of eigenvalues to (3.32) less than or equal to λ, then it holds that

$$A(\lambda) = \begin{cases} \frac{\lambda}{4\pi}\,|\Omega|\,(1+o(1)) & (n=2) \\ \frac{\lambda^{3/2}}{16\pi^2}\,|\Omega|\,(1+o(1)) & (n=3) \end{cases}$$

as $\lambda \to +\infty$. If $\lambda_1(\Omega) > 0$ denotes the first eigenvalue of (3.32), then *Faber-Krahn's isoperimetric inequality* assures that

$$\lambda_1(\Omega) \geq \lambda_1(\Omega^*),$$

where Ω^* denotes the ball (or disc) with the same volume as Ω. On the other hand, *Polyá-Szegö-Weinberger's isoperimetric inequality* is indicated as

$$\mu_2(\Omega) \leq \mu_2(\Omega^*),$$

where $\mu_2(\Omega)$ denote the second eigenvalue of (3.34). In this case, the first eigenvalue is always zero. In both cases, the equality holds if and only if $\Omega = \Omega^*$. On the other hand, *Kac's problem* is stated as to determine Ω by eigenvalues.

Exercise 3.20 Seek eigenvalues and eigenfunctions for (3.32) and (3.34) for the rectangle domain $\Omega = (0,a)\times(0,b) \subset \mathbf{R}^2$ by the method of separation of variables: $\psi_k(x,y) = A_k(x)B_k(y)$. Then, confirm that multiple eigenvalues can arise.

Exercise 3.21 Confirm (3.36) and (3.39) in the two- and three- dimensional polar coordinates, respectively. That is,

$$\Delta = \frac{\partial^2}{\partial r^2} + \frac{1}{r}\frac{\partial}{\partial r} + \frac{1}{r^2}\frac{\partial^2}{\partial \varphi^2}$$

for $n=2$ with $x_1 = r\cos\varphi$, $x_2 = r\sin\varphi$ in $\varphi \in [0,2\pi)$ and

$$\Delta = \frac{\partial^2}{\partial r^2} + \frac{2}{r}\frac{\partial}{\partial r} + \frac{1}{\sin\theta}\frac{\partial}{\partial\theta}\left(\sin\theta\frac{\partial}{\partial\theta}\cdot\right) + \frac{1}{\sin^2\theta}\frac{\partial^2}{\partial\varphi^2}$$

for $n=3$ with $x_1 = r\cos\theta$, $x_2 = r\sin\theta\cos\varphi$, $x_3 = r\sin\theta\sin\varphi$ in $\theta \in [0,\pi]$, $\varphi \in [0,2\pi)$, where $r = |x|$.

3.3.2 *Gel'fand Triple*

Problems (3.32) and (3.34) are weakly formulated as to find $u \in V$ satisfying

$$a(u,v) = \lambda b(u,v)$$

for any $v \in V$ with $V = H_0^1(\Omega)$ and $V = H^1(\Omega)$, respectively, where

$$a(u,v) = \int_\Omega \nabla u \cdot \nabla v dx \qquad \text{and} \qquad b(u,v) = \int_\Omega uv dx.$$

The spaces $H_0^1(\Omega)$ and $H^1(\Omega)$, to be mentioned in later chapters, are Hilbert spaces with $a : V \times V \to \mathbf{R}$ being a bounded symmetric bilinear form. It also holds that $b = b(\cdot,\cdot)$ is the inner product in $H = L^2(\Omega)$, and therefore, the following abstract theory can guarantee the generation of a complete orthonormal system in H from the eigenfunctions of (3.32) or (3.34).

Let us confirm some notions and facts described in §3.1.3. In this section, H denotes the Hilbert space with inner product $(\ ,\ )$ and norm $|\ \ |$. If it has a countable dense subset, it is said to be separable. In this case, it admits a complete ortho-normal system (composed of countable members), and henceforth this condition is supposed to be satisfied unless otherwise stated. Here, $\{\varphi\}_{i=1}^\infty \subset H$ is an ortho-normal system if and only if

$$(\varphi_i, \varphi_j) = \delta_{ij} = \begin{cases} 1 & (i = j) \\ 0 & (i \neq j) \end{cases}$$

holds, and it is said to be complete in the case that its linear hull,

$$\left\{ \sum_{i=1}^n \alpha_i \varphi_i \mid \alpha_i \in \mathbf{R},\ n = 1, 2, \cdots \right\}$$

is dense in H. Completeness of the ortho-normal system $\{\varphi_i\}_{i=1}^\infty$ is equivalent to the convergence

$$\lim_{n\to\infty} \left| f - \sum_{i=1}^n (f, \varphi_i)\, \varphi_i \right| = 0$$

for any $f \in H$. It is also equivalent to Parseval's relation

$$\sum_{i=1}^\infty |(f, \varphi_i)|^2 = |f|^2 \tag{3.42}$$

for any $f \in H$. Even if $\{\varphi_i\}_{i=1}^{\infty}$ is not necessarily complete, the left-hand side is less than or equal to the right-hand side in (3.42), in which case it is called Bessel's inequality. In the present section, we show that the complete ortho-normal system is obtained by the eigenvalue problem with symmetry and compactness.

The mapping $b = b(\ ,\ ) : H \times H \to \mathbf{R}$ is *bi-linear* if it satisfies the property that

$$\begin{aligned} b(\alpha u + \beta v, w) &= \alpha b(u,v) + \beta b(v,w) \\ b(u, \alpha v + \beta w) &= \alpha b(u,v) + \beta b(u,w), \end{aligned}$$

where $u, v, w \in H$ and $\alpha, \beta \in \mathbf{R}$. It is *symmetric* if

$$b(u,v) = b(v,u)$$

holds for any $u, v \in H$. It is *positive definite* if there is $\delta > 0$ satisfying

$$b(u,u) \geq \delta\, |u|^2$$

for any $u \in H$. Finally, it is *bounded* if there is $M > 0$ satisfying

$$|b(u,v)| \leq M\, |u| \cdot |v|$$

for any $u, v \in H$. If such $b(\ ,\ )$ is given, it provides the inner product to the vector space H, and then H becomes a Hilbert space with its topology equivalent to the original one.

Let V be another Hilbert space with inner product $((\ ,\ ))$ and norm $\|\cdot\|$, satisfying $V \subset H$ as a vector space. We suppose that this inclusion involves the topology so that there is $K > 0$ such that

$$|u| \leq K\, \|u\|$$

for any $u \in V$, and furthermore, that V is dense in H. Throughout this section, those relations are written as $V \hookrightarrow H$ in short. In this case, if dual space H' is identified with H by Riesz' representation theorem, we obtain

$$V \hookrightarrow H \cong H' \subset V' \tag{3.43}$$

by restricting $T : H \to \mathbf{R}$ on V, where V' is the dual space of V. This is called the *Gel'fand triple.*

If $a = a(\ ,\ ) : V \times V \to \mathbf{R}$ is a symmetric, positive definite, and bounded bi-linear form. As is shown later, then we can introduce the bounded linear operator $\tilde{A} : V \to V'$ by

$$a(u,v) = \left\langle \tilde{A}u, v \right\rangle_{V',V} \qquad (u, v \in V), \tag{3.44}$$

where $\langle\ ,\ \rangle_{V',V}$ denotes the paring between V' and V so that $\langle T, v\rangle_{V',V} = T(v)$ for $T \in V'$ and $v \in V$. Then the (unbounded) linear operator A on H is defined by

$$D(A) = \left\{ u \in V \mid \tilde{A}u \in H \right\}$$

and

$$Au = \tilde{A}u \qquad (u \in D(A))\,.$$

We shall show also that this A is *self-adjoint*, of which notion is described in the following paragraph. Henceforth, $D(A)$ is called the *domain* of A. Then, the *range* and the *kernel* of A are given as

$$R(A) = \{Au \mid u \in D(A)\} \qquad \text{and} \qquad N(A) = \{u \in D(A) \mid Au = 0\}\,,$$

respectively.

This paragraph is concluded by the following.

Theorem 3.11 *The linear operator $\tilde{A} : V \to V'$ defined by (3.44) is an isomorphism so that it is one-to-one, onto, and bounded with its inverse.*

Proof. The linearity of $\tilde{A}$ is obvious. Remember that the norm in V is denoted by $\|\ \|$. From the boundedness of $a(\ ,\ )$ we have

$$\left| \left\langle \tilde{A}u, v \right\rangle_{V',V} \right| = |a(u,v)| \le M \|u\| \cdot \|v\|$$

for $u, v \in V$ and hence it follows that

$$\begin{aligned} \left\| \tilde{A}u \right\|_{V'} &= \sup \left\{ \left| \left\langle \tilde{A}u, v \right\rangle_{V',V} \right| \mid v \in V,\ \|v\| = 1 \right\} \\ &\le M \|u\|\,. \end{aligned}$$

This assures that $\tilde{A} : V \to V'$ is a bounded linear operator.

On the other hand, the relation

$$\left\langle \tilde{A}u, u \right\rangle_{V',V} = a(u,u) \ge \delta \|u\|^2$$

for $u \in V$ implies that

$$\begin{aligned}\left\|\tilde{A}u\right\|_{V'} &= \sup\left\{\frac{\left|\left\langle \tilde{A}u, v\right\rangle_{V',V}\right|}{\|v\|} \mid v \in V \setminus \{0\}\right\} \\ &\geq \delta \|u\|.\end{aligned}$$

The operator $\tilde{A}$ is thus one-to-one. If $R(\tilde{A}) = V'$ is shown, then $T = \tilde{A}^{-1} : V' \to V$ satisfies

$$\|Tf\| \leq \delta^{-1} \|f\|_{V'}$$

for $f \in V'$ and hence is bounded.

In fact, replacing the inner product in V by $a(\ ,\ )$ makes it a Hilbert space with an equivalent topology. Given $f \in V'$, we can apply Riesz' representation theorem and obtain $u \in V$ such that

$$\left\langle \tilde{A}u, v\right\rangle_{V',V} = a(u,v) = \langle f, v\rangle_{V',V}$$

for any $v \in V$. This means $\tilde{A}u = f$ and the proof is complete. □

For the Dirichlet problem (3.32), we provide the inner products

$$b(u,v) = \int_\Omega uvdx \qquad (u, v \in H)$$

and

$$a(u,v) = \int_\Omega \nabla u \cdot \nabla v dx \qquad (u, v \in V)$$

to $H = L^2(\Omega)$ and $V = H_0^1(\Omega)$, respectively, and then, it is known that the relations

$$D(A) = \left\{u \in V \mid \tilde{A}u \in H\right\} = H_0^1(\Omega) \cap L^2(\Omega)$$

and $Au = -\Delta u$ for $u \in D(A)$ hold.

Exercise 3.22 Confirm (3.43), which means that any $T \in H'$ is regarded uniquely as $T \in V'$. Then show that $H' \hookrightarrow V'$ holds by *Hahn-Banach's theorem*, which means that the inclusion is continuous and dense.

Exercise 3.23 Confirm that $\tilde{A} : V \to V'$ is linear. Confirm also that $\tilde{A} : V \to V'$ is one-to-one.

3.3.3 *Self-adjoint Operator*

Let A be a linear operator with the domain $D(A)$ and the range $R(A)$ in the Hilbert space H with inner product $(\ ,\ )$ and norm $|\cdot|$, so that $D(A) \subset H$ is a linear subspace and the mapping

$$A : D(A) \subset H \to H$$

is provided with the property that

$$A(\alpha u + \beta v) = \alpha Au + \beta Av \qquad (\alpha, \beta \in \mathbf{R},\ u, v \in V).$$

In the case that $D(A)$ is dense in H, that is $\overline{D(A)} = H$, the *adjoint operator* A' is defined by $A'v = v'$ if and only if

$$(Au, v) = (u, v')$$

holds for any $u \in D(A)$. The densely defined linear operator A is called *self-adjoint* if $A = A'$ holds including their domains.

The linear operator A is said to be *closed* if it satisfies the condition that $\{u_j\} \subset D(A)$, $u_j \to u$, and $Au_j \to v$ imply $u \in D(A)$ and $Au = v$. It is known that if the linear operator A is densely defined, then its adjoint A' is closed. Therefore, a self-adjoint operator is always closed. For a densely defined closed operator between Banach spaces, we have the *closed range theorem*. In this case, it is stated as follows.

Theorem 3.12 *If H is a Hilbert space and $A : D(A) \subset H \to H$ is a closed linear operator satisfying $\overline{D(A)} = H$, then it holds that*

$$\overline{R(A)} = N(A')^{\perp} \qquad \text{and} \qquad \overline{R(A')} = N(A)^{\perp}.$$

Furthermore, $R(A)$ is closed if and only if $R(A')$ is so.

Here and henceforth,

$$L^{\perp} = \{u \in H \mid (u, v) = 0 \ \text{ for any } \ v \in L\}$$

for $L \subset H$. In the case that L is a linear subspace, it is known that Hahn-Banach's theorem guarantees that $\overline{L} = H$ if and only if $L^{\perp} = \{0\}$. On the other hand, $R(A)^{\perp} = N(A')$ and $R(A')^{\perp} = N(A)$ are easier to prove.

A self-adjoint operator A is said to be *positive definite* if there is $\delta > 0$ such that

$$(Au, u) \geq \delta |u|^2$$

holds for any $u \in D(A)$. In this case, we have

$$|Au| \geq \delta |u| \qquad (u \in D(A)) \tag{3.45}$$

and hence $N(A) = \{0\}$ follows. This implies also the closedness of $R(A)$. From Theorem 3.12, we have $R(A) = H$ and therefore,

$$A^{-1} : H \to H$$

is well-defined as a closed linear operator. Because the *open mapping theorem* of Banach guarantees that a closed operator with the domain of the whole space is bounded, any positive definite self-adjoint operator on a Hilbert space has a bounded inverse.

The self-adjoint operator takes the *spectral decomposition*, and from this fact a *non-negative* self-adjoint operator admits its *fractional powers*, although details are not described here.

Exercise 3.24 Confirm that if $D(A)$ is dense then the adjoint operator A' of A is well-defined.

Exercise 3.25 A densely defined linear operator A is said to be *symmetric* if it satisfies $A \subset A'$, that is, $D(A) \subset D(A')$ and $A'u = Au$ holds for $u \in D(A)$. Confirm that this relation is equivalent to

$$(Au, v) = (u, Av) \qquad (u, v \in D(A)).$$

Exercise 3.26 Show that a linear operator A on H is closed if and only if its *graph*

$$G(A) = \{u \oplus Au \mid u \in D(A)\}$$

is closed in $H \times H$. Then, confirm that the inverse operator of a closed linear operator is closed if it exists.

Exercise 3.27 Confirm that if the linear operator A is densely defined, then its adjoint A' is always closed.

Exercise 3.28 Confirm that $R(A)^\perp = N(A')$ and $R(A')^\perp = N(A)$ hold for a densely defined linear operator A on H.

Exercise 3.29 Confirm that (3.45) implies the closedness of $R(A)$ because of the completeness H and the closedness of $A = A'$.

3.3.4 *Symmetric Bi-linear Form*

We recall the notations of §3.3.2. Let

$$V \hookrightarrow H \hookrightarrow V'$$

be a Gel'fand triple, where V and H are Hilbert spaces provided with the inner products $(\ ,\ )$ and $((\ ,\ ))$ and with the norms $|\cdot|$ and $\|\cdot\|$, respectively. The bi-linear form

$$a = a(\ ,\ ) : V \times V \to \mathbf{R}$$

is symmetric, positive definite, and bounded. Then, the bounded linear operator $\tilde{A} : V \to V'$ and the linear operator $A : D(A) \subset H \to H$ are defined by

$$\left\langle \tilde{A}u, v \right\rangle_{V',V} = a(u,v) \qquad (u, v \in H)$$

and

$$A = \tilde{A}\Big|_{D(A)}, \qquad D(A) = \left\{ u \in V \mid \tilde{A}u \in H \right\},$$

respectively.

The purpose of this paragraph is to show the following.

Theorem 3.13 *The operator A given above is self-adjoint in H.*

Proof. First, we prove that $D(A)$ is dense in V. This implies that it is also dense in H. For this purpose, from Hahn-Banach's theorem it suffices to show that if $f \in V'$ satisfies

$$\langle f, v \rangle_{V',V} = 0$$

for any $v \in D(A)$, then it follows that $f = 0$. In fact, by Theorem 3.11, there is $u \in V$ satisfying $\tilde{A}u = f$. This implies that

$$a(u,v) = \langle f, v \rangle_{V',V} = 0$$

for any $v \in V$. Letting $v = u$, we have

$$0 = a(u,u) \geq \delta \left| u \right|^2.$$

Hence $u = 0$ and $f = 0$ follow in turn.

Using $R(\tilde{A}) = V'$ and $D(A) = \left\{u \in V \mid \tilde{A}u \in H\right\}$, we see that $R(A) = H$ holds. Because A is one-to-one, the operator

$$T = A^{-1} : H \to H$$

is bounded by the open mapping theorem. Because $a = a(\ ,\ ) : V \times V \to \mathbf{R}$ is symmetric, we have for $v = Aw \in R(A) = H$ and $u \in H$ that

$$\begin{aligned}(Tu, v) &= (Tu, Aw) = (Aw, Tu) = a(w, Tu) = a(Tu, w)\\ &= (A \cdot Tu, w) = (u, w) = (u, Tv).\end{aligned}$$

Thus, T is bounded and symmetric and hence is self-adjoint. From the following fact and the relation $T^{-1} = A$ including the domains, we see that A is self-adjoint in H. □

In the following, T may not be bounded.

Theorem 3.14 *If T is a self-adjoint operator on a Hilbert space H and is one-to-one, then T^{-1} is also self-adjoint.*

Proof. First, we prove that $D(T^{-1}) = R(T)$ is dense in H by showing that $R(T)^\perp = \{0\}$. In fact, if $u \in R(T)^\perp$, then it holds that

$$(Tv, u) = 0$$

for any $v \in D(T)$. This implies $u \in D(T)$ and $(v, Tu) = 0$ because T is self-adjoint. We now make use of $\overline{D(A)} = H$ and $N(T) = \{0\}$ to get that $Tu = 0$ and $u = 0$ in turn. Thus, $\overline{R(T)} = H$ is proven, and $(T^{-1})'$ is well-defined.

If $v \in D\left((T^{-1})'\right)$, we have

$$(u, v) = (T^{-1}Tu, v) = \left(Tu, (T^{-1})' v\right)$$

for any $u \in D(T)$. Because $T' = T$, we have

$$(T^{-1})' v \in D(T) \qquad \text{and} \qquad T(T^{-1})' v = v.$$

This means $v \in D(T^{-1})$ and $T^{-1}v = (T^{-1})' v$. Conversely, if $v \in D(T^{-1})$ and $w = T^{-1}v$, we have $w \in D(T)$ and $v = Tw$. It holds that

$$(u, w) = (TT^{-1}u, w) = (T^{-1}u, Tw) = (T^{-1}u, v)$$

for any $u \in D(T^{-1})$. This means $v \in D\left((T^{-1})'\right)$ and $w = (T^{-1})' v$. We have $(T^{-1})' = T^{-1}$ including the domains and the proof is complete. □

It is obvious that the self-adjoint operator A in Theorem 3.13 is positive definite. In connection with its fractional powers, it holds that $D(A^{1/2}) = V$ and

$$a(u,v) = \left(A^{1/2}u, A^{1/2}v\right) \qquad (u, v \in V).$$

Exercise 3.30 Confirm that $D(A)$ is dense in H using the fact that it is dense in V in Theorem 3.13.

3.3.5 *Compact Operator*

We continue to suppose that H is a Hilbert space with inner product $(\, , \,)$ and norm $| \cdot |$. A bounded linear operator $T : H \to H$ is said to be *compact* if it maps bounded sets to relatively compact set, that is, $\{Tu_n\}$ contains a converging subsequence if $\{u_n\}$ is bounded in H. As is indicated in Exercise 2.18, any bounded sequence in H admits a weakly converging subsequence so that $T : H \to H$ is compact if and only if $u_n \rightharpoonup u$ implies $Tu_n \to Tu$ in H. Here and henceforth, $u_n \rightharpoonup u$ and $u_n \to u$ indicate the weak and the strong convergences so that $\lim_{n\to\infty}(u_n, v) = (u, v)$ for any $v \in H$ and $\lim_{n\to\infty} |u_n - u| = 0$, respectively. This property, referred to as the *completely continuity*, is not equivalent to the compactness for the nonlinear operator.

In the Gel'fand triple $V \hookrightarrow H \hookrightarrow V'$, we say that $V \hookrightarrow H$ is compact if the imbedding (that is, identity) mapping $i : V \to H$ is compact. In the case that the symmetric, positive definite, and bounded bi-linear form $a = a(\, , \,) : V \times V \to \mathbf{R}$ exists, the inner product $((\, , \,))$ of V is replaced by $a(\, , \,)$. Therefore, this condition is equivalent to saying that if $u_n, u \in V$ $(n = 1, 2, \cdots)$ satisfies $a(u_n - u, v) \to 0$ holds for any $v \in V$, then it holds that $|u_n - u| \to 0$. We continue to write $((\, , \,))$ and $\| \cdot \|$ for inner product and norm in V.

First, we show the following.

Theorem 3.15 *If the Gel'fand triple $V \hookrightarrow H \hookrightarrow V'$ with the compact inclusion $V \hookrightarrow H$ and the symmetric, positive definite, and bounded bi-linear form $a = a(\, , \,) : V \times V \to \mathbf{R}$ are given, then the bounded linear* *operator $T = A^{-1} : H \to H$ is compact, where A is the self-adjoint operator introduced in Theorem 3.14.*

Proof. Let $\{u_n\} \subset H$ be a bounded sequence. Then, it holds that

$$\begin{aligned} \|Tu_n\|^2 &= a(Tu_n, Tu_n) = (u_n, Tu_n) \\ &\leq |u_n| \cdot |Tu_n| \leq K\,|u_n| \cdot \|Tu_n\| \end{aligned}$$

and hence

$$\|Tu_n\| \leq K\,|u_n|$$

follows. Thus $\{Tu_n\} \subset D(A) \subset V$ is bounded in V and hence has a converging subsequence in H. □

For the moment, T denotes a linear operator on H not necessarily bounded. Given $\lambda \in \mathbf{R}$, the set of solutions to

$$Tv = \lambda v \tag{3.46}$$

forms a linear subspace of H, called the *eigenspace* associated with the eigenvalue λ. If the dimension of this space is greater than 1, that is, (3.46) admits a non-trivial solution $0 \neq v \in D(T)$, we say that λ and v are an *eigenvalue* and an *eigenfunction* of T, respectively. It is easy to see that if T is self-adjoint, then the eigenfunctions u, v associated with different eigenvalues are orthogonal so that $(u, v) = 0$ holds in H.

This paragraph is concluded by the following.

Theorem 3.16 *If $T : H \to H$ is a compact self-adjoint bounded linear operator, then $\|T\|$ or $-\|T\|$ is an eigenvalue of T, where $\|T\|$ denotes the operator norm of T.*

Proof. We may suppose that $\lambda = \|T\| \neq 0$. The linear operator $S = \lambda^2 - T^2 : H \to H$ is bounded, self-adjoint, and non-negative, so that $(Sv, v) \geq 0$ holds for any $v \in H$. Therefore, $[u, v] = (Su, v)$ provides a non-negative and bounded bi-linear form on $H \times H$. Similarly to the proof of the abstract Schwarz inequality of Theorem 3.1, we obtain

$$|[u, v]|^2 \leq [u, u] \cdot [v, v]$$

for $u, v \in H$. Letting $v = Su$ here, we get that

$$\begin{aligned} |Su|^4 &\leq (Su, u) \cdot (S^2 u, Su) \leq (Su, u) \cdot |S^2 u| \cdot |Su| \\ &\leq (Su, u) \cdot |Su|^2 \cdot \|S\| \end{aligned}$$

or equivalently,

$$|Su| \le \|S\|^{1/2} (Su, u)^{1/2} \qquad (u \in H). \tag{3.47}$$

Now, from $\|T\| = \{|Tu| \mid u \in H,\ |u| = 1\}$, we have $\{u_n\} \subset H$ with $|u_n| = 1$ and $|Tu_n| \to \lambda$. In this case, it holds that

$$\left(\left(\lambda^2 - T^2 \right) u_n, u_n \right) = \lambda^2 - |Tu_n|^2 \to 0$$

and hence

$$\left(\lambda^2 - T^2 \right) u_n \to 0 \tag{3.48}$$

follows from (3.47).

Because T is compact, $\{u_n\}$ admits a subsequence, denoted by the same symbol, and $w \in H$ such that

$$Tu_n \to w. \tag{3.49}$$

This implies $\lambda = |w| \neq 0$ and $T^2 u_n \to Tw$. Therefore, $\lambda^2 u_n \to Tw$ follows from (3.48), and then, $\lambda^2 Tu_n \to T^2 w$ follows so that

$$\lambda^2 w = T^2 w$$

holds by (3.49).

Therefore, $v = (\lambda + T)w$ satisfies $(\lambda - T)v = 0$ so that if $v \neq 0$, then $\lambda = \|T\|$ is an eigenvalue of T. Otherwise, $v = 0$ and hence $-\|T\|$ is an eigenvalue of T by $|w| = \lambda$. □

3.3.6 *Eigenfunction Expansions*

In this paragraph, the eigenfunction v is normalized as $|v| = 1$, and an ortho-normal basis is taken to each eigenspace. Namely, we show the following.

Theorem 3.17 *If H is a Hilbert space and $T : H \to H$ is a compact self-adjoint bounded linear operator, then its eigenvalues are discrete and converge to 0, and each eigenspace is of finite dimension. The family composed of ortho-normal basis of those spaces forms a basis of $R(T)$.*

Proof. We show that linearly independent $\{u_n\}_{n=1}^{\infty} \subset H$ with $|u_n| = 1$ cannot satisfy $Tu_n = \lambda_n u_n$ with $\lambda_n \to \lambda \neq 0$. This assures that eigenvalues of T is discrete and can accumulate only to 0.

In fact, if $\{u_n\}$ are linearly independent and M_n denotes the linear hull of $\{u_1, \cdots, u_n\}$, then there is $v_n \in M_n$ satisfying

$$|v_n| = 1 \qquad \text{and} \qquad \operatorname{dist}(v_n, M_{n-1}) \equiv \inf_{v \in M_{n-1}} |v_n - v| = 1. \tag{3.50}$$

We show that in this case $\{\lambda_n^{-1} T v_n\}$ cannot accumulate in H, which contradicts to the compactness of T.

For this purpose, we take $m < n$ and note that

$$\lambda_n^{-1} T v_n - \lambda_m^{-1} T v_m = v_n - \left(\lambda_m^{-1} T v_m - \lambda_n^{-1} (T - \lambda_n) v_n\right) \tag{3.51}$$

holds true. We have $v_m \in M_{n-1}$ and hence $Tv_m \in M_{n-1}$ holds. On the other hand, $(T - \lambda_n) u_n = 0$ is valid and therefore, $(T - \lambda_n) v_n \in M_{n-1}$ holds. The second term of the right-hand side of (3.51) is in M_{n-1}. We obtain

$$\left|\lambda_n^{-1} T v_n - \lambda_m^{-1} T v_m\right| \geq \operatorname{dist}(v_n, M_{n-1}) = 1.$$

Thus, $\{\lambda_n^{-1} T v_n\}$ has no subsequence converging in H.

We proceed to the latter part. From Theorem 3.16, there are $v_1 \in H$ and $\lambda_1 \in \mathbf{R}$ satisfying $|v_1| = 1$, $|\lambda_1| = \|T\|$, and $Tv_1 = \lambda_1 v_1$. Then, we take $H_1 = \{v_1\}^\perp$ and note that $T_1 = T|_{H_1} : H_1 \to H_1$ is a compact self-adjoint bounded linear operator. There are $v_2 \in H_1$ and $\lambda_2 \in \mathbf{R}$ satisfying $|v_2| = 1$, $|\lambda_2| = \|T_1\| \leq \|T\|$, and $Tv_2 = \lambda_2 v_2$. We repeat the argument, taking $H_2 = \{v_1, v_2\}^\perp$ and in this way obtain a sequence of eigenvalues denoted by $\{\lambda_j\}$.

Let us consider the case that those λ_j becomes eventually 0. There is n such that $\lambda_n = 0$ and $\lambda_j \neq 0$ for $1 \leq j \leq n-1$. In this case, we have $H_{n-1} \subset N(T)$ and $Tv_j = 0$ for $j \geq n$. On the other hand, $v_j \in R(T)$ for $j < n$ and hence $R(T)^\perp \subset \{v_1, \cdots, v_{n-1}\}^\perp = H_{n-1}$. Therefore, recalling $N(T) = R(T)^\perp$, we get that $H_{n-1} = N(T)$ and $R(T)$ is a linear hull of $\{v_1, \cdots, v_{n-1}\}$.

In the other case, we have $\lambda_n \neq 0$ for any n. Because of the fact proven before, it holds that $|\lambda_1| \geq |\lambda_2| \geq \cdots \to 0$. We show that any $w \in R(T)$ has the property that

$$\lim_{n\to\infty} \left| w - \sum_{i=1}^{n} (w, v_j) v_j \right| = 0. \tag{3.52}$$

In fact, in this case we have $w = Tu$ with some $u \in H$. Letting $b_j = (w, u_j)$ and $c_j = (u, v_j)$, we have

$$\begin{aligned} b_j &= (w, v_j) = (Tu, v_j) = (u, Tv_j) \\ &= \lambda_j(u, v_j) = \lambda_j c_j \end{aligned}$$

and hence

$$T(c_j v_j) = b_j v_j$$

follows. This implies

$$w - \sum_{j=1}^{n} b_j v_j = T\left(u - \sum_{j=1}^{n} c_j v_j\right) = T_n\left(u - \sum_{j=1}^{n} c_j v_j\right),$$

and therefore, by $\|T_n\| = |\lambda_{n+1}|$ and

$$\left|u - \sum_{j=1}^{n} c_j v_j\right| \leq |u| \tag{3.53}$$

we obtain

$$\left|w - \sum_{j=1}^{n} b_j v_j\right| \leq |\lambda_{n+1}| \cdot |u| \to 0.$$

This means (3.52) and the proof is complete. □

Exercise 3.31 Confirm that the eigenvalues $\{\lambda\}$ of T in $|\lambda| > \varepsilon$ is finite for each $\varepsilon > 0$ and that they are actually discrete.

Exercise 3.32 Show that there is $v_n \in M_n$ satisfying (3.50) in the proof of Theorem 3.17.

Exercise 3.33 Confirm that $T_1 = T|_{H_1} : H_1 \to H_1$ is a compact self-adjoint bounded linear operator in the proof of Theorem 3.17.

Exercise 3.34 Confirm (3.53) in use of the argument developed in §3.2.2.

3.3.7 Mini-Max Principle

Let $V \hookrightarrow H \hookrightarrow V'$ be the Gel'fand triple with compact $V \hookrightarrow H$ and $b = b(\ ,\ ) : H \times H \to \mathbf{R}$ and $a = a(\ ,\ ) : V \times V \to \mathbf{R}$ be symmetric, positive definite, and bounded bilinear forms. Replacing the inner product in H by $b(\ ,\ )$, we take the self-adjoint operator A associated with $a(\ ,\ )$ so that $\tilde{A} : V \to V'$ is defined by $\tilde{A}u = f$ if and only if

$$a(u, v) = b(f, v)$$

for any $v \in V$, and $A : D(A) \subset H \to H$ is the restriction of $\tilde{A}$ to

$$D(A) = \left\{ u \in V \mid \tilde{A}u \in H \right\}.$$

Then, $T = A^{-1} : H \to H$ is a compact self-adjoint bounded operator, and the eigenfunctions associated to the non-zero eigenvalue to

$$Tv = \lambda v$$

provides an ortho-normal basis $\{v_i\}$ of $R(T) = D(A)$ such that

$$b(v_i, v_j) = \delta_{ij}.$$

Thus, any $v \in D(A)$ admits the relation

$$\lim_{n \to \infty} \left| v - \sum_{i=1}^{n} \alpha_i v_i \right| = 0 \qquad \text{for} \quad \alpha_i = b(v, v_i). \tag{3.54}$$

This relation is valid for any $v \in H$, because $\overline{D(A)} = H$. Theorem 3.17 now implies the following.

Theorem 3.18 *Suppose that H is separable and of infinite dimension, and that the inclusion $V \hookrightarrow H$ is compact in the Gel'fand triple of Hilbert spaces $V \hookrightarrow H \hookrightarrow V'$. Suppose, furthermore, that $b = b(\ ,\ ) : H \times H \to \mathbf{R}$ and $a = a(\ ,\ ) : V \times V \to \mathbf{R}$ are symmetric, positive definite, and bounded bilinear forms. Then, the eigenvalue problem*

$$v \in V, \qquad a(v, w) = \mu b(v, w) \qquad \text{for any} \qquad w \in V \tag{3.55}$$

provides a countably many eigenvalues with the finite multiplicity, denoted by

$$0 < \mu_1 \leq \mu_2 \leq \cdots \to +\infty,$$

and the associated normalized system of eigenfunctions $\{v_i\}_{i=1}^{\infty}$ *forms a complete ortho-normal system in* H *in the sense that*

$$b(v_i, v_j) = \delta_{ij}$$

and (3.54) holds for any $v \in H$.

Here and henceforth, dimension of eigenspace associated with the eigenvalue in consideration is called *multiplicity.* In the above theorem, the eigenvalues are labeled according to their multiplicities and each eigenvalue takes one normalized eigenfunction. Furthermore, those eigenfunctions with the same eigenvalue are so arranged to be orthogonal by the orthogonalization of Schmidt.

Under those circumstances, it holds that

$$a(v_i, v_j) = \mu_j \delta_{ij} \tag{3.56}$$

$$b(v, v) = \sum_{k=1}^{\infty} c_k^2, \quad c_k = b(v, v_j) \qquad (v \in H) \tag{3.57}$$

$$a(v, v) = \sum_{k=1}^{\infty} \mu_k c_k^2 \qquad (v \in V). \tag{3.58}$$

Then, it is easy to see that

$$\mu_1 = \inf \{a(v, v)/b(v, v) \mid v \in V \setminus \{0\}\} \tag{3.59}$$

holds. Henceforth,

$$R[v] = a(v, v)/b(v, v)$$

is called the *Rayleigh quotient,* and (3.59) the *Rayleigh principle.*

A generalization of the Rayleigh principle is given by the following, referred to as the *mini-max principle.*

Theorem 3.19 *It holds that*

$$\begin{aligned} \mu_k &= \min_{L_k} \max_{v \in L_k \setminus \{0\}} R[v] \qquad (3.60) \\ &= \max_{V_k} \min_{v \in V_k \setminus \{0\}} R[v], \qquad (3.61) \end{aligned}$$

where $\{L_k\}$ *and* $\{V_k\}$ *denote the families of all subspaces of* H *with the dimension* k *and the codimension* $k - 1$*, respectively.*

Proof. To prove the first equality, we set

$$\Lambda_k = \inf_{L_k} \max_{v \in L_k \setminus \{0\}} R[v]$$

and

$$\begin{aligned} H_n &= \{v_1, \cdots, v_n\}^{\perp} \\ &= \{v \in H \mid b(v, v_i) = 0 \ (1 \le i \le n)\}. \end{aligned}$$

Then, for any L_k we have $\tilde{v} \in L_k \setminus \{0\}$ in $\tilde{v} \in H_{k-1}$ and it follows from (3.58) that

$$\max_{v \in L_k \setminus \{0\}} R[v] \ge R[\tilde{v}] \ge \mu_k.$$

This implies

$$\Lambda_k \ge \mu_k.$$

On the other hand, if L_k is taken as the linear hull of $\{v_1, \cdots, v_k\}$, then again from (3.58) we obtain

$$\max_{v \in L_k \setminus \{0\}} R[v] = \mu_k.$$

Thus, Λ_k is attained and is equal to μ_k.

To prove the second equality, we put

$$\Sigma_k = \sup_{V_k} \inf_{v \in V_k \setminus \{0\}} R[v].$$

Let W_n be the linear hull of $\{v_1, \cdots, v_n\}$. Then, any V_k satisfies $V_k \cap W_k \neq \{0\}$, and (3.58) guarantees that

$$\inf_{v \in V_k \setminus \{0\}} R[v] = \min_{v \in V_k \setminus \{0\}} R[v] \le \mu_k,$$

and hence $\Sigma_k \le \mu_k$ follows. On the other hand, for $V_k = H_{k-1}$ we have

$$\inf_{v \in V_k \setminus \{0\}} R[v] = \min_{v \in V_k \setminus \{0\}} R[v] = \mu_k.$$

Thus, Σ_k is attained, and is equal to μ_k. The proof is complete. □

Exercise 3.35 Confirm that (3.54) holds for any $v \in H$.

Exercise 3.36 Confirm that (3.58) and (3.59) hold true.

3.4 Distributions

3.4.1 *Dirac's Delta Function*

Distributional derivative is mentioned in §2.3.3. The notion of *distribution* is a generalization of that of function, and in this framework several calculations are admitted rigorously.

In §5.2.4, we shall introduce the Gaussian kernel

$$G(x,t) = \left(\frac{1}{4\pi t}\right)^{n/2} e^{-|x|^2/4t} \qquad (x = (x_1, \cdots, x_n) \in \mathbf{R}^n,\ t > 0) \quad (3.62)$$

and the operation

$$[G(\cdot,t) * u_0](x) = \int_{\mathbf{R}^n} G(x-y,t)u_0(y)dy.$$

There, it will be shown that $u(\cdot,t) = G(\cdot,t) * u_0$ solves

$$\frac{\partial u}{\partial t} = \Delta u \qquad (x \in \mathbf{R}^n,\ t > 0)$$

with

$$\lim_{t \downarrow 0} \|u(t) - u_0\|_p = 0$$

if $u_0 \in L^p(\mathbf{R}^n)$ with $p \in [1,\infty)$, where $L^p(\mathbf{R}^n)$ denotes the set of p-integrable functions. Then, what is

$$\delta(x) = \lim_{t \downarrow 0} G(x,t) \qquad ?$$

Because of

$$\int_{\mathbf{R}^n} G(x,t)dx = 1$$

it must hold that

$$\int_{\mathbf{R}^n} \delta(x)dx = 1, \qquad \delta(x) = \begin{cases} +\infty & (x = 0) \\ 0 & (x \neq 0). \end{cases}$$

However, such an object is not obviously a function any more. Actually, it is called *Dirac's delta function*, but is contained in the category of *distribution*, regarded as a natural extension of the notion of functions. This section is closely related to the objects treated in §5.2. The reader can skip this, but the program is as follows.

Let $\Omega \subset \mathbf{R}^n$ be a domain, and $C_0^\infty(\Omega)$ be a set of functions differentiable arbitrarily many times with compact supports contained in Ω. It is topologized as the *inductive limit* of $\mathcal{D}_{K_n}(\Omega)$ as $n \to \infty$, where $\{K_n\}_{n=1}^\infty$ is a family of compact sets in Ω satisfying $\cup_{n=1}^\infty K_n = \Omega$, and

$$\mathcal{D}_K = \{f \in C_0^\infty(\Omega) \mid \text{supp } f \subset K\}.$$

This space $\mathcal{D}_K$ can be provided with the distance if $K \subset \Omega$ is compact, although it is impossible to give norm here. It is complete in this distance and belongs to the category *Frechét space.* The vector space $C_0^\infty(\Omega)$ topologized in this way is denoted by $\mathcal{D}(\Omega)$. Then, we take the dual space

$$\mathcal{D}'(\Omega) = \{T : \mathcal{D}(\Omega) \to \mathbf{R} \mid \text{continuous linear}\}$$

and each element in $\mathcal{D}'(\Omega)$ is called the distribution on Ω.

A remarkable *theorem of Bourbaki* assures that the linear mapping $T : \mathcal{D}(\Omega) \to \mathbf{R}$ is in $\mathcal{D}'(\Omega)$ if and only if for any compact set $K \subset \Omega$ and any sequence $\{\varphi_j\} \subset \mathcal{D}_K(\Omega)$ satisfying $\partial^\alpha \varphi_j \to 0$ uniformly on K for any multi-index α, it holds that $T(\varphi_j) \to 0$. Remember that $\alpha = (\alpha_1, \cdots, \alpha_n)$ with non-negative integers $\alpha_1, \cdots, \alpha_n$ is indicated as the multi-index, and

$$\partial^\alpha = \left(\frac{\partial}{\partial x_1}\right)^{\alpha_1} \cdots \left(\frac{\partial}{\partial x_n}\right)^{\alpha_n}.$$

If a measurable function $f = f(x)$ is summable on any compact set $K \subset \Omega$, it is called *locally summable.* The set of such functions is denoted by $L_{loc}^1(\Omega)$. For such f, the linear mapping $T : \mathcal{D}(\Omega) \to \mathbf{R}$ is defined by

$$T(\varphi) = \int_{\mathbf{R}^n} f(x)\varphi(x)dx \qquad (\varphi \in \mathcal{D}(\Omega)). \tag{3.63}$$

It is not difficult to see that this T, denoted by T_f, is in $\mathcal{D}'(\Omega)$.

The following theorem assures that the function in $L_{loc}^1(\Omega)$ is identified with a distribution in Ω in this way.

Theorem 3.20 *The mapping* $f \in L_{loc}^1(\Omega) \mapsto T_f \in \mathcal{D}'(\Omega)$ *is one-to-one.*

Proof. Given $f \in L_{loc}^1(\Omega)$, we shall show that if

$$\int_\Omega f(x)\varphi(x)dx = 0 \tag{3.64}$$

holds for any $\varphi \in \mathcal{D}(\Omega)$, then it follows that $f = 0$ almost everywhere.

Let $\mu(dx)$ be the n-dimensional *Lebesgue measure* and $\{K_m\}_{m=1}^{\infty}$ be a sequence of compact sets in Ω monotone increasing and satisfies $\cup_m K_m = \Omega$. Taking

$$A_{m,n}^{\pm} = \left\{ x \in K_m \mid \pm f(x) > \frac{1}{n} \right\},$$

we shall show that

$$\mu(A_{m,n}^{\pm}) = 0 \qquad \text{for any} \qquad m, n = 1, 2, \cdots. \tag{3.65}$$

Then, it holds that $\mu(B) = 0$ for

$$B = \cup_{n,m=1}^{\infty} \left[A_{m,n}^{+} \cup A_{m,n}^{-} \right].$$

Because $B = \{x \in \Omega \mid f(x) \neq 0\}$ holds, this implies that $f = 0$ almost everywhere and the conclusion follows.

To confirm (3.65), let us assume the contrary that $\mu(A) \neq 0$ for some $A = A_{m,n}^{\pm}$. Then, from a property of the Lebesgue measure any $\varepsilon > 0$ admits compact K and open G in $K \subset A \subset G \subset \overline{G} \subset \Omega$ with $\overline{G}$ compact such that $\mu(G \setminus K) < \varepsilon$. In this case, there is $\varphi \in C_0^{\infty}(\Omega)$ with the value in $[0,1]$ such that

$$\varphi = \begin{cases} 1/\mu(K) & (\text{on } K) \\ 0 & (\text{on } G^c). \end{cases}$$

Thus, we obtain

$$\begin{aligned} \pm \int_{\Omega} f(x)\varphi(x)dx &= \pm \int_{G\setminus K} f(x)\varphi(x)dx \pm \int_K f(x)\varphi(x)dx \\ &\geq -\frac{1}{\mu(K)} \int_{G\setminus K} |f(x)|\, dx + \frac{1}{n}. \end{aligned} \tag{3.66}$$

Because of $f \in L_{loc}^1(\Omega)$, we have

$$\int_{G\setminus K} |f(x)|\, dx < \frac{\mu(K)}{2n}$$

if $\varepsilon > 0$ is small enough. Then the right-hand side of (3.66) is greater than $\frac{1}{2n}$, but this contradicts to (3.64) for any $\varphi \in \mathcal{D}(\Omega)$. □

If $T \in \mathcal{D}'(\Omega)$, then $S(\varphi) = (-1)^{|\alpha|}T(\partial^{\alpha}\varphi)$ for $\varphi \in \mathcal{D}(\Omega)$ is in $\mathcal{D}'(\Omega)$. We set $S = \partial^{\alpha} T$ and in this sense the distribution is differentiable arbitrarily many times and the order of differentiation can be changed arbitrarily.

If $T_j, T \in \mathcal{D}'(\Omega)$ satisfy

$$\lim_{j\to\infty} T_j(\varphi) = T(\varphi)$$

for any $\varphi \in \mathcal{D}(\Omega)$, then we say that $T_j \to T$ in $\mathcal{D}'(\Omega)$.

Exercise 3.37 Show that $T = T_f$ defined by (3.63) is a distribution. Show also that if f is a C^1 function on Ω, then

$$\frac{\partial}{\partial x_j} T_f = T_{\frac{\partial f}{\partial x_j}}$$

for $j = 1, \cdots, n$.

3.4.2 *Locally Convex Spaces*

Let X be a linear space over $\mathbf{R}$ or C. In the rest of this section, we take the case that X is over C. Then the mapping $p : X \to \mathbf{R}$ is said to be a *semi-norm* if it satisfies the axioms that

$$\begin{aligned} &p(x+y) \le p(x) + p(y) \quad (x, y \in X) \\ &p(\alpha x) = |\alpha|\, p(x) \quad\quad\quad (x \in X,\ \alpha \in \mathbf{C}). \end{aligned} \tag{3.67}$$

From those relations we can derive that

$$\begin{aligned} &p(0) = 0 \\ &|p(x) - p(y)| \le p(x-y) \quad (x, y \in X). \end{aligned} \tag{3.68}$$

In particular, $p(x) \ge 0$ holds for any $x \in X$.

We say that $V \subset X$ is *absolutely convex* if $\alpha x + \beta y \in V$ follows from $x, y \in V$ and $|\alpha| + |\beta| \le 1$, where $\alpha, \beta \in \mathbf{C}$. We say that $V \subset X$ is *absorbing* if any $x \in V$ admits $\alpha > 0$ such that $\alpha^{-1} x \in V$. The proof of the following theorem is left to the reader.

Theorem 3.21 *If p is a semi-norm on X, then*

$$V_p = \{x \in X \mid p(x) \le 1\}$$

is absolutely convex and absorbing. Conversely, if $V \subset X$ is absolutely convex and absorbing, then

$$p_V(x) = \inf\left\{\alpha > 0 \mid \alpha^{-1} x \in V\right\}$$

is a semi-norm on X.

Henceforth, p_V is referred to as the *Minkowski functional* induced by V.

A family $\mathcal{P}$ of semi-norms on X is said to satisfy the *axiom of separation* if $x \in X$ with $p(x) = 0$ for any $p \in \mathcal{P}$ implies $x = 0$. In this case, X is provided with the *Hausdorff topology* by defining the convergence of net $\{x_\nu\} \subset X$ to x by $p(x_\nu - x) \to 0$ for any $p \in \mathcal{P}$. It is referred to as the $\mathcal{P}$ topology of X. Actually, in this topology, the fundamental neighborhood system of $0 \in X$ is given by

$$\mathcal{U} = \{U_{n;p_1,\cdots,p_n;\varepsilon_1,\cdots,\varepsilon_n} \mid n = 1, 2, \cdots;\ p_1, \cdots, p_n \in \mathcal{P};\ \varepsilon_1, \cdots, \varepsilon_n > 0\},$$

where $U_{n;p_1,\cdots,p_n;\ \varepsilon_1,\cdots,\varepsilon_n} = \{x \in X \mid p_j(x) \leq \varepsilon_j \ \ (1 \leq j \leq n)\}$, and then, the fundamental neighborhood system of $x_0 \in X$ is $\{x_0 + U \mid U \in \mathcal{U}\}$. Under the $\mathcal{P}$ topology, the linear operations $(x, y) \mapsto x+y$ and $(\alpha, x) \mapsto \alpha x$ are continuous in X, and in this sense, X becomes a topological vector space.

The vector space X is said to be a *locally convex space* if it is provided with some $\mathcal{P}$ topology. Then, $\mathcal{P}$ is said to be the semi-norm system determining the topology of X. The proof of the following theorem is also left to the reader.

Theorem 3.22 *If q is a semi-norm on X, then it is continuous in $\mathcal{P}$ topology if and only if there are some n, $p_1, \cdots, p_n \in \mathcal{P}$, and $c_1, \cdots, c_n > 0$ such that*

$$q(x) \leq c_1 p_1(x) + \cdots c_n p_n(x) \qquad (x \in X).$$

Admitting the above theorem, we see that if q is a continuous semi-norm on X with respect to $\mathcal{P}$ topology, then $\mathcal{P}$ topology is equivalent to $\mathcal{P}'$ topology, where $\mathcal{P}' = \mathcal{P} \cup \{q\}$. Generally, we say that a family of semi-norms $\mathcal{P}'$ defines the topology of X if $\mathcal{P}'$ topology is equivalent to the original one, the $\mathcal{P}$ topology. From those considerations, we see that a fundamental neighborhood system of 0 in X is taken as

$$\{V_p \mid p \text{ is a continuous semi-norms on } X\}.$$

Exercise 3.38 Derive (3.68) from (3.67).

Exercise 3.39 Give the proof of Theorem 3.21.

Exercise 3.40 Confirm that X becomes a topological vector space under the $\mathcal{P}$ topology.

Exercise 3.41 Show that a semi-norm q on X is continuous if and only if $V_q = \{x \in X \mid q(x) \le 1\}$ is a neighborhood of 0.

Exercise 3.42 Given an absolutely convex and absorbing set $V \subset X$, show that

$$V \subset V_p \subset \cap_{\lambda>1} \lambda V$$

holds for $p = p_V$. Then, prove that it is a neighborhood of 0 if and only if its Minkowski functional p_V is continuous.

Exercise 3.43 Mimicking the case of normed spaces, show that if X and Y are locally convex spaces and $T : X \to Y$ is linear, then T is continuous if and only if any continuous semi-norm q on Y admits a continuous semi-norm p on X such that $q(Tx) \le p(x)$ holds for any $x \in X$.

3.4.3 *Fréchet Spaces*

Locally convex space X is said to be *normizable* if there is a family of semi-norms $\mathcal{P}$ determining its topology consisting of one element. Locally convex space X is said to be *metrizable* if there is a distance $d(\ ,\ )$ which provides an equivalent topology to X. Now, we show the following.

Theorem 3.23 *Locally convex space X is metrizable if and only if there is a family of semi-norms $\mathcal{P}$ determining its topology which consists of a countable number of elements.*

Proof. If $\mathcal{P}$ is countable as $\mathcal{P} = \{p_j \mid j = 1, 2, \cdots\}$, then

$$d(x, y) = \sum_{j=1}^{\infty} \frac{1}{2^j} \cdot \frac{p_j(x-y)}{1 + p_j(x-y)} \tag{3.69}$$

is a metric provided with the required properties. Conversely, if $d(\ ,\ )$ provides the metric equivalent to the original topology to X, then $\{V_j\}_{j=1}^{\infty}$ is a fundamental neighborhood system of 0, where $V_j = \left\{x \in X \mid d(x, 0) \le \frac{1}{j}\right\}$. Then, there exists a continuous semi-norm p_j such that $V_{p_j} \subset V_j$, and $\mathcal{P} = \{p_j\}$ provides the equivalent topology to X. □

A net $\{x_\nu\} \subset X$ is said to be *Cauchy* if $p(x_\nu - x_\mu) \to 0$ holds for any $p \in \mathcal{P}$ as $\nu, \mu \to \infty$. Then, locally convex space X is said to be *complete* if any Cauchy net $\{x_\nu\} \subset X$ admits $x \in X$ such that $x_\nu \to x$ as $\nu \to \infty$. *Fréchet space* indicates a complete metrizable locally convex space. In this case, convergence of Cauchy sequences is sufficient for the completeness to establish.

We have put that $\mathcal{D}_K = \{\varphi \in C_0^\infty(\Omega) \mid \text{supp}\, \varphi \subset K\}$ if $\Omega \subset \mathbf{R}^n$ is a domain and $K \subset \Omega$ is compact. Then,

$$p_m(f) = \sup_{x \in K,\ |\alpha| \leq m} |\partial^\alpha f(x)|$$

is a semi-norm on $\mathcal{D}_K(\Omega)$ for $m = 0, 1, 2, \cdots$, and $\mathcal{D}_K(\Omega)$ becomes a Fréchet space by $\mathcal{P} = \{p_m \mid m = 0, 1, 2, \cdots\}$.

Another example is $\mathcal{E}(\Omega) = C^\infty(\Omega)$. In fact,

$$p_{m,K}(f) = \sup_{x \in K,\ |\alpha| \leq m} |\partial^\alpha f(x)|$$

is a semi-norm on $\mathcal{E}(\Omega)$ for $m = 0, 1, \cdots$ and a compact set $K \subset \Omega$. Then, $\mathcal{E}(\Omega)$ becomes a Fréchet space by

$$\mathcal{P} = \{p_{m,K} \mid m = 0, 1, 2, \cdots; K \subset \Omega \text{ :compact}\}.$$

Exercise 3.44 Show that $d(\ ,\ )$ defined by (3.69) is a metric on X. Show also that the net $\{x_\nu\} \subset X$ converges to $x \in X$ in $\mathcal{P}$ topology if and only if $d(x_\nu, x) \to 0$ holds.

Exercise 3.45 If X is a locally convex space with the family $\mathcal{P}$ of semi-norms determining its topology, then $B \subset X$ is said to be *bounded* if

$$\sup_{x \in B} p(x) < +\infty$$

holds for any $p \in \mathcal{P}$. Show that X is normizable if and only if there is a bounded neighborhood of 0.

Exercise 3.46 Show that $\mathcal{D}_K(\Omega)$ is a Fréchet space.

Exercise 3.47 Show that $\mathcal{E}(\Omega)$ is a Fréchet space.

3.4.4 *Inductive Limit*

Let $X_1 \subset X_2 \subset \cdots \subset X_n \subset X_{n+1} \subset \cdots$ be a sequence of locally convex spaces and suppose that the original topology of X_n is equivalent to the one induced by X_{n+1} for $n = 1, 2, \cdots$. Henceforth, we shall write $X_n \hookrightarrow X_{n+1}$ for this situation. Our purpose is to make $X = \cup_{n=1}^{\infty} X_n$ a locally convex space satisfying $X_n \hookrightarrow X$ for any n. First, we note the following.

Lemma 3.2 *If X and Y are locally convex spaces satisfying $X \hookrightarrow Y$, then any absolutely convex neighborhood of 0 in X denoted by U admits an absolutely convex neighborhood W of 0 in Y such that $U = W \cap X$.*

Proof. From the assumption, there is an absolutely convex neighborhood V of 0 in Y such that $X \cap V \subset U$. We shall show that the convex hull of $U \cup V$ is provided with the required properties:

$$W = \{\lambda u + (1-\lambda)v \mid u \in U,\ v \in V,\ 0 \le \lambda \le 1\}.$$

First, to show $U = W \cap X$ we note that $U \subset W \cap X$ is obvious. If $w \in W \cap X$, then it is written as $\lambda u + (1-\lambda)v$. We have

$$(1-\lambda)v = w - \lambda u \in X.$$

If $\lambda \neq 1$, then $v \in X \cap V \subset U$ and hence $w = \lambda u + (1-\lambda)v \in U$ follows. In the case of $\lambda = 1$, it holds that $w = u \in U$. In any case, we get $w \in U$ and hence $W \cap X \subset U$ follows.

Next, we show that $W \subset Y$ is absolutely convex. For this purpose, we take $|\alpha| + |\beta| \le 1$, $u_1, u_2 \in U$, $v_1, v_2 \in V$, and $0 \le \lambda_1, \lambda_2 \le 1$. Then, it holds that

$$\begin{aligned}
&(\alpha\lambda_1 u_1 + \alpha(1-\lambda_1)v_1) + (\beta\lambda_2 u_2 + \beta(1-\lambda_2)v_2) \\
&= \frac{|\alpha|\lambda_1 + |\beta|\lambda_2}{|\alpha| + |\beta|} \cdot \frac{\alpha\lambda_1 u_1 + \beta\lambda_2 u_2}{(|\alpha|\lambda_1 + |\beta|\lambda_2)/(|\alpha| + |\beta|)} \\
&\quad + \frac{|\alpha|(1-\lambda_1) + |\beta|(1-\lambda_2)}{|\alpha| + |\beta|} \\
&\quad \cdot \frac{\alpha(1-\lambda_1)v_1 + \beta(1-\lambda_2)v_2}{(|\alpha|(1-\lambda_1) + |\beta|(1-\lambda_2))/(|\alpha| + |\beta|)}.
\end{aligned}$$

The right-hand side is expressed as $\lambda_3 u_3 + (1-\lambda_3)v_3$ with $u_3 \in U$, $v_3 \in V$, and $0 \le \lambda_3 \le 1$, and therefore, in W.

We finally note that $W \subset Y$ is a neighborhood of 0 because $V \subset W$, and the proof is complete. □

Analytic expression of the previous lemma is given as follows.

Lemma 3.3 *If X and Y are locally convex spaces satisfying $X \hookrightarrow Y$, then any continuous semi-norm p on X admits a continuous semi-norm $\tilde{p}$ on Y such that $\tilde{p}|_X = p$.*

Proof. We can apply Lemma 3.2 to $V_p = \{x \in X \mid p(x) \leq 1\}$. Then, we get an absolutely convex neighborhood W of 0 in Y, such that $W \cap X = V_p$. Letting $\tilde{p} = p_W$, the Minkowski functional of $W \subset Y$, we obtain the conclusion. In fact, $\tilde{p}$ is a continuous semi-norm on Y. Furthermore, if $x \in X$ then it holds that

$$\begin{aligned} p_W(x) &= \inf\left\{\alpha > 0 \mid \alpha^{-1}x \in W\right\} \\ &= \inf\left\{\alpha > 0 \mid \alpha^{-1}x \in W \cap X\right\} \\ &= \inf\left\{\alpha > 0 \mid \alpha^{-1}x \in V_p\right\} \\ &= \inf\left\{\alpha > 0 \mid \alpha^{-1}p(x) \leq 1\right\} = p(x). \end{aligned}$$

The proof is complete. □

Now, we show the following

Theorem 3.24 *Let $X_1 \hookrightarrow X_2 \hookrightarrow \cdots \hookrightarrow X_n \hookrightarrow X_{n+1} \hookrightarrow \cdots$ be a sequence of locally convex spaces, $X = \cup_{n=1}^{\infty} X_n$, and*

$$\mathcal{P} = \left\{p : \textit{semi-norm on } X \mid p|_{X_n} \quad \textit{is continuous for any } n\right\}.$$

Then, $\mathcal{P}$ admits the axiom of separation, and induces $\mathcal{P}$ topology to X.

Proof. Let $x \in X$ satisfy $p(x) = 0$ for any $p \in \mathcal{P}$.

There is X_n such that $x \in X_n$. Let $\mathcal{P}_n$ be the family of semi-norms determining the topology of X_n. Applying Lemma 3.3 successively, we see that any $q \in \mathcal{P}_n$ admits a semi-norm $\tilde{q}$ on X such that $\tilde{q}|_{X_n} = q$ and $\tilde{q}|_{X_m}$ is continuous on X_m for any $m \geq n$. This guarantees $\tilde{q} \in \mathcal{P}$ and hence $\tilde{q}(x) = q(x) = 0$ from the assumption. Because $q \in \mathcal{P}_n$ is arbitrary, we see that $x \in X_n$ is equal to 0. This means the conclusion. □

Because $\mathcal{P}$ induces topology to any $Y \subset X$, the following fact is to be noted.

Theorem 3.25 *The $\mathcal{P}$ topology introduced to X in the previous theorem induces the topology to $X_n \subset X$ equivalent to the original one.*

Proof. The assertion follows from the definition of $\mathcal{P}$ and the fact that any $q \in \mathcal{P}_n$ admits $\tilde{q} \in \mathcal{P}$ such that $\tilde{q}|_{X_n} = q$. □

The following fact indicates that $\mathcal{P}$ topology of X is independent of the choice of $\{X_n\}$, and the proof is left to the reader.

Theorem 3.26 *Let $\hat{X}_1 \hookrightarrow \hat{X}_2 \hookrightarrow \cdots \hookrightarrow \hat{X}_m \hookrightarrow \hat{X}_{m+1} \hookrightarrow \cdots$ be a sequence of locally convex spaces equivalent to $X_1 \hookrightarrow X_2 \hookrightarrow \cdots \hookrightarrow X_n \hookrightarrow X_{n+1} \hookrightarrow \cdots$, which means that any n admits m satisfying $X_n \hookrightarrow \hat{X}_m$ and any m admits n satisfying $\hat{X}_m \hookrightarrow X_n$. Then, $\mathcal{P}$ and $\hat{\mathcal{P}}$ topologies introduced to X through $\{X_n\}$ and $\{\hat{X}_m\}$, respectively, are equivalent to each other.*

The locally convex space X defined in this way is called the *inductive limit* of $\{X_n\}$ and is written as $X = \lim_{n\to} X_n$. The only if part of the following theorem is obvious by Theorem 3.25.

Theorem 3.27 *If $X = \lim_{n\to} X_n$, Y is a locally convex space, and $T : X \to Y$ is linear, then T is continuous if and only if $T|_{X_n} : X_n \to Y$ is continuous for any X_n.*

Proof. To prove the if part, we shall apply the criterion given by Exercise 3.43. Thus, given continuous semi-norm q on X, we show the existence of a continuous semi-norm p on Y such that $q(Tx) \leq p(x)$ for any $x \in X$. In fact, $p(x) = q(Tx)$ is a semi-norm on X and $p|_{X_n}$ is continuous for each X_n from the assumption. This implies $p \in \mathcal{P}$ and the proof is complete. □

If $\Omega \subset \mathbf{R}^n$ is a domain, the locally convex space $\mathcal{D}(\Omega) = C_0^\infty(\Omega)$ is defined by $\mathcal{D}(\Omega) = \lim_{n\to} \mathcal{D}_{K_n}(\Omega)$, where $\{K_n\}$ denotes a monotone increasing family of compact sets in Ω satisfying $\cup_n K_n = \Omega$. However, $\mathcal{D}(\Omega)$ does not satisfy the *first axiom of countability*.

Exercise 3.48 Let $X = \cup_{n=1}^\infty X_n$ and $\mathcal{P}$ be as in Theorem 3.5, and $U \subset X$ be absolutely convex. Then, show that U is a neighborhood of 0 in X if and only if $U|_{X_n}$ is a neighborhood of 0 in X_n for any n.

Exercise 3.49 Prove Theorem 3.26 by showing that any $p \in \mathcal{P}$ satisfies $p \in \tilde{\mathcal{P}}$.

Exercise 3.50 Show that the topology of $\mathcal{D}(\Omega)$ is independent of the choice of the monotone increasing family of compact sets $\{K_n\}$ satisfying $\cup_n K_n = \Omega$.

3.4.5 Bounded Sets

If X is a locally convex space with the family of semi-norms $\mathcal{P}$ determining its topology, the set $B \subset X$ is said to be bounded if $\{p(x) \mid x \in B\}$ is bounded for any $p \in \mathcal{P}$. In the case of $X = \lim_{n\to} X_n$, if $B \subset X$ is bounded then we have $B \subset X_n$ for some X_n. This is a key structure to make the definition of distributions, $\mathcal{D}'(\Omega)$, much simpler. We note the following.

Lemma 3.4 *If $X \hookrightarrow Y$ are locally convex spaces, $x_0 \in Y \setminus X$, and U is an absolutely convex neighborhood of 0 in X, then there is an absolutely convex neighborhood of 0 in Y, denoted by W, such that $U = W \cap X$ and $x_0 \notin W$.*

Proof. We shall only give the outline. First, we take an absolutely convex neighborhood V of 0 in Y such that $V \cap X \subset U$ and $(x_0 + V) \cap U = \emptyset$. Then, the desired set is given by

$$W = \{\lambda u + (1-\lambda)v \mid u \in U,\ v \in V,\ 0 \leq \lambda \leq 1\}.$$

□

Now, we give the following.

Theorem 3.28 *If $X = \lim_{n\to} X_n$, then $B \subset X$ is bounded if and only if there is X_n such that $B \subset X_n$ is bounded.*

Proof. We show that if $B \subset X$ is bounded then $B \subset X_n$ holds with some X_n because the other cases are obvious. In fact, if this is not the case, there are $X_{n_1} \hookrightarrow X_{n_2} \hookrightarrow \cdots$ and $\{u_k\} \subset B$ in $u_k \in X_{n_{k+1}} \setminus X_{n_k}$ for $k = 1, 2, \cdots$. By Lemma 3.4, we have an absolutely convex neighborhood W_k of 0 in X_{n_k} satisfying

$$W_{k+1} \cap X_{n_k} = W_k \qquad \text{and} \qquad \frac{u_k}{k} \notin W_{k+1}$$

for $k = 1, 2, \cdots$. Then $W = \cup_{k=1}^{\infty} W_k$ is absolutely convex in X and is a neighborhood of 0 there by Exercise 3.48.

If $\frac{u_k}{k} \in W$, then $\frac{u_k}{k} \in W \cap X_{n_{k+1}} = W_{k+1}$. This is impossible and hence

$$\frac{u_k}{k} \notin W$$

holds for $k = 1, 2, \cdots$. However, $\{u_k\} \subset B$ is bounded and hence $\frac{u_k}{k} \to 0$ follows as $k \to \infty$. This is a contradiction, and the proof is complete. □

The following fact is a direct consequence of Theorem 3.28 because any Cauchy sequence is bounded. Actually $X = \lim_{n\to} X_n$ is a complete uniform space in that case, although X may not be a Fréchet space in that case.

Theorem 3.29 *If each X_n is a Fréchet space, then $X = \lim_{n\to} X_n$ is sequentially complete.*

Exercise 3.51 Complete the proof of Lemma 3.4.

Exercise 3.52 Introduce a suitable topology to the set of holomorphic functions on a domain $\Omega \subset \mathbf{C}$ and make it a (sequentially) complete locally convex space.

3.4.6 *Definition and Examples*

If $\Omega \subset \mathbf{R}^n$ is a domain, then each element of

$$\mathcal{D}'(\Omega) = \{T : \mathcal{D}(\Omega) \to \mathbf{C} \mid \text{continuous linear}\}$$

is called the *distribution* on Ω. From Theorem 3.27, if $T : \mathcal{D}(\Omega) \to \mathbf{C}$ is linear, $T \in \mathcal{D}'(\Omega)$ if and only if

$$T|_{\mathcal{D}_K(\Omega)} : \mathcal{D}_K(\Omega) \quad \to \quad \mathcal{C}$$

is continuous for any compact $K \subset \Omega$. This condition is equivalent to saying that any compact $K \subset \Omega$ and integer $m \geq 0$ admits $C > 0$ such that

$$|T(\varphi)| \leq C \sup_{x\in K,\ |\alpha|\leq m} |\partial^\alpha \varphi(x)|$$

for any $\varphi \in \mathcal{D}_K(\Omega)$, or that for any compact $K \subset \Omega$ and for any sequence $\{\varphi_j\} \subset \mathcal{D}_K(\Omega)$ satisfying $\partial^\alpha \varphi_j \to 0$ uniformly on K it holds that $T(\varphi_j) \to 0$. This fits the rough definition in §3.4.1 and by Theorem 3.20 any locally summable function on Ω is regarded as a distribution there.

A *Radon measure* $\mu(dx)$ on Ω is a *Borel measure* such that $\mu(K) < +\infty$ for any compact $K \subset \Omega$. Then, it is easy to see that

$$T_\mu(\varphi) = \int_\Omega \varphi(x)\mu(dx) \qquad (\varphi \in \mathcal{D}_K(\Omega))$$

defines a distribution on Ω. Furthermore, $\mu \mapsto T_\mu$ is shown to be one-to-one, and in this sense the Radon measure is regarded as a distribution. In particular,

$$\delta_p(\varphi) = \varphi(p) \qquad (\varphi \in \mathcal{D}(\Omega))$$

defines a distribution for $p \in \Omega$. It is called the Dirac's delta function and is denoted by $\delta = \delta_0 \in \mathcal{D}'(\mathbf{R}^n)$ for simplicity. Also, sometimes it is written as $\delta(dx)$ or $\delta(x)$.

In the one-dimensional case,

$$\left(\text{P.f. } \frac{1}{x}\right)(\varphi) = \lim_{\varepsilon \downarrow 0} \int_{|x| \geq \varepsilon} \frac{\varphi(x)}{x} dx \qquad (\varphi \in \mathcal{D}(\mathbf{R}))$$

defines a distribution P.f. $\frac{1}{x}$. In fact, we have

$$\begin{aligned} \int_{|x|\geq\varepsilon} \frac{\varphi(x)}{x} dx &= \int_{-\infty}^{-\varepsilon} \frac{\varphi(x)}{x} dx + \int_{\varepsilon}^{\infty} \frac{\varphi(x)}{x} dx \\ &= \int_{\varepsilon}^{\infty} \frac{\varphi(x) - \varphi(-x)}{x} dx \end{aligned}$$

and hence

$$\left(\text{P.f. } \frac{1}{x}\right)(\varphi) = \int_0^\infty \frac{\varphi(x) - \varphi(-x)}{x} dx \tag{3.70}$$

follows for $\varphi \in \mathcal{D}(\Omega)$. If $m \in \mathbf{C} \setminus \{-1, -2, \cdots\}$ we take an integer $k > 0$ in $\Re\, m + k + 1 > -1$ and set

$$\left(\text{P.f. } (x^m)_{x>0}\right)(\varphi) = \frac{(-1)^k}{(m+1)\cdots(m+k+1)} \int_0^\infty x^{m+k+1} \varphi^{(k)}(x) dx \tag{3.71}$$

for $\varphi \in \mathcal{D}(\mathbf{R})$.

If $T \in \mathcal{D}'(\Omega)$, then

$$S(\varphi) = (-1)^{|\alpha|} T(\partial^\alpha \varphi) \qquad (\varphi \in \mathcal{D}(\Omega))$$

defines $S \in \mathcal{D}'(\Omega)$, where α is a multi-index. It is denoted by $S = \partial^\alpha T$. In §3.4.1 we note that $\partial^\alpha T_f = T_{\partial^\alpha f}$ if $f = f(x)$ is a C^m function in Ω for $m = |\alpha|$. It also holds that $\partial^\alpha \delta(\varphi) = (-1)^{|\alpha|} \partial^\alpha \varphi(0)$ for $\varphi \in \mathcal{D}(\Omega)$. In this way, the distribution can *differentiable* any times and the order of differentiations can be changed arbitrarily.

In the one-dimensional case, the locally summable function

$$Y(x) = \begin{cases} 1 & (x \geq 0) \\ 0 & (x < 0) \end{cases}$$

is called the *Heaviside function*. Then it holds that

$$\frac{d}{dx}Y(x) = \delta(x) \tag{3.72}$$

as a distribution on $\mathbf{R}$. In the n-dimensional case, the locally summable function

$$\Gamma(x) = \begin{cases} \frac{1}{\omega_n(2-n)} \cdot \frac{1}{|x|^{n-2}} & (n \geq 3) \\ \frac{1}{\omega_2} \log |x| & (n = 2) \end{cases}$$

satisfies

$$\Delta\Gamma = \delta(dx), \tag{3.73}$$

where $\Delta = \sum_{j=1}^{n} \frac{\partial^2}{\partial x_j^2}$ denotes the *Laplacian* and ω_n is the area of n-dimensional unit ball. The cases of $n = 2$ and $n = 3$ are called the *Newton potential* and the *logarithmic potential*, respectively.

Generally, the linear differential operator $L(D)$ with constant coefficients, $T \in \mathcal{D}'(\mathbf{R}^n)$ is called the *fundamental solution* if $L(D)T = \delta(dx)$ holds. The *Malgrange-Ehrenpreis* theorem guarantees the existence of the fundamental solution for each $L(D)$. The Gaussian kernel $G(x,t)$ defined by (3.62) is also called the fundamental solution to $\partial_t - \Delta$, because it satisfies that

$$(\partial_t - \Delta)G(x,t) = 0 \qquad (x,t) \in \mathbf{R}^n \times (0,\infty)$$

and

$$\lim_{t \downarrow 0} G(x,t) = \delta(dx).$$

Exercise 3.53 Show that $\mu(dx) \in \mathcal{M}(\Omega) \mapsto T_\mu \in \mathcal{D}'(\Omega)$ is one-to-one, where $\mathcal{M}(\Omega)$ denotes the set of Radon measures on Ω.

Exercise 3.54 Confirm that P.f. $\frac{1}{x} \in \mathcal{D}'(\Omega)$ is well-defined by (3.70).

Exercise 3.55 Show that P.f. $(x^m)_{x>0}$ defined by (3.71) is a distribution on $\mathbf{R}$. In use of the partial integral and take the convergent part of $\int_0^\infty x^m \varphi(x)dx$ and in that way define P.f. $(x^m)_{x>0}$ for $m = -2, -3, \cdots$ as in the case of $m = -1$.

Exercise 3.56 Confirm (3.72).

Exercise 3.57 Show that (3.73) holds in use of the argument in §5.3.6.

Exercise 3.58 Seek fundamental solutions to Δ^2 in $\mathbf{R}^2$ and $\mathbf{R}^3$ in the form of $K(|x|)$ in use of the logarithmic and Newton potentials, respectively, and the representation in polar coordinate of $x = |x|\,\omega$.

Exercise 3.59 Writing $z = x + \imath y \in \mathbf{R}^2 \cong \mathbf{C}$ and $\overline{z} = x - \imath y$, we have

$$x = \frac{z + \overline{z}}{2} \qquad \text{and} \qquad y = \frac{z - \overline{z}}{2\imath}$$

and hence

$$\frac{\partial}{\partial \overline{z}} = \frac{1}{2}\left(\frac{\partial}{\partial x} + \imath \frac{\partial}{\partial y}\right) \qquad \text{and} \qquad \frac{\partial}{\partial z} = \frac{1}{2}\left(\frac{\partial}{\partial x} - \imath \frac{\partial}{\partial y}\right)$$

follow. First, confirm that f is holomorphic if and only if $\partial f/\partial \overline{z} = 0$ and then it holds that $\partial f/\partial z = f'(z)$ in use of Cauchy-Riemann's relation. Second, prove that

$$\frac{1}{\pi}\frac{\partial}{\partial \overline{z}}\left(\frac{1}{z}\right) = \delta$$

holds as $\mathcal{D}'(\mathbf{R}^2)$.

3.4.7 *Fundamental Properties*

In this paragraph, we shall try to perform some calculations. First, we show the following.

Theorem 3.30 *It holds that*

$$\frac{d}{dx}(\log|x|) = P.f.\ \frac{1}{x} \tag{3.74}$$

as $\mathcal{D}'(\mathbf{R})$.

Proof. In fact, we have $\log|x| \in L^1_{loc}(\mathbf{R})$. Then, from the definition we get for $\varphi \in \mathcal{D}(\mathbf{R})$ that

$$\begin{aligned}\left(\frac{d}{dx}\log|x|\right)(\varphi) &= -\int_{-\infty}^{\infty}(\log|x|)\,\varphi'(x)dx \\ &= -\lim_{\varepsilon\downarrow 0}\left(\int_{\varepsilon}^{\infty}+\int_{-\infty}^{-\varepsilon}\right)(\log|x|)\,\varphi'(x)dx \\ &= \lim_{\varepsilon\downarrow 0}\int_{|x|\geq\varepsilon}\frac{\varphi(x)}{x}dx + [\log|x|\,\varphi(x)]_{x=-\varepsilon}^{x=\varepsilon}.\end{aligned}$$

Here, we have

$$\begin{aligned}[\log|x|\,\varphi(x)]_{x=-\varepsilon}^{x=\varepsilon} &= (\log\varepsilon)\,(\varphi(\varepsilon)-\varphi(-\varepsilon)) \\ &= 2\varepsilon\log\varepsilon\cdot\frac{\varphi(\varepsilon)-\varphi(-\varepsilon)}{2\varepsilon}\to 0\end{aligned}$$

as $\varepsilon\downarrow 0$ and hence (3.74) follows as

$$\left(\frac{d}{dx}\log|x|\right)(\varphi) = \lim_{\varepsilon\downarrow 0}\int_{|x|\geq\varepsilon}\frac{\varphi(x)}{x}dx.$$

The proof is complete. □

Now, we show the following.

Theorem 3.31 *If $T \in \mathcal{D}'(\mathbf{R})$ satisfies*

$$\frac{d}{dx}T = 0,$$

then $T = c$ with some $c \in \mathbf{C}$.

Proof. We take $\varphi_0 \in \mathcal{D}(\mathbf{R})$ in $\int_{-\infty}^{\infty}\varphi_0(x)dx = 1$. If $\varphi \in \mathcal{D}(\mathbf{R})$, then

$$\psi(x) = \varphi(x) - \varphi_0(x)\int_{-\infty}^{\infty}\varphi(x)dx$$

satisfies that

$$\int_{-\infty}^{\infty}\psi(x)dx = 0 \quad \text{and} \quad \psi \in C_0^{\infty}(\mathbf{R}).$$

We have

$$\Psi(x) = \int_{-\infty}^{x}\psi(y)dy \in C_0^{\infty}(\mathbf{R}),$$

and from the assumption it follows that

$$T(\psi) = T(\Psi') = -T'(\Psi) = 0.$$

This means that

$$T(\varphi) = \int_{-\infty}^{\infty} \varphi(x)dx \cdot T(\varphi_0)$$

or equivalently, $T = T(\varphi_0) \in C$, and the proof is complete. □

Given an absolutely continuous function $f = f(x)$, we distinguish its classical derivative f' and its distributional derivative $\frac{df}{dx}$ for the moment. However, those two notions coincide as follows.

Theorem 3.32 *If $f = f(x)$ is locally absolutely continuous on $\mathbf{R}$, then it holds that $\frac{df}{dx} = f'$ as distributions. Conversely, if f and $\frac{df}{dx}$ are locally summable functions, then $f = f(x)$, modified on a set of Lebesgue measure 0, becomes locally absolutely continuous on $\mathbf{R}$ and it holds that $f' = \frac{df}{dx}$ almost everywhere.*

Proof. The first part is easier to prove, as under the assumption it holds for $\varphi \in \mathcal{D}(\mathbf{R})$ that

$$\begin{aligned} T_{\frac{df}{dx}}(\varphi) &= -T_f(\varphi') = -\int_{-\infty}^{\infty} f(x)\varphi'(x)dx \\ &= \int_{-\infty}^{\infty} f'(x)\varphi(x)dx. \end{aligned}$$

To prove the latter part, we also take $\varphi \in \mathcal{D}(\mathbf{R})$ and note that

$$\begin{aligned} \int_{-\infty}^{\infty} \left(\frac{df}{dx}\right) \varphi dx &= T_{\frac{df}{dx}}(\varphi) = -T_f(\varphi') \\ &= -\int_{-\infty}^{\infty} f(x)\varphi'(x)dx. \end{aligned} \tag{3.75}$$

Let $\chi \in C_0^{\infty}(\mathbf{R})$ be such that $0 \leq \chi = \chi(x) \leq 1$ and

$$\chi(x) = \begin{cases} 1 & (|x| \leq 1) \\ 0 & (|x| \geq 2). \end{cases}$$

Given $a < b$, we take

$$\varphi_n(x) = \begin{cases} \chi(nx + 1 - nb) & x \in (b, \infty) \\ 1 & x \in [a, b] \\ \chi(nx - 1 - na) & x \in (-\infty, a), \end{cases}$$

where $n = 1, 2, \cdots$. Then, it holds that $\chi(x) = 0$ if $x < a - n^{-1}$ or $x > b + n^{-1}$, and relation (3.75) gives

$$\int_{-\infty}^{\infty} \left(\frac{df}{dx} \right) \varphi_n(x) dx = - \int_{-\infty}^{\infty} f(x) \varphi_n'(x) dx.$$

Here, the left-hand side converges to

$$\int_a^b \frac{df}{dx} dx$$

as $n \to \infty$ from the dominated convergence theorem. On the other hand, the right-hand side is equal to

$$- \int_{-\infty}^{a} f(x) \varphi_n'(x) dx - \int_b^{\infty} f(x) \varphi_n'(x) dx.$$

The first term is treated as

$$\begin{aligned} \left| \int_{-\infty}^{a} f(x) \varphi_n'(x) dx - f(a) \right| &= \left| \int_{a-n^{-1}}^{a} (f(x) - f(a)) \varphi_n'(x) dx \right| \\ &\leq \sup_{a - n^{-1} \leq x \leq a} |\varphi_n'(x)| \cdot \int_{a-n^{-1}}^{a} |f(x) - f(a)| \, dx \\ &\leq \|\chi'\|_{\infty} \cdot n \int_{a-n^{-1}}^{a} |f(x) - f(a)| \, dx, \end{aligned}$$

and therefore, it converges to 0 when a is the Lebesgue point of f. The second term is treated similarly, and thus we obtain

$$\int_a^b \left(\frac{df}{dx} \right) dx = f(b) - f(a)$$

for almost every $(a, b) \in \mathbf{R}^2$. This implies the conclusion, and the proof is complete. □

Exercise 3.60 Put

$$Y_m = \begin{cases} \frac{1}{\gamma(m)}\text{P.f. } \left(x^{m-1}\right)_{x>0} & (m \neq 0, -1, -2, \cdots) \\ \delta^{(\ell)}(x) & (m = -\ell; \ell = 0, 1, 2, \cdots), \end{cases}$$

and show that $Y_m(\varphi)$ is a continuous function of m for any $\varphi \in \mathcal{D}'(\mathbf{R})$ and that

$$\frac{d}{dx}Y_m = Y_{m-1}$$

holds true, where

$$\gamma(x) = \int_0^\infty e^{-s}s^{x-1}ds \qquad (x > 0)$$

denotes the *Gamma function.*

Exercise 3.61 Confirm that if $g(x)$ is summable on $[a, b]$, then

$$f(x) = \int_a^x g(x)dx$$

is absolutely continuous on $[a, b]$.

Exercise 3.62 Let $\Omega \subset \mathbf{R}^n$ be a domain and $f \in C^\infty(\Omega)$ and $T \in \mathcal{D}'(\Omega)$. Then, $S \in \mathcal{D}'(\Omega)$ is defined by

$$S(\varphi) = T(f\varphi) \qquad (\varphi \in \mathcal{D}(\Omega)),$$

and is denoted by $S = fT$. Show that the *Leibniz formula* (or *chain rule*)

$$\frac{\partial}{\partial x_j}(fT) = \left(\frac{\partial}{\partial x_j}f\right)T + f\frac{\partial T}{\partial x_j}$$

holds for $j = 1, \cdots, n$.

3.4.8 *Support*

Let $\Omega \subset \mathbf{R}^n$ be a domain, and $T \in \mathcal{D}'(\Omega)$. If $U \subset \Omega$ is open, then we say that T vanishes on U, or $T|_U = 0$ in short, if $T(\varphi) = 0$ for any $\varphi \in C_0^\infty(U)$. The following fact will be obvious.

Theorem 3.33 *If $\{U_\alpha\}$ be a family of open sets in Ω and $T|_{U_\alpha} = 0$ for each α, then it holds that $T|_{\cup_\alpha U_\alpha} = 0$.*

Therefore, supp T of $T \in \mathcal{D}'(\Omega)$ is defined as the maximal closed set F in Ω such that T vanishes on $\Omega \backslash F$. In §3.4.3, we have introduced the space $\mathcal{E}(\Omega) = C^\infty(\Omega)$. We have $\mathcal{D}(\Omega) \subset \mathcal{E}(\Omega)$ including their topologies, and hence any $T \in \mathcal{E}'(\Omega)$ restricted to $\mathcal{D}(\Omega)$ is regarded as an element in $\mathcal{D}'(\Omega)$, denoted by T_0. Furthermore, the mapping $T \in \mathcal{E}'(\Omega) \mapsto T_0 \in \mathcal{D}'(\Omega)$ is one-to-one because the inclusion $C_0^\infty(\Omega) \subset \mathcal{E}(\Omega)$ is dense. Thus, we obtain

$$\mathcal{E}'(\Omega) \subset \mathcal{D}'(\Omega).$$

In fact, we have the following.

Theorem 3.34 *It holds that*

$$\mathcal{E}'(\Omega) = \{T \in \mathcal{D}'(\Omega) \mid \textit{supp}\, T: \ \textit{compact}\}.$$

Proof. In fact, if $T \in \mathcal{E}'(\Omega)$, there is a compact set $K \subset \Omega$, an integer $m \geq 0$, and a constant $C > 0$ such that

$$|T(\varphi)| \leq C \sup_{|\alpha| \leq m,\ x \in K} |\partial^\alpha \varphi(x)|$$

for any $\varphi \in \mathcal{E}(\Omega)$. This implies $T(\varphi) = 0$ for $\varphi \in C_0^\infty(\Omega)$ satisfying supp $\varphi \subset \Omega \setminus K$. In particular, supp $T \subset K$ follows.

Conversely, if the support of $T \in \mathcal{D}'(\Omega)$ is compact, denoted by $K \subset \Omega$, then we take $\psi \in C_0^\infty(\Omega)$ in $\psi(x) = 1$ for $x \in K$. Then, taking $T_0(\varphi) = T(\psi\varphi)$ for $\varphi \in C^\infty(\Omega)$, we have $T_0|_{C_0^\infty(\Omega)} = T$ and $T_0 \in \mathcal{E}'(\Omega)$. The proof is complete. □

The proof of the following two theorems is not described here. In fact, for the first theorem, the representation theorem of Riesz is made use of, and then the second theorem is a consequence of the first one.

Theorem 3.35 *Any $T \in \mathcal{E}'(\Omega)$ admits an integer $m \geq 0$ and $f_\alpha \in C(\overline{\Omega})$ with $|\alpha| \leq m$ such that supp $T \subset$ supp f_α and*

$$T = \sum_{|\alpha| \leq m} \partial^\alpha f_\alpha.$$

Theorem 3.36 *If $T \in \mathcal{D}(\mathbf{R}^n)$ satisfies supp $T \subset \{0\}$, then for some integer $m \geq 0$ and constants C_α it holds that*

$$T = \sum_{|\alpha| \leq m} C_\alpha \partial^\alpha \delta.$$

Exercise 3.63 Confirm Theorem 3.33.

Exercise 3.64 Show that supp $\partial^\alpha T \subset$ supp T for $T \in \mathcal{D}'(\Omega)$.

Exercise 3.65 Confirm that $C_0^\infty(\Omega) \subset \mathcal{E}(\Omega)$ is dense.

3.4.9 *Convergence*

First, we develop the abstract theory. Let X, Y be locally compact spaces, and set

$$L(X, Y) = \{T : X \to Y \mid \text{continuous linear}\}.$$

Several topologies can be introduced to $L(X, Y)$. Henceforth, we put

$$P_{q,F}(T) = \sup_{x \in F} q(Tx),$$

where q is a semi-norm on Y, $F \subset X$, and $T \in L(X, Y)$. If q is continuous and F is bounded, then this value is finite, and $P_{q,F}$ becomes a semi-norm on $L(X, Y)$. Then, each family

$$\begin{aligned}
\mathcal{P}_b &= \{P_{q,F} \mid q : \text{continuous semi-norm on } Y, \quad F \subset X : \text{bounded}\} \\
\mathcal{P}_c &= \{P_{q,F} \mid q : \text{continuous semi-norm on } Y, \quad F \subset X : \text{compact}\} \\
\mathcal{P}_s &= \{P_{q,F} \mid q : \text{continuous semi-norm on } Y, \quad F \subset X : \text{finite}\}
\end{aligned}$$

enjoys the axiom of separation and induces a topology to $L(X, Y)$. Those topologies by $\mathcal{P}_b$, $\mathcal{P}_c$, and $\mathcal{P}_s$ induce for the net on $L(X, Y)$ the convergence of uniform on each bounded set, that of uniform on each compact set, and that of pointwise, respectively. A subset B of $L(X, Y)$ is called *equi-continuous* if any continuous semi-norm q on Y admits a continuous semi-norm p on X such that

$$\sup_{T \in B} q(Tx) \leq p(x)$$

for any $x \in X$. The following theorem is referred to as that of *Banach-Steinhaus.*

Theorem 3.37 *If X is a Fréchet space or an inductive limit of Fréchet spaces and Y is a locally convex space, then for $B \subset L(X, Y)$ the following items are equivalent to each other.*

(1) B is bounded in $\mathcal{P}_s$ topology.
(2) B is bounded in $\mathcal{P}_c$ topology.
(3) B is bounded in $\mathcal{P}_b$ topology.
(4) B is equi-continuous.

In fact, the inclusions (4) $\Rightarrow$ (3) $\Rightarrow$ (2) $\Rightarrow$ (1) are obvious. The inclusion (1) $\Rightarrow$ (4) is called the *resonance theorem* in the case that X is a Fréchet space. The general case of X is treated by this fact, although details are not described here.

The proof of the following fact is left to the reader.

Theorem 3.38 *If X, Y are locally convex spaces, $\{T_j\} \subset L(X, Y)$ is an equi-continuous sequence, $T \in L(X, Y)$, and $T_j \to T$ holds in $\mathcal{P}_s$, then, this convergence is in $\mathcal{P}_c$.*

Now, we show the following.

Theorem 3.39 *For X, Y as in Theorem 3.37, if a sequence $\{T_j\} \subset L(X, Y)$ attains $Tx = \lim_{j\to\infty} T_j x$ for any $x \in X$, then T is in $L(X, Y)$ and $T_j \to T$ holds in $\mathcal{P}_c$.*

Proof. In fact, $\{T_j\}$ is bounded in $\mathcal{P}_s$ topology in this case, and hence is equi-continuous by Theorem 3.37. Any continuous semi-norm q on Y attains a continuous semi-norm p on X such that

$$q(T_j x) \leq p(x) \qquad (x \in X)$$

for $j = 1, 2, \cdots$. This implies

$$q(Tx) \leq p(x) \qquad (x \in X)$$

and hence $T \in L(X, Y)$ follows. We get the conclusion by Theorem 3.38 $\square$

Thus, the following fact is obtained as a corollary.

Theorem 3.40 *If $\{T_j\} \subset \mathcal{D}'(\Omega)$ is a sequence attaining*

$$T(\varphi) = \lim_{j \to \infty} T_j(\varphi)$$

for each $\varphi \in \mathcal{D}(\Omega)$, then $T \in \mathcal{D}'(\Omega)$ and $T_j \to T$ uniformly on each bounded set in $\mathcal{D}(\Omega)$.

Proof. We have $T \in \mathcal{D}'(\Omega)$ by Theorem 3.39 and the convergence $T_j \to T$ is uniform on each compact set of $\mathcal{D}(\Omega)$. However, any bounded set $B \subset \mathcal{D}(\Omega)$ is relatively compact. In fact, in this case we have a compact $K \subset \Omega$ such that $B \subset \mathcal{D}_K(\Omega)$ is bounded. Therefore, Ascoli'-Arzelá's theorem applies. The proof is complete. □

Exercise 3.66 Confirm Theorem 3.38.

Exercise 3.67 Prove an analogous result to Theorem 3.40.

Chapter 4

Random Motion of Particles

The underlying structure of chemotaxis is the movement of many particles controlled by the other species. This chapter describes the way to derive dynamical partial differential equations from the statistic model.

4.1 Process of Diffusion

4.1.1 *Master Equation*

Random walk on lattice induces the equation of diffusion. In this section, we describe mostly one-dimensional lattice $\mathcal{L}$, but n-dimensional lattice $\mathcal{L}^n$ is treated similarly. Also, we restrict our considerations to the one-step jump process with continuous time.

Thus, we identify $\mathcal{L}$ with

$$\mathcal{Z} = \{\cdots, -n-1, -n, -n+1, \cdots, -1, 0, 1, \cdots, n-1, n, n+1, \cdots\}.$$

Let $p_n(t) \in [0,1]$ be the conditional probability that the walker stayed on site $n = 0$ at time $t = 0$ is on site $n = n$ at time $t = t$. Then, it holds that

$$\frac{\partial p_n}{\partial t} = \hat{T}^+_{n-1} p_{n-1} + \hat{T}^-_{n+1} p_{n+1} - (\hat{T}^+_n + \hat{T}^-_n) p_n, \tag{4.1}$$

where $\hat{T}^\pm_n$ denotes the transition rates that the walker staying on site n jumps to site $n \pm 1$ in the unit time. Sometimes, equation (4.1) is called the *master equation*. Because it is regarded as an ordinary differential equation on the infinite dimensional space, the reader may skip over the following exercise first.

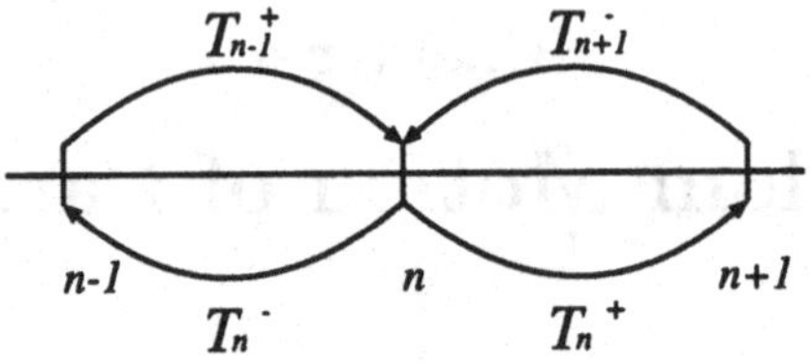

Fig. 4.1

Exercise 4.1 Suppose that $\hat{T}_n^{\pm}$'s are constants in $[0, 1]$ and study (4.1) in the following way. First, formulate the equation as an abstract linear ordinary differential equation in the Banach space ℓ^1, the set of absolutely convergent sequences as

$$\frac{d}{dt}P = TP, \tag{4.2}$$

where T is a bounded linear operator on ℓ^1 and $P = (p_n(t)) \in C^1([0, T], \ell^1)$. Then, show the unique existence the solution globally in time for give initial values $\{p_n(0)\}$ in $p_n(0) \geq 0$ and $\sum_n p_n(0) = 1$, by means of the semigroup $\{e^{tT}\}_{t\geq 0}$ defined by

$$e^{tT} = \sum_{k=0}^{\infty} \frac{t^k T^k}{k!}.$$

Then, show that $p_n(t) \geq 0$ and $\sum_n p_n(t) = 1$ hold. Finally, generalize those results to the case that $\hat{T}_n^{\pm}$'s are given continuous functions in t with the values in $[0, 1]$, replacing (4.2) by

$$P(t) = P(0) + \int_0^t T(s)P(s)ds.$$

In the following, we take the case that those transient rates $\hat{T}_n^{\pm}$ are controlled by the other species living in sub-lattice $\hat{\mathcal{L}}$, of which mesh size is a half of that of $\mathcal{L}$. Let the density of that species be

$$w = \left(\cdots, w_{-n-1/2}, w_{-n}, w_{-n+1/2}, \cdots, w_{-1/2},\right.$$
$$\left. w_0, w_{1/2}, \cdots, w_{n-1/2}, w_n, w_{n+1/2}, \cdots\right).$$

4.1.2 *Local Information Model*

If the transition probabilities depend only on the density of the control species at that site, it holds that $\hat{T}_n^{\pm} = \hat{T}(w_n)$ and hence (4.1) is written as

$$\frac{\partial p_n}{\partial t} = \hat{T}(w_{n-1})p_{n-1} + \hat{T}(w_{n+1})p_{n+1} - 2\hat{T}(w_n)p_n. \tag{4.3}$$

Therefore, writing $x = nh$ by the mesh size h of the lattice, we obtain

$$\frac{\partial p}{\partial t} = h^2 \frac{\partial^2}{\partial x^2}\left(\hat{T}(w)p\right) + O(h^4) \tag{4.4}$$

because

$$f(x+h) + f(x-h) - 2f(x) = h^2 f''(x) + O(h^4)$$

holds by Taylor's expansion.

If we have the scaling $t' = \lambda t$, then we can take $\hat{T}(w) = \lambda T(w)$. Under the assumption $\lim_{h\downarrow 0} \lambda h^2 = D > 0$, it follows formally that

$$\frac{\partial p}{\partial t} = D\frac{\partial^2}{\partial x^2}(T(w)p)$$

by replacing t' by t. It is formal because the forth derivatives of $T(w)p$ is supposed to be bounded a priori. Thus, the response function $T(w)$ represents the microscopic mechanism of the jump process.

Taking $\mathcal{L}^n$, we get the n-dimensional model

$$\frac{\partial p}{\partial t} = D\Delta(T(w)p), \tag{4.5}$$

where $\Delta = \nabla \cdot \nabla = \sum_{i=1}^{n} \frac{\partial^2}{\partial x_i^2}$ denotes the Laplacian. We take non-negative p defined on the bounded domain Ω with smooth boundary $\partial\Omega$. Usually, we impose the boundary condition

$$\frac{\partial}{\partial \nu}T(w)p = 0 \qquad \text{on} \qquad \partial\Omega, \tag{4.6}$$

where

$$\frac{\partial f}{\partial \nu} = \nu \cdot \nabla f = \frac{d}{ds}f(\cdot + s\nu)\Big|_{s=0}$$

indicates the direction derivative of f toward ν. Then, by the divergence formula of Gauss it holds that

$$\begin{aligned}\frac{d}{dt}\int_\Omega p(x,t)dx &= \int_\Omega p_t dx = \int_\Omega D\nabla\cdot\nabla(T(w)p)dx \\ &= \int_{\partial\Omega}\frac{\partial}{\partial\nu}(T(w)p)\,dS = 0.\end{aligned}$$

This implies the conservation of total mass,

$$\int_\Omega p(x,t)dx = \int_\Omega p(x,0)dx \equiv \lambda.$$

Equation (4.5) is written as

$$\frac{\partial p}{\partial t} + \nabla\cdot j = 0 \tag{4.7}$$

for $j = -D\nabla\,(T(w)p)$. This form is referred to as the *equation of continuity*. It describes the conservation of mass, as

$$\frac{d}{dt}\int_\omega pdx = -\int_{\partial\omega}\nu\cdot jdS \tag{4.8}$$

holds for any $\omega\subset\Omega$ with $\overline{\omega}\subset\Omega$ and smooth boundary $\partial\omega$ by the divergence formula of Gauss. In fact, equality (4.8) means that the vector field j indicates the *flux* of p, by which the particles flow in the unit time. Thus, $v = j/p$ represents the *average particle velocity*, and (4.6) indicates the zero flux boundary condition.

In this case of (4.5), we have

$$\begin{aligned} j &= -D\nabla\cdot(T(w)p) \\ &= -DT(w)\nabla p - DpT'(w)\nabla w \\ &= -DT(w)\nabla p + p\chi(w)\nabla w,\end{aligned}$$

where $\chi(w) = -DT'(w)$ denotes the *chemotactic sensitivity*. Here, the first term of the right-hand side is ∇p times a negative scalar determined by w. Because $T(w) > 0$, this vector is parallel to the direction that p decreases mostly. This term of flux indicates that the particles are subject to the *diffusion*. On the other hand, the second term $p\chi(w)\nabla w$ indicates that p is carried under the flow subject to the vector field $\chi(w)\nabla w$. Therefore, if $\chi(w) > 0$ the particles are attracted to the place where the concentration of

w is higher. Similarly, if $\chi(w) < 0$ the particles are repulsive in there. Those cases are referred to as the *positive* and *negative* chemotaxis, respectively.

Presence of the chemotactic sensitivity induces the non-constant equilibrium state. In fact, in this case $T(w)p$ is independent of t. It is harmonic and satisfies the homogeneous Neumann boundary condition by (4.5) and (4.6), respectively. This means that it is a constant in Ω. In fact, generally,

$$\Delta f = 0 \quad \text{in} \quad \Omega \qquad \text{with} \qquad \frac{\partial f}{\partial \nu} = 0 \quad \text{on} \quad \partial\Omega$$

implies

$$\int_\Omega |\nabla f|^2 \, dx = 0$$

from the divergence formula of Gauss (1.25), and hence f is a constant in Ω.

Thus, we obtain

$$p(x) = \frac{\lambda}{T(w)(x)} \left(\int_\Omega \frac{dx}{T(w)(x)} \right)^{-1} \tag{4.9}$$

for

$$\lambda = \|p\|_1 = \int_\Omega p(x) dx$$

and $p(x)$ is not a constant unless $T(w)$ is.

Here and henceforth, $\|\cdot\|_q$ indicates the standard L^q norm on Ω so that $\|v\|_q = \left\{\int_\Omega |v(x)|^q \, dx\right\}^{1/q}$ for $q \in [1, \infty)$ and $\|v\|_\infty = \text{ess. sup}_{x \in \Omega} |v(x)|$, of which details are described in the next chapter.

If this stationary solution is stable, then the solution, with the initial value close to it, keeps to stay near from it. Then it can happen that

$$\|p_0\|_\infty < \liminf_{t\to\infty} \|p(t)\|_\infty \leq \limsup_{t\to\infty} \|p(t)\|_\infty < +\infty \tag{4.10}$$

or

$$\limsup_{t\to\infty} \|p(t)\|_\infty < \|p_0\|_\infty \,. \tag{4.11}$$

The cases (4.10) and (4.11) are called the *aggregation* and the *collapse*, respectively.

In the actual model of biology, the variables p and w are coupled, so that w is subject to an equation involving p. In this case, (4.9) becomes a functional equation on p, and the stability of the solution must be examined from the viewpoint of the dynamical system made by those coupled equations.

Exercise 4.2 Prove (4.9).

Exercise 4.3 Investigate the sign of chemotaxis for the case of linear response, $T(w) = \alpha + \beta w$, and that of saturating response at large w, $T(w) = \alpha + \beta w/(\gamma + w)$.

4.1.3 *Barrier Model*

In the barrier model, the transient rate at site n is determined by the densities of the control species at site $n \pm 1/2$. Therefore, the control species which govern the jump process makes barrier to the particle. We have

$$\hat{T}_n^{\pm}(w) = \hat{T}(w_{n\pm 1/2})$$

and the master equation (4.3) is now reduced to

$$\begin{aligned}\frac{\partial p_n}{\partial t} &= \hat{T}(w_{n-1/2})p_{n-1} + \hat{T}(w_{n+1/2})p_{n+1} \\ &\quad - \left(\hat{T}(w_{n+1/2}) + \hat{T}(w_{n-1/2})\right) p_n.\end{aligned}$$

Here, the right-hand side is equal to

$$\begin{aligned}&\hat{T}(w_{n+1/2})\,(p_{n+1} - p_n) + \hat{T}(w_{n-1/2})\,(p_{n-1} - p_n) \\ &\quad = h\left(\hat{T}(w_{n+1/2}) - \hat{T}(w_{n-1/2})\right)\left(\frac{\partial p}{\partial x} + o(1)\right) \\ &\quad = h^2\left\{\frac{\partial}{\partial x}\left(\hat{T}(w)\frac{\partial p}{\partial x}\right) + o(1)\right\}\end{aligned}$$

and we obtain

$$\frac{\partial p}{\partial t} = D\frac{\partial}{\partial x}\left(T(w)\frac{\partial p}{\partial x}\right) \tag{4.12}$$

under the same scaling $\lim_{h\downarrow 0} \lambda h^2 = D > 0$. In n space dimension, we have

$$\frac{\partial p}{\partial t} = D\nabla \cdot (T(w)\nabla p)$$

and the average particle velocity is given by

$$v = -DT(w)\frac{\nabla p}{p}.$$

This model has only diffusion part and any stationary solution is a constant in Ω. However, aggregation can happen transiently.

4.1.4 *Renormalization*

We note that the mean waiting time of the particle at site n is given by $\left(\hat{T}_n^+ + \hat{T}_n^-\right)^{-1}$. In the case that it is independent of w and n, it holds that

$$\hat{T}_n^+(w) + \hat{T}_n^-(w) = 2\lambda,$$

where $\lambda > 0$ is a constant. If the barrier model is adopted here, then $\hat{T}_n^\pm(w) = \hat{T}(w_{n\pm1/2})$ follows. Those relations imply

$$\hat{T}_n^\pm(w) = 2\lambda \cdot \frac{\hat{T}(w_{n\pm1/2})}{\hat{T}(w_{n+1/2}) + \hat{T}(w_{n-1/2})}.$$

Renormalization is the procedure of introducing a new jump process by replacing the right-hand side as

$$\hat{T}_n^\pm(w) = 2\lambda \cdot \frac{T(w_{n\pm1/2})}{T(w_{n+1/2}) + T(w_{n-1/2})}$$

with some $T(w)$.

Writing

$$N^+(w_{n+1/2}, w_{n-1/2}) = \frac{T(w_{n+1/2})}{T(w_{n+1/2}) + T(w_{n-1/2})}$$

$$N^-(w_{n-1/2}, w_{n+1/2}) = \frac{T(w_{n-1/2})}{T(w_{n-1/2}) + T(w_{n+1/2})},$$

we have

$$\hat{T}_n^\pm(w) = \begin{cases} 2\lambda N^+(w_{n+1/2}, w_{n-1/2}) \\ 2\lambda N^-(w_{n-1/2}, w_{n+1/2}) \end{cases}$$

and the master equation (4.3) is reduced to

$$\frac{1}{2\lambda}\frac{\partial p_n}{\partial t} = N^+(w_{n-1/2}, w_{n-3/2})p_{n-1} + N^-(w_{n+1/2}, w_{n+3/2})p_{n+1}$$

$$- \left(N^+(w_{n+1/2}, w_{n-1/2}) + N^-(w_{n-1/2}, w_{n+1/2})\right) p_n. \tag{4.13}$$

In this model, the sub-lattice is assumed to be homogeneous so that $N^\pm$ is independent of n. Letting $N(u,v) = N^+(u,v)$, we have $N^-(v,u) = 1 - N(u,v)$ and

$$N(u,v) = \frac{T(u)}{T(u) + T(v)}. \tag{4.14}$$

Putting $x = nh$, we examine the right-hand side of (4.13). In fact, we have

$$\begin{aligned} N^+(w_{n-1/2}, w_{n-3/2})p_{n-1} &- N^+(w_{n+1/2}, w_{n-1/2})p_n \\ = N^+ &(w(x - h/2), w(x - 3h/2))\, p(x - h) \\ &-N^+ (w(x + h/2), w(x - h/2))\, p(x). \end{aligned}$$

This term vanishes at $h = 0$. Differentiating the right-hand side in h, we have

$$\begin{aligned} -\frac{1}{2}N_u^+ &(w(x - h/2), w(x - 3h/2))\, w_x(x - h/2)p(x - h) \\ &-\frac{3}{2}N_v^+ (w(x - h/2), w(x - 3h/2))\, w_x(x - 3h/2)p(x - h) \\ &-N^+ (w(x - h/2, w(x - 3h/2))\, p_x(x - h) \\ &-\frac{1}{2}N_u^+ (w(x + h/2), w(x - h/2))\, w_x(x + h/2)p(x) \\ &+\frac{1}{2}N_v^+ (w(x + h/2), w(x - h/2))\, w_x(x - h/2)p(x). \end{aligned} \tag{4.15}$$

If $h = 0$, this term is equal to

$$\begin{aligned} -N_u^+ (w(x), w(x))\, w_x(x)p(x) &- N_v^+ (w(x), w(x))\, w_x(x)p(x) \\ &-N^+ (w(x), w(x))\, p_x(x). \end{aligned}$$

Similarly, we have

$$\begin{aligned} N^- \left(w_{n+1/2}, w_{n+3/2}\right) p_{n+1} &- N^- \left(w_{n-1/2}, w_{n+1/2}\right) p_n \\ = N^- &(w(x + h/2), w(x + 3h/2))\, p(x + h) \\ &-N^- (w(x - h/2), w(x + h/2))\, p(x) \end{aligned}$$

and this term vanishes at $h = 0$. Differentiating in h, we have

$$\frac{1}{2}N_u^- (w(x + h/2), w(x + 3h/2))\, w_x(x + h/2)p(x + h)$$

$$
\begin{aligned}
&+\frac{3}{2}N_v^-\left(w(x+h/2),w(x+3h/2)\right)w_x(x+3h/2)p(x+h)\\
&+N^-\left(w(x+h/2),w(x+3h/2)\right)p_x(x+h)\\
&+\frac{1}{2}N_u^-\left(w(x-h/2),w(x+h/2)\right)w_x(x-h/2)p(x)\\
&-\frac{1}{2}N_v^-\left(w(x-h/2),w(x+h/2)\right)w_x(x+h/2)p(x), \qquad (4.16)
\end{aligned}
$$

which is equal to

$$
\begin{aligned}
&N_u^-\left(w(x),w(x)\right)w_x(x)p(x)+N_v^-\left(w(x),w(x)\right)w_x(x)p(x)\\
&+N^-\left(w(x),w(x)\right)p_x(x)
\end{aligned}
$$

at $h=0$.

We shall write $N^\pm = N^\pm(w(x),w(x))$ and $N = N(w(x),w(x))$ for simplicity. Then, we see that $O(h)$ term of the right-hand side of (4.13) vanishes as

$$
\begin{aligned}
&\left(N_u^- - N_u^+\right)w_xp+\left(N_v^- - N_v^+\right)w_xp+\left(N^- - N^+\right)p_x\\
&=\left(-N_v-N_u\right)w_xp+\left(-N_u-N_v\right)w_xp+\left(1-2N\right)p_x\\
&=\left((1-2N)p\right)_x=0.
\end{aligned}
$$

In fact, $N(w,w)=1/2$ holds by (4.14).

Now, we differentiate (4.15) in h and put $h=0$. This gives that

$$
\begin{aligned}
&2N_{uv}^+w_x^2p+2N_{vv}^+w_x^2p\\
&\quad+2N_v^+w_{xx}p+N_u^+w_xp_x+3N_v^+w_xp_x+N^+p_{xx}.
\end{aligned}
$$

Under the same operation to (4.16), we have

$$
\begin{aligned}
&2N_{uv}^-w_x^2p+2N_{vv}^-w_x^2p\\
&\quad+2N_v^-w_{xx}p+N_u^-w_xp_x+3N_v^-w_xp_x+N^-p_{xx}.
\end{aligned}
$$

Therefore, $O(h^2)$ term of the right-hand side of (4.13) is a half of the following:

$$
\begin{aligned}
&(N^++N^-)p_{xx}+2(N^++N^-)_{uv}w_x^2p\\
&\quad+2(N^++N^-)_{vv}w_x^2p+2(N^++N^-)_vw_{xx}p\\
&\quad+(N^++N^-)_uw_xp_x+3(N^++N^-)_vw_xp_x. \qquad (4.17)
\end{aligned}
$$

Here, we have $N^+(u,v) = N(u,v)$, $N^-(u,v) = 1 - N(v,u)$, and hence it follows that

$$\begin{aligned}
&(N^+ + N^-)_{uv} = 0\\
&N^+ + N^- = 1\\
&(N^+ + N^-)_{vv} = N_{vv} - N_{uu}\\
&(N^+ + N^-)_v = N_v - N_u\\
&(N^+ + N^-)_u = N_u - N_v
\end{aligned}$$

at $u = v = w$, where $N_u = N_u(w,w)$, $N_{uu} = N_{uu}(w,w)$, and so forth.

We see that the quantity defined by (4.17) is equal to

$$\begin{aligned}
&p_{xx} + 2\,(N_{vv} - N_{uu})\,w_x^2 p + 2\,(N_v - N_u)\,w_{xx}p - 2\,(N_u - N_v)\,w_x p_x\\
&\quad = p_{xx} - 2\,(p\,(N_u - N_v)\,w_x)_x
\end{aligned}$$

so that (4.17) has the form

$$\frac{1}{\lambda}\frac{\partial p}{\partial t} = h^2 \frac{\partial}{\partial x}\left(\frac{\partial p}{\partial x} - 2p\,(N_u - N_v)\,\frac{\partial w}{\partial x}\right) + o(h^2).$$

Under the scaling $\lim_{h\downarrow 0} \lambda h^2 = D > 0$ we obtain

$$\frac{\partial p}{\partial t} = D\frac{\partial}{\partial x}\left(\frac{\partial p}{\partial x} - 2p\,(N_u - N_v)\,\frac{\partial w}{\partial x}\right). \tag{4.18}$$

Here, we have

$$N_u(w,w) = \left.\frac{T(v)T'(u)}{(T(u)+T(v))^2}\right|_{u=v=w} = \frac{1}{4}\,(\log T(w))'$$

and $N_v(w,w) = -N_u(w,w)$ by (4.14). Equation (4.18) is written as

$$\frac{\partial p}{\partial t} = D\frac{\partial}{\partial x}\left(\frac{\partial p}{\partial x} - p\frac{\partial}{\partial x}\log T(w)\right).$$

In n space dimension, we have

$$\frac{\partial p}{\partial t} = D\nabla\cdot(\nabla p - p\nabla \log T(w)).$$

Therefore, the chemotactic sensitivity function is $\chi(w) = D\,(\log T(w))'$ and the average particle velocity is $\boldsymbol{v} = -D\nabla \log p + D\,(\log T(w))'\,\nabla w$. This chemotactic profile is positive if $T'(w) > 0$.

In the renormalized *nearest neighbor model*, one takes

$$N_n^{\pm}(w) = \frac{T(w_{n\pm1})}{T(w_{n+1}) + T(w_{n-1})}.$$

In this case, it follows that

$$\frac{\partial p}{\partial t} = D\frac{\partial}{\partial x}\left(\frac{\partial p}{\partial x} - 2p\left(\log T(w)\right)'\frac{\partial w}{\partial x}\right).$$

Thus, the chemotactic sensitivity function becomes twice of that of the renormalized barrier model.

In the case that the organism searches local environment before it decides the movement, the transient rate depends on the difference between the state at the present position and that of the nearest neighbor in the direction of the movement. That is,

$$\begin{aligned}\hat{T}_{n-1}^{+}(w) &= \alpha + \beta\left(\tau(w_n) - \tau(w_{n-1})\right)\\ \hat{T}_{n+1}^{-}(w) &= \alpha + \beta\left(\tau(w_n) - \tau(w_{n+1})\right),\end{aligned} \tag{4.19}$$

called the gradient-based model.

On the other hand, movement of w is under the influence of p. In this model derived from cellular automaton, it is given by the ordinary differential equation such as

$$\begin{aligned}\frac{dw}{dt} &= p - \mu w\\ \frac{dw}{dt} &= (p-\mu)w\\ \frac{dw}{dt} &= \frac{pw}{1+\gamma w} - \mu w + \frac{\gamma p}{1+p}.\end{aligned}$$

Those cases describe *linear*, *exponential*, and *saturating* growths, respectively. In *biological field*, the equation for w is given by the ordinary differential equation in those ways.

Exercise 4.4 Derive

$$\frac{\partial p}{\partial t} = D\frac{\partial}{\partial x}\left(\alpha\frac{\partial p}{\partial x} - 2\beta p\frac{\partial}{\partial x}\tau(w)\right)$$

from the gradient-based model (4.19). Derive, also, the limiting equation for the renormalized model to this case.

4.2 Kinetic Model

4.2.1 *Transport Equation*

This section studies the self-interaction of many particles. *Boltzmann equation* is a description of macroscopic motion of gas particles by the atomic probability of entropy. Then the *transport equation* arises as the linearization around the *Maxwell distribution*. However, the latter also comes from the following physical situation.

Imagine that two families of particles, gas and medium, are interacting. Suppose that each of them is provided with so many particles as to make the statistical description of their motions possible, that the number of particles of medium is much more than that of gas, that the interaction occurs mostly between gas and medium, and that the macroscopic state of the medium is free from the influence of interaction.

Let $f(\boldsymbol{x}, \boldsymbol{v}, t)$ be the distribution function of gas particles, so that

$$\int\int_G f(\boldsymbol{x}, \boldsymbol{v}, t) d\boldsymbol{x} d\boldsymbol{v}$$

denotes their number staying in the domain $G \subset \mathbf{R}^3 \times \mathbf{R}^3$ in $\boldsymbol{x} - \boldsymbol{v}$ space at the time t. The gas particles are labelled by j and $\boldsymbol{x}^j = \boldsymbol{x}^j(t)$, $\boldsymbol{v}^j = \boldsymbol{v}^j(t)$ denote the position and the velocity of the j-th particle. If mass of each gas particle is put to be one and $\boldsymbol{F}(\boldsymbol{x})$ denotes the outer force acting on gas particles, then $\{\boldsymbol{x}^j, \boldsymbol{v}^j\}$ are subject to the Newton equation as

$$\frac{d\boldsymbol{x}^j}{dt} = \boldsymbol{v}^j \quad \text{and} \quad \frac{d\boldsymbol{v}^j}{dt} = \boldsymbol{F}(\boldsymbol{x}^j). \tag{4.20}$$

Henceforth, the set of infinitely many differentiable function with compact support contained in a domain Ω is denoted by $C_0^\infty(\Omega)$. Then, we take arbitrary $\varphi \in C_0^\infty(\mathbf{R}^3 \times \mathbf{R}^3)$ and put

$$\begin{aligned} I(t) &= \sum_j \varphi\left(\boldsymbol{x}^j(t), \boldsymbol{v}^j(t)\right) \\ &= \int\int_{\mathbf{R}^3 \times \mathbf{R}^3} f(\boldsymbol{x}, \boldsymbol{v}, t) \varphi(\boldsymbol{x}, \boldsymbol{v}) d\boldsymbol{x} d\boldsymbol{v}, \end{aligned} \tag{4.21}$$

where the second equality comes from

$$f(\boldsymbol{x}, \boldsymbol{v}, t) d\boldsymbol{x} d\boldsymbol{v} = \sum_j \delta_{\boldsymbol{x}^j(t)}(d\boldsymbol{x}) \otimes \delta_{\boldsymbol{v}^j(t)}(d\boldsymbol{v}). \tag{4.22}$$

See Chapter 3.4.9 for *Dirac's delta function* used in the right-hand side of (4.22).

If $|\Delta t| \ll 1$, then $\Delta I = I(t+\Delta t) - I(t)$ is decomposed into the contributions of gas particles without collision between medium and those with more than once. Those quantities are denoted by $\Delta I_1(t)$ and $\Delta I_2(t)$, respectively. Then, we set

$$I_1(t) = \sum_{j \in \Lambda'} \varphi\left(\boldsymbol{x}^j(t), \boldsymbol{v}^j(t)\right)$$

and

$$I_2(t) = \sum_{j \in \Lambda''} \varphi\left(\boldsymbol{x}^j(t), \boldsymbol{v}^j(t)\right).$$

First, we have

$$\begin{aligned}\Delta I_1(t) &= \sum_{j \in \Lambda'} \left\{\varphi\left(\boldsymbol{x}^j(t+\Delta t), \boldsymbol{v}^j(t+\Delta t)\right) - \varphi\left(\boldsymbol{x}^j(t), \boldsymbol{v}^j(t)\right)\right\} \\ &= \sum_{j \in \Lambda'} \left\{\frac{d\boldsymbol{x}^j(t)}{dt} \cdot \nabla_x \varphi\left(\boldsymbol{x}^j(t), \boldsymbol{v}^j(t)\right) + \frac{d\boldsymbol{v}^j(t)}{dt} \cdot \nabla_v \varphi\left(\boldsymbol{x}^j(t), \boldsymbol{v}^j(t)\right)\right\} \\ &\quad \cdot \Delta t + o(\Delta t).\end{aligned}$$

Then, equation (4.20) gives that

$$\Delta I_1 = \sum_{j \in \Lambda'} \left(\boldsymbol{v}^j \cdot \nabla_x \varphi(\boldsymbol{x}^j, \boldsymbol{v}^j) + \boldsymbol{F}(\boldsymbol{x}^j) \cdot \nabla_v \varphi\right) \Delta t + o(\Delta t).$$

Because the number of particles without collision is much more than that with collision, $\sum_{j \in \Lambda'}$ is approximated by $\sum_j$. In this case it holds that

$$\begin{aligned}\Delta I_1 = \Delta t \cdot \int\!\!\int_{\mathbf{R}^3 \times \mathbf{R}^3} &\left(\boldsymbol{v} \cdot \nabla_x \varphi(\boldsymbol{x}, \boldsymbol{v}) + \boldsymbol{F}(\boldsymbol{x}) \cdot \nabla_v \varphi(\boldsymbol{x}, \boldsymbol{v})\right) \\ &\cdot f(\boldsymbol{x}, \boldsymbol{v}, t) d\boldsymbol{x} d\boldsymbol{v} + o(\Delta t).\end{aligned}$$

Let us proceed to ΔI_2. The possibility that one gas particle interacts to the medium in Δt is proportional to Δt. If $\sigma = \sigma(\boldsymbol{x}, \boldsymbol{v}, t)$ denotes its rate, then $\Delta t \cdot \sigma(\boldsymbol{x}, \boldsymbol{v}, t) f(\boldsymbol{x}, \boldsymbol{v}, t)$ indicates the distribution of gas particles that interact to the medium during that time. When collision occurs, the gas particle changes the velocity although the position is unchanged. If $k_1 = k_1(\boldsymbol{x}, \boldsymbol{v}, \boldsymbol{v}', t)$ denotes the conditional probability that the velocity changes from $\boldsymbol{v}'$ to $\boldsymbol{v}$, then the distribution of gas particles that increases by the collision (at time t, position $\boldsymbol{x}$, and velocity $\boldsymbol{v}$) is approximated by

$$\Delta t \cdot \int_{\mathbf{R}^3} k(\boldsymbol{x}, \boldsymbol{v}, \boldsymbol{v}', t) f(\boldsymbol{x}, \boldsymbol{v}', t) d\boldsymbol{v}',$$

where $k = k_1 \cdot \sigma$. In this way, the change of distribution function of gas particles with collision is approximated by

$$\Delta t \cdot \left\{ -\sigma(\boldsymbol{x}, \boldsymbol{v}, t) f(\boldsymbol{x}, \boldsymbol{v}, t) + \int_{\mathbf{R}^3} k(\boldsymbol{x}, \boldsymbol{v}, \boldsymbol{v}', t) f(\boldsymbol{x}, \boldsymbol{v}', t) d\boldsymbol{v}' \right\}$$

and it holds that

$$\begin{aligned} \Delta I_2 = \Delta t \int\int_{\mathbf{R}^3 \times \mathbf{R}^3} & \{ -\sigma(\boldsymbol{x}, \boldsymbol{v}, t) f(\boldsymbol{x}, \boldsymbol{v}, t) \\ & + \int_{\mathbf{R}^3} k(\boldsymbol{x}, \boldsymbol{v}, \boldsymbol{v}', t) f(\boldsymbol{x}, \boldsymbol{v}', t) d\boldsymbol{v}' \} \varphi(\boldsymbol{x}, \boldsymbol{v}) \cdot d\boldsymbol{x} d\boldsymbol{v} + o(\Delta t) . \end{aligned}$$

We can summarize as

$$\begin{aligned} \frac{\Delta I}{\Delta t} = \int\int_{\mathbf{R}^3 \times \mathbf{R}^3} & \{ \boldsymbol{v} \cdot \nabla_x \varphi(\boldsymbol{x}, \boldsymbol{v}) f(\boldsymbol{x}, \boldsymbol{v}, t) \\ & + \boldsymbol{F}(\boldsymbol{x}) \cdot \nabla_v \varphi(\boldsymbol{x}, \boldsymbol{v}) f(\boldsymbol{x}, \boldsymbol{v}, t) - \sigma(\boldsymbol{x}, \boldsymbol{v}, t) f(\boldsymbol{x}, \boldsymbol{v}, t) \varphi(\boldsymbol{x}, \boldsymbol{v}) \\ & + \int_{\mathbf{R}^3} k(\boldsymbol{x}, \boldsymbol{v}, \boldsymbol{v}', t) f(\boldsymbol{x}, \boldsymbol{v}', t) d\boldsymbol{v}' \varphi(\boldsymbol{x}, \boldsymbol{v}) \} d\boldsymbol{x} d\boldsymbol{v} + o(1). \end{aligned}$$

This implies

$$\begin{aligned} \int\int_{\mathbf{R}^3 \times \mathbf{R}^3} & \{ \boldsymbol{v} \cdot \nabla_x \varphi(\boldsymbol{x}, \boldsymbol{v}) f(\boldsymbol{x}, \boldsymbol{v}, t) + \boldsymbol{F}(\boldsymbol{x}) \cdot \nabla_v \varphi(\boldsymbol{x}, \boldsymbol{v}) f(\boldsymbol{x}, \boldsymbol{v}, t) \\ & - \sigma(\boldsymbol{x}, \boldsymbol{v}, t) f(\boldsymbol{x}, \boldsymbol{v}, t) \varphi(\boldsymbol{x}, \boldsymbol{v}) \\ & + \int_{\mathbf{R}^3} k(\boldsymbol{x}, \boldsymbol{v}, \boldsymbol{v}', t) f(\boldsymbol{x}, \boldsymbol{v}', t) d\boldsymbol{v}' \cdot \varphi(\boldsymbol{x}, \boldsymbol{v}') \} d\boldsymbol{x} d\boldsymbol{v} \\ = \int\int_{\mathbf{R}^3 \times \mathbf{R}^3} & f_t(\boldsymbol{x}, \boldsymbol{v}, t) \varphi(\boldsymbol{x}, \boldsymbol{v}) d\boldsymbol{x} d\boldsymbol{v} \end{aligned}$$

by (4.21). Because $\varphi \in C_0^\infty(\mathbf{R}^3 \times \mathbf{R}^3)$ is arbitrary, we get that

$$\frac{\partial f}{\partial t} = -\boldsymbol{v} \cdot \nabla_x f - \boldsymbol{F} \cdot \nabla_v f - \sigma f + \int_{\mathbf{R}^3} k(\cdot, \cdot, \boldsymbol{v}', t) f(\cdot, \boldsymbol{v}', t) d\boldsymbol{v}'. \quad (4.23)$$

This is linear Boltzmann or transport equation.

4.2.2 *Boltzmann Equation*

The last two terms of the right-hand side of (4.23) are called the *collision term*, totally denoted by

$$\left(\frac{\partial f}{\partial t}\right)_c.$$

The original Boltzmann equation was derived when the same kind of particles are interacting. There, collision between more than three particles at once is ignored and it is assumed that mass, momentum, and kinetic energy are preserved when the collision between two particles occurs. Because the mass of particles is preserved we set it to be one as before.

Thus we have

$$\left(\frac{\partial f}{\partial t}\right)_c = Q[f,f](\boldsymbol{x},\boldsymbol{v},t)$$

with

$$\begin{aligned} Q[f,f](\boldsymbol{v}) = \int_{\mathbf{R}^3} d\boldsymbol{v}_1 \int\int_{\mathbf{R}^3\times\mathbf{R}^3} w(\boldsymbol{v},\boldsymbol{v}_1;\boldsymbol{v}',\boldsymbol{v}_1') \\ \cdot \left(f(\boldsymbol{v}')f(\boldsymbol{v}_1') - f(\boldsymbol{v})f(\boldsymbol{v}_1)\right) d\boldsymbol{v}'d\boldsymbol{v}_1'. \end{aligned} \tag{4.24}$$

Here, $(\boldsymbol{v},\boldsymbol{v}_1) \mapsto (\boldsymbol{v}',\boldsymbol{v}_1')$ indicates the change of velocities in the pair of collision particles. Namely,

$$\left(\frac{\partial f}{\partial t}\right)_c (\boldsymbol{v})$$

decreases and increases proportionally to $f(\boldsymbol{v})f(\boldsymbol{v}_1)$ and to $f(\boldsymbol{v}')f(\boldsymbol{v}_1')$, respectively, with the rate $w(\boldsymbol{v},\boldsymbol{v}_1;\boldsymbol{v}',\boldsymbol{v}_1') \geq 0$.

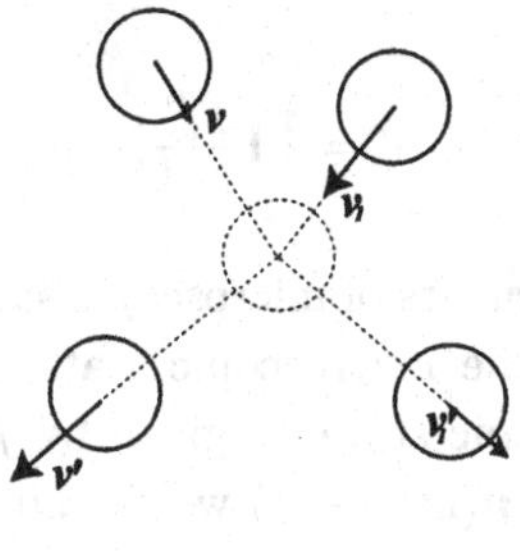

Fig. 4.2

Because the momentum and the kinetic energy are preserved at the collision, we have

$$\boldsymbol{v} + \boldsymbol{v}_1 = \boldsymbol{v}' + \boldsymbol{v}_1' \tag{4.25}$$

and

$$\frac{1}{2}\left(|\boldsymbol{v}|^2 + |\boldsymbol{v}_1|^2\right) = \frac{1}{2}\left(|\boldsymbol{v}'|^2 + |\boldsymbol{v}_1'|^2\right). \tag{4.26}$$

Therefore, the *correlation function* (or correlation measure, precisely) w has the support where (4.25) and (4.26) hold. Furthermore, because the collision is symmetric and reversible we have

$$\begin{aligned} w(\boldsymbol{v}, \boldsymbol{v}_1; \boldsymbol{v}', \boldsymbol{v}_1') &= w(\boldsymbol{v}_1, \boldsymbol{v}; \boldsymbol{v}_1', \boldsymbol{v}') \\ &= w(\boldsymbol{v}', \boldsymbol{v}_1'; \boldsymbol{v}, \boldsymbol{v}_1) = w(\boldsymbol{v}_1', \boldsymbol{v}'; \boldsymbol{v}_1, \boldsymbol{v}). \end{aligned} \tag{4.27}$$

Those requirements make $Q[f,f]$ in more detailed form.

The Boltzmann equation

$$\frac{\partial f}{\partial t} = -\boldsymbol{v} \cdot \nabla_x f - \boldsymbol{F}(\boldsymbol{x}) \cdot \nabla_v f + Q[f,f] \tag{4.28}$$

makes the second law of thermodynamics to be underlined by the intermolecular source. In fact, Boltzmann's formula says that entropy is given by

$$S = k_B \log W,$$

where k_B and W denote the Boltzmann constant and the multiplicity of thermodynamical states.

In the classical theory, microscopic states in the same macroscopic state are not distinguished. If the macroscopic states are labelled by $i = 1, 2, \cdots$, then we have

$$W = \prod_i \frac{g_i^{n_i}}{n_i!},$$

where g_i and n_i are the numbers of microscopic states and that of particles, respectively, involved by the macroscopic state i. The mean number of particles in those microscopic states is given by $f_i = n_i/g_i$, and in use of Starling's formula $\log n! \approx n(\log n - 1)$ we obtain

$$S = k_B \sum_i (n_i \log g_i - \log n_i!) \approx -k_B \sum_i g_i f_i \, (\log f_i - 1). \tag{4.29}$$

Here, replacing g_i by constant times $\Delta \boldsymbol{x} \cdot \Delta \boldsymbol{v}$ gives the representation of entropy,

$$S = -k \int\int_{\mathbf{R}^3 \times \mathbf{R}^3} f\ (\log f - 1)\ d\boldsymbol{x} d\boldsymbol{v},$$

where k is a physical constant.

Based on this formula, Boltzmann's *H-function* is defined by

$$H(\boldsymbol{x}, t) = \int_{\mathbf{R}^3} f\ (\log f - 1)\ d\boldsymbol{v}.$$

It follows formally that

$$\frac{\partial H}{\partial t} = \int_{\mathbf{R}^3} \frac{\partial}{\partial t} \{f\ (\log f - 1)\}\ d\boldsymbol{v} = \int_{\mathbf{R}^3} f_t \cdot \log f\ d\boldsymbol{v}$$

$$\int_{\mathbf{R}^3} (\nabla_x f) \log f d\boldsymbol{x} = \int_{\mathbf{R}^3} \nabla_x \left(f (\log f - 1)\right) d\boldsymbol{x} = 0$$

$$\int_{\mathbf{R}^3} (\nabla_v f) \log f d\boldsymbol{v} = \int_{\mathbf{R}^3} \nabla_v \left(f (\log f - 1)\right) d\boldsymbol{v} = 0.$$

Because $f = f(\boldsymbol{x}, \boldsymbol{v}, t)$ satisfies the Boltzmann equation we have

$$\frac{d\overline{H}}{dt} = \int\int_{\mathbf{R}^3 \times \mathbf{R}^3} Q[f, f] \log f\ d\boldsymbol{x} d\boldsymbol{v}, \tag{4.30}$$

where

$$\overline{H}(t) = \int_{\mathbf{R}^3} H(\boldsymbol{x}, t) d\boldsymbol{x}.$$

Here, the right-hand side of (4.30) is treated as

$$\begin{aligned}
&\int_{\mathbf{R}^3} Q[f, f](\boldsymbol{v}) \log f(\boldsymbol{v}) d\boldsymbol{v} \\
&\quad = \int\int_{\mathbf{R}^3 \times \mathbf{R}^3} \log f(\boldsymbol{v}) d\boldsymbol{v}_1 d\boldsymbol{v} \\
&\quad \cdot \int\int_{\mathbf{R}^3 \times \mathbf{R}^3} w(\boldsymbol{v}, \boldsymbol{v}_1; \boldsymbol{v}', \boldsymbol{v}_1') \left(f(\boldsymbol{v}')f(\boldsymbol{v}_1') - f(\boldsymbol{v})f(\boldsymbol{v}_1)\right) d\boldsymbol{v}' d\boldsymbol{v}_1' \\
&\quad = -\frac{1}{4} \int\int_{\mathbf{R}^3 \times \mathbf{R}^3} d\boldsymbol{v}_1 d\boldsymbol{v} \\
&\quad \cdot \int\int_{\mathbf{R}^3 \times \mathbf{R}^3} w(\boldsymbol{v}, \boldsymbol{v}_1; \boldsymbol{v}', \boldsymbol{v}_1') \left\{\log\left(f(\boldsymbol{v})f(\boldsymbol{v}_1)\right) - \log\left(f(\boldsymbol{v}')f(\boldsymbol{v}_1')\right)\right\} \\
&\quad \cdot \left\{f(\boldsymbol{v})f(\boldsymbol{v}_1) - f(\boldsymbol{v}')f(\boldsymbol{v}_1')\right\} d\boldsymbol{v}' d\boldsymbol{v}_1'
\end{aligned} \tag{4.31}$$

by (4.27). This term is non-negative because $(\log z - \log w)(z - w) \geq 0$ holds for $z, w > 0$. This means the *second law of thermodynamics*, the decreasing of H-function.

In the equilibrium state we have $\frac{dH}{dt} = 0$. This implies $f(\boldsymbol{v})f(\boldsymbol{v}_1) = f(\boldsymbol{v}')f(\boldsymbol{v}_1')$ and hence

$$\int_{\mathbf{R}^3} \{\log f(\boldsymbol{x}, \boldsymbol{v}) + \log f(\boldsymbol{x}, \boldsymbol{v}_1)\}\, d\boldsymbol{x} = \int_{\mathbf{R}^3} \{\log f(\boldsymbol{x}, \boldsymbol{v}') + \log f(\boldsymbol{x}, \boldsymbol{v}_1')\}\, d\boldsymbol{x}$$

follows. This means the conservation of $\int_{\mathbf{R}^3} \log f d\boldsymbol{x}$ at the occasion of collision. Such a quantity must be a linear combination of mass, momentum, and kinetic energy from the physical point of view. In the case that f is uniform in $\boldsymbol{x}$, it holds that

$$\log f(\boldsymbol{v}) = \alpha - \frac{\beta}{2} |\boldsymbol{v}|^2 + \boldsymbol{\gamma} \cdot \boldsymbol{v}$$

with some $\alpha, \beta \in \mathbf{R}$ and $\boldsymbol{\gamma} \in \mathbf{R}^3$, and in this way the Maxwell distribution

$$f_0(\boldsymbol{v}) = n(2\pi k_B T)^{-3/2} \exp\left(-\frac{|\boldsymbol{v}|^2}{2k_B T}\right)$$

is regarded as an equilibrium state, where n and T stand for the particle density and the temperature, respectively.

Exercise 4.5 Confirm the last equality of (4.31).

Exercise 4.6 In Fermi and Bose statistics, multiplicities of the state are given by

$$\prod_i \frac{g_i!}{n_i!(g_i - n_i)!} \quad \text{and} \quad \prod_i \frac{(g_i + n_i - 1)!}{n_i!(g_i - 1)!},$$

respectively. In those cases the entropies are given by

$$S = -k \int\int_{\mathbf{R}^3 \times \mathbf{R}^3} (f \log f + (1 \pm f) \log(1 \pm f))\, d\boldsymbol{x} d\boldsymbol{v}.$$

Derive them by Starling's formula.

4.3 Semi-Conductor Device Equation

4.3.1 *Modelling*

Transportation of electrons and positrons inside semi-conductor devices is governed by the Boltzmann-Poisson system. If $f(\boldsymbol{x}, \boldsymbol{k}, t)$ denotes the distribution function of electrons or positrons at the position $\boldsymbol{x}$, the wave number $\boldsymbol{k}$, and the time t, then the Boltzmann equation is given by

$$f_t + \boldsymbol{u} \cdot \nabla_x f + \frac{\boldsymbol{F}}{\hbar} \cdot \nabla_k f = \left(\frac{\partial f}{\partial t} \right)_c, \tag{4.32}$$

where $\boldsymbol{F}$, $\hbar$, and $\boldsymbol{u}$ stand for the outer force, the Planck constant, and the carrier velocity, respectively. Here, mass of particles is not put to be one. Wave number, adopted for velocity as an independent variable, represents the momentum of particles. The Planck constant arises as the rate for $\Delta \boldsymbol{x} \cdot \Delta \boldsymbol{v}$ to be replaced by g_i, from the principle of the quantum mechanics.

The reader can skip this section first. In the theory of semi-conductor devices, the carrier velocity

$$\boldsymbol{u} = \boldsymbol{u}(\boldsymbol{x}, \boldsymbol{k}, t)$$

of electron or positron is associated with the energy band $\mathcal{E}$ of crystal in such a way as

$$\boldsymbol{u} = \frac{1}{\hbar} \nabla_k \mathcal{E}.$$

Furthermore, similarly to the transport equation, the effective collision is taken only between the particle and medium. Therefore, the collision term is given as $\left(\frac{\partial f}{\partial t} \right)_c = Q[f, f]$, where

$$\begin{aligned} Q[f, f](\boldsymbol{k}) = \int_{\mathbf{R}^3} \{ & f(\boldsymbol{k}') \, (n - f(\boldsymbol{k})) \, P(\boldsymbol{k}', \boldsymbol{k}) \\ & - f(\boldsymbol{k}) \left(n - f(\boldsymbol{k}') \right) P(\boldsymbol{k}, \boldsymbol{k}') \} \, d\boldsymbol{k}'. \end{aligned} \tag{4.33}$$

Here, $P(\boldsymbol{k}', \boldsymbol{k})$ denotes the scattering probability of $\boldsymbol{k}' \mapsto \boldsymbol{k}$ at the collision, and $n(\boldsymbol{x}, t) = \int_{\mathbf{R}^3} f(\boldsymbol{x}, \boldsymbol{k}, t) d\boldsymbol{k}$ is the carrier density.

Outer force $\boldsymbol{F}$ comes from the electric field $\boldsymbol{E}$ in such a way as

$$\boldsymbol{F} = q\boldsymbol{E} = -q\nabla\varphi, \tag{4.34}$$

where q denotes the elementary electric charge of the particle and φ is the electric field potential. The carrier velocity $\boldsymbol{u}$ is the sum of the drift velocity $\boldsymbol{v}(\boldsymbol{x},t)$ and the heat velocity $\mathbf{c}(\boldsymbol{x},\boldsymbol{k},t)$:

$$\boldsymbol{u}(\boldsymbol{x},\boldsymbol{k},t) = \boldsymbol{v}(\boldsymbol{x},t) + \mathbf{c}(\boldsymbol{x},\boldsymbol{k},t).$$

From the randomness of the latter it follows that

$$\int_{\mathbf{R}^3} \mathbf{c}(\boldsymbol{x},\boldsymbol{k},t) f(\boldsymbol{x},\boldsymbol{k},t) d\boldsymbol{k} = 0$$

and hence we obtain

$$\boldsymbol{v}(\boldsymbol{x},t) = \frac{1}{n(\boldsymbol{x},t)} \int_{\mathbf{R}^3} \boldsymbol{u}(\boldsymbol{x},\boldsymbol{k},t) f(\boldsymbol{x},\boldsymbol{k},t) d\boldsymbol{k}.$$

Henceforth $\langle A \rangle$ denotes the mean value of the physical quantity A:

$$\langle A \rangle (\boldsymbol{x},t) = \int_{\mathbf{R}^3} A(\boldsymbol{x},\boldsymbol{k},t) f(\boldsymbol{x},\boldsymbol{k},t) d\boldsymbol{k}.$$

Relations (4.32) and (4.33) imply

$$\frac{\partial}{\partial t} \langle A \rangle + \nabla_x \cdot \langle \boldsymbol{u} A \rangle - \frac{\boldsymbol{F}}{\hbar} \cdot \langle \nabla_{\boldsymbol{k}} A \rangle = n(\boldsymbol{x},t) C_A, \tag{4.35}$$

where

$$C_A = \frac{1}{n} \int_{\mathbf{R}^3} A(\boldsymbol{k}) Q[f,f](\boldsymbol{k}) d\boldsymbol{k}.$$

Putting $A = q$, we take the zeroth moment, where q denotes the elementary electric charge of particles. In this case, we have

$$\begin{aligned} \langle q \rangle &= qn \\ \langle \boldsymbol{u} q \rangle &= \langle q \boldsymbol{v} \rangle = qn\boldsymbol{v} \\ C_q &= \frac{q}{n} \int_{\mathbf{R}^3} Q[f,f](\boldsymbol{k}) d\boldsymbol{k} = 0, \end{aligned}$$

and hence

$$q\frac{\partial n}{\partial t} + \nabla \cdot (qn\boldsymbol{v}) = 0$$

follows. Here, $\boldsymbol{J} = qn\boldsymbol{v}$ is the electric current density and we obtain the equation of continuity for the electric current,

$$\frac{\partial n}{\partial t} = -\frac{1}{q}\nabla \cdot \boldsymbol{J}. \tag{4.36}$$

Next, we take the first moment, setting $A = \boldsymbol{u}$. In this case, it holds that

$$\langle \boldsymbol{u} \rangle = \langle \boldsymbol{v} \rangle = n\boldsymbol{v}$$
$$\langle \boldsymbol{u} \otimes \boldsymbol{u} \rangle = n\boldsymbol{v} \otimes \boldsymbol{v} + \langle \mathbf{c} \otimes \mathbf{c} \rangle .$$

Here we assume that the distribution function $f(\boldsymbol{x}, \boldsymbol{k}, t)$ is isotropic, so that the non-diagonal part of the second tensor

$$\boldsymbol{v} \otimes \boldsymbol{v} = \left(v^i v^j\right)_{1 \le i,j \le 3}$$

of $\boldsymbol{v} = {}^t(v^1, v^2, v^3)$ is small. It is approximated by $\left(\frac{1}{3}\boldsymbol{v} \cdot \boldsymbol{v}\delta_{ij}\right)_{1 \le i,j \le 3}$, which comes from

$$\frac{1}{3}\mathrm{Tr}\,(\boldsymbol{v} \otimes \boldsymbol{v}) = \frac{1}{3}\boldsymbol{v} \cdot \boldsymbol{v}.$$

Writing this process simply as

$$\boldsymbol{v} \otimes \boldsymbol{v} = \frac{1}{3}\boldsymbol{v} \cdot \boldsymbol{v},$$

similarly we have

$$\langle \mathbf{c} \otimes \mathbf{c} \rangle = \frac{1}{3}\mathrm{Tr}\,\langle \mathbf{c} \otimes \mathbf{c} \rangle .$$

Temperature tensor $\boldsymbol{T}$ and scalar temperature T are given as

$$nk_B\boldsymbol{T} = m\,\langle \mathbf{c} \otimes \mathbf{c} \rangle$$

and

$$T = \frac{1}{3}\mathrm{Tr}(\boldsymbol{T}) = \frac{m}{3nk_B}\mathrm{Tr}\,\langle \mathbf{c} \otimes \mathbf{c} \rangle ,$$

where k_B and m denote the Boltzmann constant and the effective mass, respectively. This gives

$$\langle \mathbf{c} \otimes \mathbf{c} \rangle = \frac{nk_B}{m}T$$

and hence

$$\langle \boldsymbol{u} \otimes \boldsymbol{u} \rangle = \frac{2}{3} n \cdot \frac{E}{m}$$

follows for

$$E = \frac{1}{2} m \boldsymbol{v} \cdot \boldsymbol{v} + \frac{3}{2} k_B T.$$

Here, E is the sum of the kinetic energy $\frac{1}{2} m \boldsymbol{v} \cdot \boldsymbol{v}$ and the heat energy $\frac{3}{2} k_B T$. Compared with the latter, we can ignore the former in the ordinary temperature. Letting

$$E = \frac{3}{2} k_B T,$$

we obtain

$$\langle \boldsymbol{u} \otimes \boldsymbol{u} \rangle = \frac{1}{m} n k_B T.$$

The inverse effective mass tensor is given as $\boldsymbol{M}^{-1} = \frac{1}{\hbar} \nabla_k \boldsymbol{u}$. Between the effective mass m, we have the relation

$$\frac{1}{3} \mathrm{Tr} \left\langle \boldsymbol{M}^{-1} \right\rangle = \frac{n}{m}.$$

Because the density function is isotropic, it follows that

$$\begin{aligned} \frac{\boldsymbol{F}}{\hbar} \cdot \langle \nabla_k \boldsymbol{u} \rangle &= \boldsymbol{F} \cdot \left\langle \boldsymbol{M}^{-1} \right\rangle = \boldsymbol{F} \cdot \frac{1}{3} \mathrm{Tr} \left\langle \boldsymbol{M}^{-1} \right\rangle \\ &= \boldsymbol{F} \frac{n}{m}. \end{aligned}$$

Finally, we apply the relaxation time approximation as

$$n C_u = -n \frac{\boldsymbol{v}}{\tau_u},$$

where τ_u is the kinetic relaxation time. Thus, under those physical assumptions we have

$$\frac{\partial}{\partial t} (n \boldsymbol{v}) + \frac{1}{m} \nabla (n k_B T) - \boldsymbol{F} \frac{n}{m} = -n \frac{\boldsymbol{v}}{\tau_u}.$$

Multiplying the elementary charge q, we get that

$$\boldsymbol{J} + \tau_u \frac{\partial}{\partial t} \boldsymbol{J} = -\frac{q \tau_u}{m} \nabla (n k_B T) + \frac{q \tau_u n}{m} \boldsymbol{F}. \tag{4.37}$$

4.3.2 Drift-Diffusion (DD) Model

If $\boldsymbol{J}$ is stationary, equation (4.37) is reduced to

$$\boldsymbol{J} = -\frac{q\tau_u}{m}\nabla\left(nk_BT\right) + \frac{q\tau_v}{m}n\boldsymbol{F}.$$

Furthermore, if the carrier velocity is quasi-uniform in the space, we get that

$$\boldsymbol{J} = -\frac{q\tau_u}{m}k_BT\nabla n + \frac{q\tau_u}{m}n\boldsymbol{F}. \tag{4.38}$$

The elementary electric charge of electron is $q = -e$. Its mobility μ indicates the rate between the velocity of the electric field and is determined by the kinetic relaxation time τ_u and the effective mass m, associated with the scattering and the band, respectively, in such a way as

$$\mu = \frac{e\tau_u}{m}.$$

Then, Einstein's relation is expressed as

$$D = \mu\frac{k_BT}{e},$$

where D is the diffusion coefficient. In use of (4.34) and (4.38), we obtain

$$\begin{aligned}\boldsymbol{J} &= \mu k_BT\nabla n + \mu ne\boldsymbol{E}\\ &= eD\nabla n + \mu e\boldsymbol{E}.\end{aligned}$$

Because the above relation is concerned with the electron, we shall write it as

$$\boldsymbol{J}_n = e\mu_n\boldsymbol{E} + eD_n\nabla n.$$

As for the positron, the elementary electric charge and the outer force are given by $q = e$ and $\boldsymbol{F} = -\boldsymbol{E}$, respectively. It follows that

$$\boldsymbol{J}_p = e\mu_p\boldsymbol{E} - eD_p\nabla p,$$

where p denotes the carrier density of positron.

Now we re-formulate the problem. In fact, the equation of continuity on the electric current is given by

$$\frac{\partial n}{\partial t} = \frac{1}{e}\nabla\cdot\boldsymbol{J}_n + r$$

$$\frac{\partial p}{\partial t} = -\frac{1}{e}\nabla \cdot \boldsymbol{J}_p + r, \tag{4.39}$$

where r is the generation-recombination term associated with the band structure. On the other hand, the Poisson equation holds to the electric field $\boldsymbol{E} = -\nabla\varphi$ as

$$-\varepsilon\Delta\varphi = -e(n - p - c), \tag{4.40}$$

where φ and $\rho = -e(n - p - c)$ are the dielectric constant and the charge density, respectively. The term c is the difference between the concentrations of the ionized donors and adapters:

$$c = N_D^+ - N_A^-.$$

Remember that the current densities of electron and positron are given as

$$\begin{aligned} \boldsymbol{J}_n &= -e\mu_n n\nabla\varphi + eD_n\nabla n \\ \boldsymbol{J}_p &= -e\mu_p p\nabla\varphi - eD_p\nabla p. \end{aligned} \tag{4.41}$$

4.3.3 *Mathematical Structure*

We take the case that $c = r = 0$ and put one for every physical constant. Then, DD model is given as

$$\begin{aligned} &\frac{\partial n}{\partial t} = \nabla \cdot (\nabla n - n\nabla\varphi) \\ &\frac{\partial p}{\partial t} = \nabla \cdot (\nabla p + p\nabla\varphi) \qquad \text{in} \qquad \Omega \times (0,T) \\ &\Delta\varphi = n - p, \end{aligned} \tag{4.42}$$

where $\Omega \subset \mathbf{R}^n$ is a bounded domain indicating the device. Let the boundary $\partial\Omega$ be Lipschitz continuous. For (4.42) to determine the unique solution, side conditions are necessary. For simplicity, we take the homogeneous Dirichlet boundary condition for the potential of electric field φ, and zero flux boundary conditions for the electric current densities n, p of electrons and positrons:

$$\begin{aligned} &\frac{\partial n}{\partial \nu} - n\frac{\partial \varphi}{\partial \nu} = 0 \\ &\frac{\partial p}{\partial \nu} + p\frac{\partial \varphi}{\partial \nu} = 0 \qquad \text{on} \qquad \partial\Omega \times (0,T) \\ &\varphi = 0. \end{aligned} \tag{4.43}$$

Initial conditions are imposed only for p, n because the third equation of (4.42) is elliptic:

$$n|_{t=0} = n_0(\boldsymbol{x}), \qquad p|_{t=0} = p_0(\boldsymbol{x}). \tag{4.44}$$

If the initial values is taken in a suitable function space X, then the unique existence of the solution follows in an associated function space in space-time variables. Thus, well-posedness of problem (4.42), (4.43), (4.44) will be assured.

A very important feature of the solution is the positivity preserving:

$$n_0,\ p_0 \geq 0 \quad \text{in} \quad \Omega \qquad \Rightarrow \qquad n(t),\ p(t) \geq 0.$$

This is a consequence of the maximum principle in the linear parabolic equation. On the other hand, we have

$$\frac{d}{dt}\int_\Omega p(\boldsymbol{x},t)d\boldsymbol{x} = \frac{d}{dt}\int_\Omega n(\boldsymbol{x},t)d\boldsymbol{x} = 0$$

by (4.42) and (4.43). Hence we obtain conservation of mass,

$$\|n(t)\|_1 = \|n_0\|_1, \quad \|p(t)\|_1 = \|p_0\|_1 \qquad (0 \leq t < T). \tag{4.45}$$

Here and henceforth $\|\cdot\|_p$ denotes the standard L^p norm on Ω for $p \in [1,\infty]$:

$$\|v\|_p = \begin{cases} \left\{\int_\Omega |v(\boldsymbol{x})|^p\, d\boldsymbol{x}\right\}^{1/p} & (p \in [1,\infty)) \\ \text{ess. sup}_{\boldsymbol{x}\in\Omega} |v(\boldsymbol{x})| & (p = \infty). \end{cases}$$

The first step to study long time behavior of the solution is to classify stationary solutions:

$$\begin{aligned} &\nabla\cdot(\nabla n - n\nabla\varphi) = 0 \\ &\nabla\cdot(\nabla p + p\nabla n) = 0 \qquad \text{in} \qquad \Omega\times(0,T) \\ &\Delta\varphi = n - p \end{aligned} \tag{4.46}$$

with

$$\begin{aligned} &\frac{\partial n}{\partial\nu} - n\frac{\partial\varphi}{\partial\nu} = 0 \\ &\frac{\partial p}{\partial\nu} + p\frac{\partial\varphi}{\partial\nu} = 0 \qquad \text{on} \qquad \partial\Omega\times(0,T) \\ &\varphi = 0. \end{aligned} \tag{4.47}$$

For this purpose, we note that the chemical potential $\varphi_n = \varphi - \log n$ of electron is associated with the current density as

$$J_n = \nabla n - n\nabla\varphi = -n\nabla\varphi_n.$$

In the non-trivial stationary state $n = n(x) > 0$ it follows from (4.46) and (4.47) that

$$\nabla \cdot (n\nabla\varphi_n) = 0 \quad \text{in} \quad \Omega, \qquad \frac{\partial}{\partial\nu}\varphi_n = 0 \quad \text{on} \quad \partial\Omega.$$

This implies

$$\int_\Omega n\left|\nabla\varphi_n\right|^2 dx = 0$$

and hence φ_n is a constant in Ω.

Taking account of the conservation of mass, (4.45), we prescribe this unknown constant in terms of $\lambda = \|n\|_1 > 0$. Then, it follows that

$$n = \frac{\lambda e^\varphi}{\int_\Omega e^\varphi dx}$$

and similarly,

$$p = \frac{\mu e^{-\varphi}}{\int_\Omega e^{-\varphi} dx},$$

where $\mu = \|p\|_1 > 0$. Thus the stationary problem is reduced to

$$\begin{aligned} \Delta\varphi &= \frac{\lambda e^\varphi}{\int_\Omega e^\varphi dx} - \frac{\mu e^{-\varphi}}{\int_\Omega e^{-\varphi} dx} && \text{in} \quad \Omega \\ \varphi &= 0 && \text{on} \quad \partial\Omega \end{aligned} \tag{4.48}$$

by means of the third relations of (4.46) and (4.47).

We have observed mass conservation and introduced the problem for equilibrium state to satisfy. Another important factor for the large time behavior of the solution is the existence of the Lyapunov function. It is reasonable because DD model is derived from the Boltzmann equation.

In fact, thermal equilibrium is the state where free energy $\mathcal{F} = E - TS$ attains minimum. Here, E, T, and S denote inner energy, temperature, and entropy, respectively. Remember that the inner energy is preserved in the classical Boltzmann equation. Because any physical constant is set to be one, we have $T = 1$,

$$E = \frac{1}{2}\int_\Omega |\nabla\varphi|^2\, dx$$

$$S = -\int_\Omega \{p(\log p - 1) + n(\log n - 1)\}\, dx$$

and hence

$$\mathcal{F} = \int_\Omega \left\{p(\log p - 1) + n(\log n - 1) + \frac{1}{2}|\nabla\varphi|^2\right\} dx$$

is obtained. In fact, if the smooth functions $p = p(x,t)$ and $n = n(x,t)$ satisfy (4.42) and (4.43), then it follows that

$$\frac{d}{dt}\mathcal{F} \le 0 \qquad (0 \le t < T)\,. \tag{4.49}$$

Exercise 4.7 Prove (4.49).

Chapter 5

Linear PDE Theory

This chapter deals with the fundamental theory of partial differential equations, well-posedness, fundamental solution, potential, and regularity. Although the materials are restricted mostly to elliptic and parabolic equations, but several basic ideas and calculations are presented, from which one can access the standard advanced monographs or papers.

5.1 Well-posedness

5.1.1 *Heat Equation*

Imagine that a domain $\Omega \subset \mathbf{R}^3$ is occupied with the heat conductor, and let $u = u(x,t)$ be the temperature at the position $\boldsymbol{x} = {}^t(x_1, x_2, x_3) \in \Omega$ and the time $t > 0$. If ρ, c, and ω denote the ratio of specific heat, the density, and a subdomain of Ω with smooth boundary $\partial\omega$, respectively, then

$$\int_\omega c\rho u(\boldsymbol{x},t)d\boldsymbol{x}$$

denotes the heat quantity put in ω. On the other hand,

$$\int_{\partial\omega} \kappa \frac{\partial u}{\partial \boldsymbol{\nu}}(\boldsymbol{x},t)dS$$

indicates the heat quantity radiated inside ω through $\partial\omega$, where $\boldsymbol{\nu}$ and dS denote the outer unit normal vector and the area element of $\partial\omega$, respectively. Therefore, it holds that

$$\frac{d}{dt}\int_\omega c\rho u(\boldsymbol{x},t)d\boldsymbol{x} = \int_{\partial\omega} \kappa \frac{\partial u}{\partial \boldsymbol{\nu}}(\boldsymbol{x},t)dS. \tag{5.1}$$

If $u(x,t)$ is smooth, the left-hand side of (5.1) is equal to

$$\int_\omega \frac{\partial}{\partial t}(c\rho u)dx,$$

while the right-hand side is

$$\int_\omega \nabla \cdot (\kappa \nabla u)dx$$

from the divergence theorem of Gauss. Because ω is arbitrary, this implies that

$$\frac{\partial}{\partial t}(c\rho u) = \nabla \cdot (\kappa \nabla u) \qquad (x \in \Omega,\ t > 0)\,, \tag{5.2}$$

which is referred to as the *heat equation*. Usually, side conditions are provided to determine $u(x,t)$, so that the *initial condition* is given as

$$u|_{t=0} = u_0(x) \qquad (x \in \Omega)\,. \tag{5.3}$$

The *boundary condition*

$$u = \alpha(\xi, t) \qquad (\xi \in \partial\Omega,\ t > 0) \tag{5.4}$$

prescribes the temperature distribution itself on the boundary and is called the *Dirichlet* or the *first kind* boundary condition. The *Neumann* or the *second kind* boundary condition

$$\kappa \frac{\partial u}{\partial \nu} = \beta(\xi, t) \qquad (\xi \in \partial\Omega,\ t > 0) \tag{5.5}$$

prescribes the heat flux distribution radiated from the boundary. Finally, the *Robin* or the *third kind* boundary condition

$$\kappa \frac{\partial u}{\partial \nu} + \mu u = \gamma(\xi, t) \qquad (\xi \in \partial\Omega,\ t > 0) \tag{5.6}$$

prescribes the heat flux distribution on the boundary proportionally subject to the temperature.

In the *direct problem*, those *parameters* c, ρ, κ, and μ, and the *initial value* u_0, and the *boundary value* α, or β, are given and it is asked to determine the *solution* $u = u(x,t)$ satisfying (5.2), (5.3), and (5.4) (or (5.5), or (5.6)). On the other hand, the *inverse problem* determines those parameters (or initial and boundary values) by some observable concerning the solution. It is expected that the direct problem is *well-posed*, which means that the solution exists uniquely for and depends continuously on the given data.

5.1.2 *Uniqueness*

We take the simplest case as $\Omega = (0, \pi) \subset \mathbf{R}$, $c = \rho = \kappa = 1$, and $\alpha = 0$ in (5.4):

$$\begin{aligned} &u_t = u_{xx} && (0 < x < \pi,\ t > 0) \\ &u\big|_{t=0} = u_0(x) && (0 \le x \le \pi) \\ &u\big|_{x=0,\pi} = 0 && (t \ge 0) \end{aligned} \tag{5.7}$$

and consider the *classical solution*, so that its differentiability and continuity are taken in the classical sense with the least regularity to make (5.7) reasonable. Namely,

$$u = u(x,t) \in C\left([0,\pi] \times [0,\infty)\right) \tag{5.8}$$

and

$$u_t, u_x, u_{xx} \in C\left((0,\pi) \times (0,\infty)\right). \tag{5.9}$$

Therefore, for the classical solution to exist, the initial value u_0 must be in $C[0,\pi]$ and satisfy $u_0|_{x=0,\pi} = 0$. They are called the *compatibility condition*, generally.

We show the following.

Theorem 5.1 *The classical solution to (5.7) is unique.*

Proof. Let u_1, u_2 be the classical solution and set $u = u_1 - u_2$. Then, it satisfies (5.8), (5.9), and

$$\begin{aligned} &u_t = u_{xx} && (0 < x < \pi,\ t > 0) \\ &u\big|_{t=0} = 0 && (0 \le x \le \pi) \\ &u\big|_{x=0,\pi} = 0 && (t \ge 0). \end{aligned}$$

We have only to derive $u \equiv 0$ from those relations. For this purpose, we show that both $u \ge 0$ and $u \le 0$ hold in $[0,\pi] \times [0,\infty)$.

In fact, $w = e^{-t}u$ is provided with the same continuity and the differentiability as those of u. Writing $u = e^t w$, we have $u_t = e^t w + e^t w_t$ and

$u_{xx} = e^t w_{xx}$ and hence

$$\begin{array}{ll} w_t + w = w_{xx} & (0 < x < \pi,\ t > 0) \\ w|_{t=0} = 0 & (0 \le x \le \pi) \\ w|_{x=0,\pi} = 0 & (t \ge 0) \end{array} \tag{5.10}$$

follows. Let us take $T > 0$ arbitrarily. Then, $w = w(x,t)$ is continuous on $[0,\pi] \times [0,T]$ and there exists $(x_0, t_0) \in [0,\pi] \times [0,T]$ such that $\mu = w(x_0, t_0)$, where

$$\mu = \max\{w(x,t) \mid 0 \le x \le \pi,\ 0 \le t \le T\}.$$

Using the initial and the boundary conditions in (5.10), it holds that $\mu \ge 0$. If $\mu > 0$, then $x_0 \in (0,\pi)$ and $t_0 \in (0,T]$ and therefore, we obtain

$$w_x(x_0,t_0) = 0, \qquad w_{xx}(x_0,t_0) \le 0, \qquad w_t(x_0,t_0) \ge 0. \tag{5.11}$$

On the other hand, the first equation of (5.10) implies

$$w_t(x_0,t_0) + \mu = w_{xx}(x_0,t_0)$$

and hence $\mu \le 0$ follows. This is a contradiction and we obtain $\mu = 0$. This means that $w = w(x,t) \le 0$ on $[0,\pi] \times [0,T]$. Because $v = -w$ is a classical solution to (5.10) and it follows that $w \ge 0$ there. This means $w = e^{-t}u = 0$ on $[0,\pi] \times [0,T]$ and hence $u \equiv 0$ follows because $T > 0$ is arbitrary. □

To prove the above theorem, we have made use of the *argument of comparison*, or the *maximum principle*. Actually, we can show the following, where

$$\|v\|_\infty = \sup_{x \in [0,\pi]} |v(x)| \tag{5.12}$$

is called the *maximum norm*, because it is attained if v is continuous on $[0,\pi]$.

Theorem 5.2 *If $u(x,t) = u(\cdot,t)$ is the classical solution to (5.7), then it holds that*

$$\|u(t)\|_\infty \le \|u_0\|_\infty$$

for any $t \ge 0$.

Proof. Letting $\lambda = \|u_0\|_\infty$, we have

$$-\lambda \le u_0(x) \le \lambda \qquad (0 \le x \le \pi).$$

Then $w = e^{-t}(u - \lambda)$ satisfies that

$$\begin{aligned} & w_t + w = w_{xx} && (0 < x < \pi,\ t > 0) \\ & w|_{t=0} \le 0 && (0 \le x \le \pi) \\ & w|_{x=0,\pi} \le 0 && (t \ge 0) \end{aligned}$$

and the same argument as in the proof of the previous theorem guarantees that

$$\max\{w(x,t) \mid 0 \le x \le \pi,\ 0 \le t \le T\} \le 0 \tag{5.13}$$

for any $T > 0$. This means $u(x,t) \le \lambda$ for $(x,t) \in [0,\pi] \times [0,T]$. The inequality $u(x,t) \ge -\lambda$ follows similarly, and the proof is complete. □

Theorem 5.2 indicates the stability of the solution in $\|\cdot\|_\infty$. But it also implies the continuous dependence of the solution on the initial data. Namely, if $u_1(\cdot,t) = u_1(x,t)$ and $u_2(\cdot,t) = u_2(x,t)$ denote the classical solutions to (5.7) with the initial values $u_0(x)$ equal to $u_1(x)$ and $u_2(x)$, respectively, then $u(x,t) = u_1(x,t) - u_2(x,t)$ solves the problem with the initial value $u_1(x) - u_2(x)$, so that we obtain

$$\|u_1(t) - u_2(t)\|_\infty \le \|u_1 - u_2\|_\infty \tag{5.14}$$

for any $t \ge 0$. Because of this, we see that if $\|u_1 - u_2\|_\infty$ is small, then so is $\|u_1(t) - u_2(t)\|_\infty$. Furthermore, Theorem 5.1 follows from this inequality as $u_1 = u_2$ implies $u_1(\cdot,t) = u_2(\cdot,t)$ for any $t \ge 0$.

Exercise 5.1 Confirm (5.11).

Exercise 5.2 Confirm (5.13).

5.1.3 *Existence*

Let us make use of the method of §3.2.1 to construct the solution. First, the principle of super position says that if $v_1(x,t), v_2(x,t), \cdots$ are the solutions to

$$\begin{aligned} & u_t = u_{xx} && (0 < x < \pi,\ t > 0) \\ & u|_{x=0,\pi} = 0 && (t \ge 0)\,, \end{aligned} \tag{5.15}$$

then, so is $v(x,t) = v_1(x,t) + v_2(x,t) + \cdots$. Second, the special solution to (5.15) is obtained by the form of separation of variables, $v(x,t) = \varphi(x)\eta(t)$. This trial leads us to

$$\frac{\varphi''(x)}{\varphi(x)} = \frac{\eta'(t)}{\eta(t)} = -\lambda \tag{5.16}$$

and hence the eigenvalue problem

$$-\varphi''(x) = \lambda\varphi(x) \quad (0 \leq x \leq \pi), \qquad \varphi|_{x=0,\pi} = 0$$

arises, where $\lambda \in \mathbf{R}$ denotes the eigenvalue. As we have seen in §3.2.1, this problem provides a complete ortho-normal system in $L^2(0,\pi)$,

$$\left\{ \frac{1}{\sqrt{\pi}} \sin nx \mid n = 1, 2, \cdots \right\},$$

with each eigenfunction $\varphi_n(x) = \frac{1}{\sqrt{\pi}} \sin nx$ corresponds to the eigenvalue $\lambda_n = n^2$.

Writing formally that

$$u(x,t) = \sum_{n=1}^{\infty} c_n(t)\varphi_n(x),$$

we have $c_n(t) = (u(t), \varphi_n)$, where (,) denotes the L^2 inner product:

$$(f,g) = \int_0^\pi f(x)g(x)dx.$$

Thus, integration by parts guarantees

$$\begin{aligned} c_n' &= (u_t, \varphi_n) = (u_{xx}, \varphi_n) \\ &= (u, \varphi_{nxx}) = \lambda_n (u, \varphi_n) = \lambda_n c_n, \end{aligned} \tag{5.17}$$

which implies that

$$c_n(t) = (u_0, \varphi_n)e^{-\lambda_n t}$$

from the initial condition to u. We obtain

$$u(x,t) = \sum_{n=1}^{\infty} c_n e^{-\lambda_n t}\varphi_n(x) \tag{5.18}$$

with $c_n = (u_0, \varphi_n)$. This sequence is not so difficult to handle except for the delicate behavior as $t \downarrow 0$, and its justification is left to the reader. Note that

$$\sum_{n=1}^{\infty} |c_n|^2 < +\infty$$

holds by $u_0 \in L^2(0,\pi)$.

Exercise 5.3 Confirm that (5.17) follows from integration by parts.

Exercise 5.4 Confirm that $u(t) = u(\cdot, t)$ defined by (5.18) satisfies the initial condition of (5.14) in the sense that

$$\lim_{t\downarrow 0} \|u(t) - u_0\|_2 = 0,$$

where $\| \ \|_2$ denotes the L^2 norm:

$$\|f\|_2 = \left(\int_0^\pi f(x)^2 dx \right)^{1/2}.$$

Exercise 5.5 Show that the right-hand side of (5.18) converges uniformly on $[0,\pi] \times [\delta, T]$ if $0 < \delta < T$. Show also that any termwise derivative has the same property and thus prove that $u(x,t)$ defined by (5.14) satisfies (5.7).

Exercise 5.6 Prove that the right-hand side of (5.7) converges uniformly on $[0,\pi] \times [0,T]$ for $T > 0$, provided that u_0 is continuously differentiable on $[0,\pi]$ and satisfies the compatibility condition $u_0|_{x=0,\pi} = 0$.

5.2 Fundamental Solutions

5.2.1 *Fourier Transformation*

The result in §3.2.2 is summarized that

$$\left\{ \frac{1}{\sqrt{2\pi}}, \frac{1}{\sqrt{\pi}} \sin nx, \frac{1}{\sqrt{\pi}} \cos mx \mid n, m = 1, 2, \cdots \right\}$$

forms a complete ortho-normal system in $L^2(0, 2\pi)$, or in $L^2(-\pi, \pi)$.

Vector spaces treated so far are over $\mathbf{R}$. In this paragraph we make use of the complex variable. Actually, the function space $L^2(-\pi,\pi)$ is regarded as the vector space over **C** if each element takes the complex value. Then, it forms the Hilbert space over **C** through the inner product

$$(f,g) = \int_{-\pi}^{\pi} f(x)\overline{g(x)}dx.$$

Using Euler's convention that

$$e^{\imath\theta} = \cos\theta + \imath\sin\theta$$

for $\theta \in \mathbf{R}$, we have

$$e^{\pm\imath nx} = \cos nx \pm \imath\sin nx.$$

This implies that

$$\left\{\frac{1}{\sqrt{2\pi}}e^{\imath nx} \mid n = 0, \pm 1, \pm 2, \cdots\right\}$$

forms a complete ortho-normal system in (complex) $L^2(-\pi,\pi)$, and the Fourier series of $f \in L^2(-\pi,\pi)$ is written as

$$f(x) = \sum_{n=-\infty}^{+\infty} \frac{1}{2\pi}\hat{f}_n e^{\imath nx} \quad \text{with} \quad \hat{f}_n = \int_{-\pi}^{\pi} f(x)e^{-\imath nx}dx. \tag{5.19}$$

Then, taking $x' = Nx$, we see that each $f \in L^2(-N\pi, N\pi)$ is expanded as

$$f(x) = \sum_{n=-\infty}^{\infty} \frac{1}{2\pi}\hat{f}_n^N e^{\imath(n/N)x}$$

with

$$\hat{f}_n^N = \frac{1}{N}\int_{-N\pi}^{N\pi} f(x)e^{-\imath(n/N)x}dx.$$

In terms of $\hat{f}(n/N) = N\hat{f}_n^N$, those relations are expressed as

$$\hat{f}(n/N) = \int_{-N\pi}^{N\pi} f(x)e^{-\imath(n/N)x}dx \qquad (n = 0, \pm 1, \pm 2, \cdots)$$

$$f(x) = \frac{1}{2\pi}\sum_{n=-\infty}^{\infty} \frac{1}{N}\hat{f}(n/N)e^{\imath(n/N)x} \qquad (-N\pi < x < N\pi).$$

Then, making $N \to \infty$ formally, we get the relation that

$$\hat{f}(\xi) = \int_{-\infty}^{\infty} f(x) e^{-\imath \xi x} dx \qquad (\xi \in \mathbf{R}) \tag{5.20}$$

and

$$f(x) = \frac{1}{2\pi} \int_{-\infty}^{\infty} \hat{f}(\xi) e^{\imath \xi x} d\xi \qquad (x \in \mathbf{R}). \tag{5.21}$$

The right-hand sides of (5.20) and (5.21) are called the *Fourier transformation* of $f(x)$ and the *inverse Fourier transformation* of $\hat{f}(\xi)$, and denoted by $(\mathcal{F}f)(\xi)$ and $(\mathcal{F}^{-1})\hat{f}(x)$, respectively. Then, equality (5.21) is referred to as the *Plancherel's inversion formula*. Justification of those relations are done in several categories.

Exercise 5.7 Confirm that the Fourier series of $f \in L^2(-\pi, \pi)$ is written as (5.19) and show that

$$\left\{ \frac{1}{\sqrt{2\pi}} e^{\imath n x} \mid n = 0, \pm 1, \pm 2, \cdots \right\}$$

forms a complete ortho-normal system in (complex) $L^2(-\pi, \pi)$.

5.2.2 *Rapidly Decreasing Functions*

Henceforth, $L^p(\mathbf{R})$ denotes the set of *p-integrable functions* on $\mathbf{R}$ for $p \in [1, \infty)$. That is, $f \in L^p(\mathbf{R})$ holds if and only if it is measurable and

$$\|f\|_p = \left(\int_{\mathbf{R}} |f(x)|^p \, dx \right)^{1/p} < +\infty$$

in the sense of Lebesgue. On the other hand, we say that $f \in L^\infty(\mathbf{R})$ if there is $M > 0$ such that $|f(x)| \leq M$ for almost every $x \in \mathbf{R}$. In this case it is said that $f(x)$ is *essentially bounded* on $\mathbf{R}$, and infimum of such M is denoted by $\|f\|_\infty$. We note that this notation adjusts with (5.12). Furthermore, it is known that $L^p(\mathbf{R})$ becomes a Banach space under the norm $\| \ \|_p$ for $p \in [1, \infty]$. Then, inequalities

$$\|f + g\|_p \leq \|f\|_p + \|g\|_p \tag{5.22}$$

and

$$\|f \cdot g\|_1 \le \|f\|_p \cdot \|g\|_{p'} \tag{5.23}$$

hold, and referred to as *Minkowski's inequality* and *Hölder's inequality*, respectively. Here and henceforth, $p' \in [1, \infty]$ denotes the *dual exponent* of $p \in [1, \infty]$, defined by

$$\frac{1}{p} + \frac{1}{p'} = 1.$$

It is obvious that if $f \in L^1(\mathbf{R})$, then $\hat{f}(\xi)$ defined by (5.20) converges for each $\xi \in \mathbf{R}$ with the property that

$$\left|\hat{f}(\xi)\right| \le \|f\|_1 \qquad (\xi \in \mathbf{R}). \tag{5.24}$$

In the L^2 setting, the right-hand sides of (5.20) and (5.21) are taken as the limits in $L^2(\mathbf{R})$ as $N \to \infty$ of

$$\int_{-N\pi}^{N\pi} f(x) e^{-\imath \xi x} dx \qquad \text{and} \qquad \frac{1}{2\pi} \int_{-N\pi}^{N\pi} \hat{f}(\xi) e^{\imath \xi x} d\xi,$$

respectively. Then, Plancherel's inversion formula holds as

$$\mathcal{F}^{-1}(\mathcal{F}f) = f \qquad \text{and} \qquad \mathcal{F}\left(\mathcal{F}^{-1}\hat{f}\right) = \hat{f}$$

in $L^2(\mathbf{R})$ for any $f \in L^2(\mathbf{R})$ and $\hat{f} \in L^2(\mathbf{R})$.

The set of arbitrarily many differentiable functions on $\mathbf{R}$ is denoted by $C^\infty(\mathbf{R})$. We say that $f \in C^\infty(\mathbf{R})$ is *rapidly decreasing* if

$$\lim_{x \to \pm\infty} |x|^m \left| f^{(k)}(x) \right| = 0$$

for any $m, k = 0, 1, 2, \cdots$. The set of such functions is denoted by $\mathcal{S}(\mathbf{R})$. Thus, $f \in \mathcal{S}(\mathbf{R})$ if and only if its any derivative decays more rapidly than any rational functions. In particular, each $f \in \mathcal{S}(\mathbf{R})$ admits $C > 0$ such that

$$|f(x)| \le C\left(1 + x^2\right)^{-1}$$

and hence $f \in L^1(\mathbf{R})$ follows. The integral of the right-hand side of (5.20) converges absolutely. It also holds that $f \in L^2(\mathbf{R})$ and hence Plancherel's inversion formula (5.21) is valid for each $x, \xi \in \mathbf{R}$. Furthermore the following property holds.

Theorem 5.3 *If $f \in \mathcal{S}(\mathbf{R})$, then $\hat{f} \in \mathcal{S}(\mathbf{R})$. It holds also that*

$$\mathcal{F}\left(f^{(k)}\right)(\xi) = (\imath\xi)^k \hat{f}(\xi) \tag{5.25}$$

and

$$\mathcal{F}\left(x^m f\right)(\xi) = (\imath\partial_\xi)^m \hat{f}(\xi) \tag{5.26}$$

for $k, m = 0, 1, 2, \cdots$, where $\partial_\xi = \partial/\partial\xi$.

Proof. To show (5.25), we note that $f^{(k)} \in \mathcal{S}(\mathbf{R})$. Then, we have by means of the integration by parts that

$$\begin{aligned}
\mathcal{F}\left(f^{(k)}\right)(\xi) &= \int_{\mathbf{R}} f^{(k)}(x) e^{-\imath\xi x} dx \\
&= (-1)^k \int_{\mathbf{R}} f(x) \cdot (\partial_x)^k \left(e^{-\imath\xi x}\right) dx \\
&= (\imath\xi)^k \int_{\mathbf{R}} f(x) e^{-\imath\xi x} dx = (\imath\xi)^k \hat{f}(\xi).
\end{aligned}$$

Equality (5.26) follows similarly, as

$$\begin{aligned}
\mathcal{F}\left(x^m f\right)(\xi) &= \int_{\mathbf{R}} x^m f(x) e^{-\imath\xi x} dx \\
&= \int_{\mathbf{R}} (\imath\partial_\xi)^m \left(f(x) e^{-\imath\xi x}\right) dx \\
&= (\imath\partial_\xi)^m \int_{\mathbf{R}} f(x) e^{-\imath\xi x} dx = (\imath\partial_\xi)^m \hat{f}(\xi).
\end{aligned}$$

Note that the dominated convergence theorem is applied to justify the above calculations.

Given $f \in \mathcal{S}(\mathbf{R})$, we have $f' \in L^1(\mathbf{R})$. This implies that

$$|\xi| \left|\hat{f}(\xi)\right| = |\mathcal{F}(f')| \leq \|f'\|_1 < +\infty$$

and hence

$$\lim_{\xi \to \pm\infty} \left|\hat{f}(\xi)\right| = 0$$

follows. Then, we have $x^m f^{(k)} \in \mathcal{S}(\mathbf{R})$ and hence

$$\lim_{\xi \to \pm\infty} |\xi|^k \left|(\partial_\xi)^m \hat{f}(\xi)\right| = \lim_{\xi \to \pm\infty} \left|\mathcal{F}\left(x^m f^{(k)}\right)(\xi)\right| = 0$$

follows for $k, m = 0, 1, \cdots$. This means $\hat{f} \in \mathcal{S}$. □

Exercise 5.8 Calculate the Fourier transform

$$\hat{f}(\xi) = \int_{\mathbf{R}} f(x) e^{-\imath \xi x} dx$$

of

$$f(x) = \begin{cases} 1 & (-1 < x < 1) \\ \frac{1}{2} & (x = \pm 1) \\ 0 & (|x| > 1). \end{cases}$$

Then, compute

$$\lim_{N \to \infty} \frac{1}{2\pi} \int_{-N\pi}^{N\pi} \hat{f}(\xi) e^{\imath \xi x} d\xi.$$

Exercise 5.9 Given $f, g \in L^1(\mathbf{R})$, show that

$$(f * g)(x) = \int_{\mathbf{R}} f(x-y) g(y) dy$$

converges almost every $x \in \mathbf{R}$ and is in $L^1(\mathbf{R})$. Show also that

$$\mathcal{F}(f * g)(\xi) = \mathcal{F}f(\xi) \cdot \mathcal{F}g(\xi) \tag{5.27}$$

holds.

5.2.3 *Cauchy Problem*

We consider the heat equation on the whole space,

$$u_t = u_{xx} \quad (x \in \mathbf{R},\ t > 0) \qquad \text{with} \qquad u|_{t=0} = u_0(x). \tag{5.28}$$

Such a problem is called the *Cauchy problem* because the initial data u_0 is prescribed. What we wish to establish is the well-posedness, so that existence, uniqueness, and continuous dependence on the initial data $u_0(x)$ of the solution $u(x,t)$ with appropriate continuity and differentiability, and also the qualitative study, that is, the properties of the solution. Whole space $\mathbf{R}$ has no boundary, and the infinite point takes its place. Therefore, we have to prescribe the behavior of the solution at infinity.

To see this, we shall show that there is smooth $u(x,t)$ for $(x,t) \in \mathbf{R} \times \mathbf{R}$ satisfying $u(x,t) = 0$ for $x \in \mathbf{R},\ t \leq 0$, $u(x,t) \not\equiv 0$ for $x \in \mathbf{R},\ t > 0$, and

$$u_t = u_{xx} \qquad (x \in \mathbf{R}, t \in \mathbf{R}).$$

Actually, this $u(x,t)$ gets to $+\infty$ as $x \to \pm\infty$ more rapidly than any polynomial if $t > 0$. Hence uniqueness of (5.28) does not follow unless the behavior at ∞ of the solution is prescribed. The reader can skip the rest of this paragraph if he is not familiar with *complex analysis*.

In fact, first we see that

$$f(t) = \begin{cases} e^{-1/t^2} & (t > 0) \\ 0 & (t \leq 0) \end{cases}$$

satisfies $\lim_{t\downarrow 0} f^{(k)}(t) = 0$ for $k = 0, 1, 2, \cdots$, so that $f \in C^\infty(\mathbf{R})$ holds. Then, $f(t)$ is extended as $f(z) = e^{-1/z^2}$, which is holomorphic in $z \in \mathbf{C} \setminus \{0\}$. Let $t > 0$ be fixed, and apply the *integration formula of Cauchy* as

$$f^{(k)}(t) = \frac{k!}{2\pi\imath} \int_\Gamma \frac{f(z)}{(z-t)^{k+1}} dz,$$

where

$$\Gamma : \ z = t + \frac{t}{2} e^{\imath\theta} \qquad \text{with} \qquad 0 \leq \theta < 2\pi.$$

Here, we have

$$|f(z)| = e^{-\mathrm{Re}(1/z^2)}$$

with $z = t + \frac{t}{2} e^{\imath\theta}$. Therefore, there is $\delta > 0$ such that

$$\mathrm{Re}\left(1/z^2\right) = \frac{1}{t^2} \mathrm{Re}\left(\frac{1}{1 + e^{\imath\theta} + \frac{1}{4} e^{2\imath\theta}} \right) \geq \frac{\delta}{t^2}$$

for any $\theta \in [0, 2\pi)$. This implies that

$$\begin{aligned} \left| f^{(k)}(t) \right| &\leq \frac{k!}{2\pi} \int_\Gamma \frac{e^{-\delta/t^2}}{|z-t|^{k+1}} |dz| \\ &= \frac{k!}{2\pi} e^{-\delta/t^2} \left(\frac{t}{2} \right)^{-k-1} \cdot 2\pi \left(\frac{t}{2} \right) = k! 2^k t^{-k} e^{-\delta/t^2}. \end{aligned} \tag{5.29}$$

Here, we note the elementary inequality

$$e^{-cy}y^p \leq \left(\frac{p}{ec}\right)^p \qquad (y \geq 0) \tag{5.30}$$

valid for $p, c > 0$. Thus, in (5.29) we have

$$t^{-k}e^{-\delta/t^2} = (t^2)^{-k/2}e^{-\delta/t^2} \leq \left(\frac{k/2}{e\delta}\right)^{k/2} = \left(\frac{k}{2e\delta}\right)^{k/2}.$$

This implies that

$$\left|f^{(k)}(t)\right| \leq k!2^k \left(\frac{k}{2e\delta}\right)^{k/2} \equiv a_k.$$

Now we show the following.

Lemma 5.1 *Any $r > 0$ admits $M > 0$ such that*

$$a_k \leq M(2k)!/r^k$$

for $k = 1, 2, \cdots$.

Proof. We shall show that

$$\lim_{k\to\infty} \frac{r^k}{(2k)!}a_k = 0.$$

In fact, letting $b_k = \frac{r^k}{(2k)!}a_k$, we have

$$\begin{aligned} c_k \equiv \frac{b_{k+1}}{b_k} &= \frac{r}{(2k+2)(2k+1)} \cdot \frac{a_{k+1}}{a_k} \\ &= \frac{r \cdot 2(k+1)}{2(k+1)(2k+1)} \cdot \left(\frac{k+1}{2e\delta}\right)^{(k+1)/2} \Big/ \left(\frac{k}{2e\delta}\right)^{k/2} \\ &= \frac{r}{2k+1}\left(\frac{k}{2e\delta}\right)^{1/2} \cdot \left(1 + \frac{1}{k}\right)^{(k+1)/2} \to 0 \end{aligned}$$

as $k \to \infty$, because

$$\lim_{k\to\infty}\left(1 + \frac{1}{k}\right)^k = e.$$

In particular, $\limsup_{k\to\infty} c_k < 1$ and hence $\lim_{k\to\infty} b_k = 0$ follows. □

We can summarize that

$$f(t) = \begin{cases} e^{-1/t^2} & (t > 0) \\ 0 & (t \leq 0) \end{cases}$$

is in $C^\infty(\mathbf{R})$ and any $r > 0$ admits $M > 0$ satisfying

$$\left| f^{(k)}(t) \right| \leq M(2k)!/r^k \qquad (t > 0, \ k = 0, 1, 2, \cdots).$$

Then,

$$u(x,t) = \sum_{k=0}^{\infty} \frac{f^{(k)}(t)}{(2k)!} x^{2k}$$

converges locally uniformly in $(x,t) \in \mathbf{R} \times \mathbf{R}$. Furthermore, form the dominated sequence theorem of Weierstrass, it is termwise differentiable so that we obtain

$$\begin{aligned} u_{xx}(x,t) &= \sum_{k=1}^{\infty} \frac{f^{(k)}(t)}{(2k)!} (2k) \cdot (2k-1) x^{2k-2} \\ &= \sum_{k=1}^{\infty} \frac{f^{(k)}(t)}{(2(k-1))!} x^{2k-2} \\ &= \sum_{k=0}^{\infty} \frac{f^{(k+1)}(t)}{(2k)!} x^{2k} = u_t \end{aligned}$$

for $(x,t) \in \mathbf{R} \times \mathbf{R}$. Then, we can see that $u(x,t) = 0$ for $x \in \mathbf{R}$, $t \leq 0$ and $u(x,t) \not\equiv 0$ for $x \in \mathbf{R}$, $t > 0$.

Exercise 5.10 Confirm (5.30).

5.2.4 *Gaussian Kernel*

Let us seek the smooth solution to (5.28), assuming that u_0, $u(\cdot,t)$, and $u_t(\cdot,t)$ belong to $\mathcal{S}(\mathbf{R})$. In fact, in this case we can take

$$\hat{u}(\xi,t) = \int_{\mathbf{R}} u(x,t) e^{-\imath x \xi} dx.$$

Then, it holds that

$$\mathcal{F}(u_t) = (\mathcal{F}u)_t = \hat{u}_t \qquad \text{and} \qquad \mathcal{F}(u_{xx}) = (\imath\xi)^2 \mathcal{F}(u) = -\xi^2 \hat{u}$$

and hence

$$\hat{u}_t = -\xi^2 \hat{u} \quad (\xi \in \mathbf{R},\ t > 0), \qquad \hat{u}|_{t=0} = \hat{u}_0(\xi) \quad (\xi \in \mathbf{R}) \tag{5.31}$$

follows.

If $\xi \in \mathbf{R}$ is fixed, then (5.31) is the Cauchy problem of the ordinary differential equation with respect to t. Thus, we obtain

$$\hat{u}(\xi, t) = e^{-t\xi^2} \hat{u}_0(\xi) \qquad (\xi \in \mathbf{R},\ t > 0).$$

Using the inverse Fourier transformation, we can recover $u(x,t)$ as

$$\begin{aligned} u(x,t) &= \frac{1}{2\pi} \int_{\mathbf{R}} \hat{u}(\xi, t) e^{\imath x\xi} d\xi = \frac{1}{2\pi} \int_{\mathbf{R}} e^{-t\xi^2} \hat{u}_0(\xi) e^{\imath x\xi} d\xi \\ &= \frac{1}{2\pi} \int_{\mathbf{R}} e^{-t\xi^2 + \imath x\xi} d\xi \int_{\mathbf{R}} u_0(y) e^{-\imath \xi y} dy \\ &= \frac{1}{2\pi} \int_{\mathbf{R}} d\xi \int_{\mathbf{R}} e^{-t\xi^2 + \imath(x-y)\xi} u_0(y) dy. \end{aligned}$$

Here, we have

$$\left| e^{-t\xi^2 + \imath(x-y)\xi} u_0(y) \right| = e^{-t\xi^2} |u_0(y)|$$

and

$$\int_{\mathbf{R}} d\xi \int_{\mathbf{R}} dy e^{-t\xi^2} |u_0(y)| = \|u_0\|_1 \cdot \int_{\mathbf{R}} e^{-t\xi^2} d\xi < +\infty$$

for $t > 0$. Using Fubini's theorem we obtain

$$u(x,t) = \frac{1}{2\pi} \int_{\mathbf{R}} \left\{ \int_{\mathbf{R}} e^{-t\xi^2 + \imath(x-y)\xi} d\xi \right\} u_0(y) dy. \tag{5.32}$$

In this way, we get the Gaussian kernel

$$\begin{aligned} G(z,t) &= \frac{1}{2\pi} \int_{\mathbf{R}} e^{-t\xi^2 + \imath z\xi} d\xi \\ &= \frac{1}{2\pi} \int_{\mathbf{R}} e^{-t\left(\xi - \frac{\imath z}{2t}\right)^2} d\xi \cdot e^{-z^2/4t} = \frac{e^{-z^2/4t}}{2\pi} \int_{\Gamma} e^{-t\zeta^2} d\zeta \end{aligned}$$

Using the path integral, where

$$\Gamma : \mathrm{Im}(\zeta) = -\frac{z}{2t}.$$

This path Γ is deformed as

$$\begin{aligned}\Gamma_R = &\left\{\xi - \frac{\imath z}{2t} \mid -\infty < \xi \le -R\right\} \cup \left\{-R + \imath\eta \mid -\frac{z}{2t} \le \eta \le 0\right\} \\ &\cup \{\xi \mid -R < \xi < R\} \cup \left\{R + \imath\eta \mid 0 \ge \eta \ge -\frac{z}{2t}\right\} \\ &\cup \left\{\xi - \frac{\imath z}{2t} \mid R \le \xi < +\infty\right\}.\end{aligned}$$

We obtain from Cauchy's integral theorem that

$$\begin{aligned}\int_\Gamma e^{-t\zeta^2} d\zeta &= \int_{-R}^{R} e^{-t\xi^2} d\xi + \int_{|\xi|>R} e^{-t\left(\xi - \frac{\imath z}{2t}\right)^2} d\xi \\ &\quad + \int_{-z/2t}^{0} e^{-t(-R+\imath\eta)^2} d\eta + \int_0^{-z/2t} e^{-t(R+\imath\eta)^2} d\eta \\ &= \int_{-R}^{R} e^{-t\xi^2} d\xi + I + II + III.\end{aligned}$$

It is not difficult to see that the terms II and III converge to zero as $R \to \infty$ if $t > 0$ is fixed. Furthermore,

$$|I| \le \int_{|\xi|>R} e^{-t\left(\xi^2 - \frac{z^2}{4t^2}\right)} d\xi \to 0$$

follows similarly. Thus, we have

$$\int_{\mathbf{R}} e^{-t\zeta^2} d\zeta = \int_{-\infty}^{\infty} e^{-t\xi^2} d\xi = \frac{1}{\sqrt{t}} \int_{-\infty}^{\infty} e^{-s^2} ds.$$

In use of

$$\int_{-\infty}^{\infty} e^{-x^2} dx = \sqrt{\pi} \tag{5.33}$$

we have

$$G(z,t) = \frac{e^{-z^2/4t}}{2\sqrt{\pi t}} = \left(\frac{1}{4\pi t}\right)^{1/2} e^{-z^2/4t}.$$

For this $G(z,t)$, equality (5.32) is written as

$$u(x,t) = \int_{\mathbf{R}} G(x-y,t) u_0(y) dy.$$

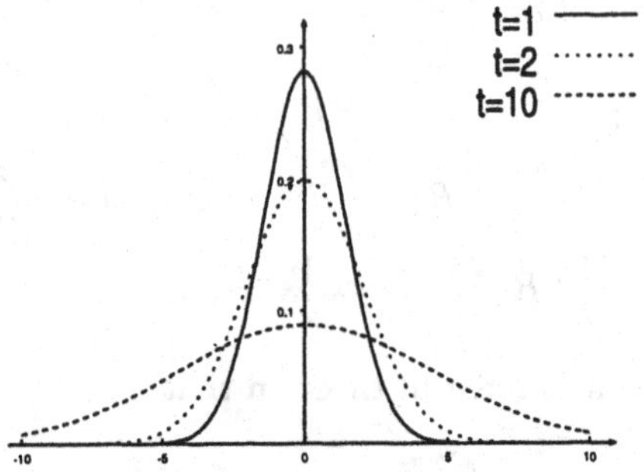

Fig. 5.1

Exercise 5.11 If the left-hand side is denoted by J, then it holds that

$$J^2 = \int_{-\infty}^{\infty}\int_{-\infty}^{\infty} e^{-(x^2+y^2)}dxdy.$$

Then, show (5.33) in use of the polar coordinate.

5.2.5 *Semi-groups*

In the n-dimensional case, the Gaussian kernel is given as

$$G(z,t) = \left(\frac{1}{4\pi t}\right)^{n/2} e^{-|z|^2/4t} \tag{5.34}$$

and the heat equation takes the form of

$$u_t = \Delta u \quad (x \in \mathbf{R}^n,\ t > 0), \qquad u|_{t=0} = u_0(x) \quad (x \in \mathbf{R}^n), \tag{5.35}$$

where $\Delta = \frac{\partial^2}{\partial x_1^2} + \cdots + \frac{\partial^2}{\partial x_n^2}$ denotes the n-dimensional Laplacian and $|z|^2 = z_1^2 + \cdots + z_n^2$ for $z = (z_1, \cdots, z_n) \in \mathbf{R}^n$. Under suitable assumptions to $u(x,t)$ and $u_0(x)$, it follows that

$$u(x,t) = \int_{\mathbf{R}^n} G(x-y,t)u_0(y)dy \qquad (x \in \mathbf{R}^n,\ t > 0). \tag{5.36}$$

In this sense, the Gaussian kernel (5.34) is called the *fundamental solution* to (5.35).

In this expression of (5.36), $G(z,t) > 0$ holds for $(z,t) \in \mathbf{R}^n \times (0,+\infty)$. Therefore, $u_0(x) \geq 0$ for $x \in \mathbf{R}^n$ implies $u(x,t) \geq 0$ for $(x,t) \in \mathbf{R}^n \times$

$(0, +\infty)$. This implies the *order preserving property* that $u_1(x) \geq u_2(x)$ for $x \in \mathbf{R}^n$ implies $u_1(x,t) \geq u_2(x,t)$ for $(x,t) \in \mathbf{R}^n \times (0,+\infty)$, where $u_1(x,t)$ and $u_2(x,t)$ denote the solution to (5.35) with the initial values $u_1(x)$ and $u_2(x)$, respectively.

More carefully, if $u_0(x) \geq 0$ with $u_0(x) \not\equiv 0$ for $x \in \mathbf{R}^n$, then $u(x,t) > 0$ follows for $x \in \mathbf{R}^n$ and $t > 0$. In particular, even if u_0 has compact support, so it is not true for $u(\cdot,t)$ with $t > 0$. This means that the heat equation (5.35) is lack of the *finite propagation* property. It also holds that $u_1(x) \geq u_2(x)$ and $u_1(x) \not\equiv u_2(x)$ for $x \in \mathbf{R}^n$ imply $u_1(x,t) > u_2(x,t)$ for $(x,t) \in \mathbf{R}^n \times (0,\infty)$. This property is referred to as the *strong order preserving*. Those properties are valid to the general *parabolic equation* appropriately posed, although details are not described here.

The following fact is referred to as *Hausdorff-Young's inequality*. The proof is left to the reader.

Theorem 5.4 *If $f \in L^1(\mathbf{R}^n)$ and $g \in L^p(\mathbf{R}^n)$, the function*

$$(f \star g)(x) = \int_{\mathbf{R}^n} f(x-y)g(y)dy$$

is well-defined for almost every $x \in \mathbf{R}^n$ and it holds that

$$\|f \star g\|_p \leq \|f\|_1 \cdot \|g\|_p, \tag{5.37}$$

where $p \in [1,\infty]$.

We have for $t > 0$ that

$$\begin{aligned} \|G(\cdot,t)\|_1 &= \int_{\mathbf{R}^n} G(x,t)dx = \left(\frac{1}{4\pi t}\right)^{n/2} \int_{\mathbf{R}^n} e^{-|x|^2/4t}dx \\ &= \left(\frac{1}{4\pi t}\right)^{n/2} \cdot (4t)^{n/2} \int_{\mathbf{R}^n} e^{-|y|^2}dy = 1 \end{aligned}$$

by

$$\int_{\mathbf{R}^n} e^{-|y|^2}dy = \left(\int_{\mathbf{R}} e^{-s^2}ds\right)^n = \pi^{n/2}.$$

Because of $u(x,t) = [G(\cdot,t) \star u_0](x)$, thus we obtain

$$\|u(\cdot,t)\|_p \leq \|u_0\|_p$$

for $p \in [1, \infty]$. Writing $u(\cdot, t) = T_t u_0$, we have

$$\|T_t\|_{L^p(\mathbf{R}^n), L^p(\mathbf{R}^n)} \leq 1,$$

where the left-hand side denotes the operator norm. Thus, $T_t : L^p(\mathbf{R}^n) \to L^p(\mathbf{R}^n)$ becomes a *contraction mapping.* Even if $u_0 \in L^p(\mathbf{R}^n)$ is not continuous, $u(x,t) = (T_t u_0)(x)$ is differentiable arbitrarily many times for $(x,t) \in \mathbf{R}^n \times (0, +\infty)$. This property is called the *smoothing effect.*

On the other hand, by (5.27) we obtain

$$T_t \circ T_s = T_{t+s} \qquad (t, s \geq 0), \tag{5.38}$$

which is called the *semi-group property* and is proven directly if the unique existence of the solution to (5.35) is established. In fact, we have

$$\hat{G}(\xi, t) = [\mathcal{F}G](\cdot, t)(\xi) = e^{-|\xi|^2 t}$$

and hence

$$\hat{G}(\xi, t) \cdot \hat{G}(\xi, s) = e^{-|\xi|^2 (t+s)} = \hat{G}(\xi, t+s)$$

follows. This implies

$$\int_{\mathbf{R}^n} G(x-y, t) G(y, s) dy = G(x, t+s) \qquad (x \in \mathbf{R}^n,\ t, s > 0) \tag{5.39}$$

and therefore, (5.38) is obtained.

Now we show the following.

Theorem 5.5 *If $p \in [1, \infty)$ and $u_0 \in L^p(\mathbf{R}^n)$, then it holds that*

$$\lim_{t \downarrow 0} \|T_t u_0 - u_0\|_p = 0. \tag{5.40}$$

Proof. Letting $G(x) = G(x, 1)$, we have

$$G(x, t) = t^{-n/2} G(x/\sqrt{t})$$

and

$$\begin{aligned} u(x,t) &= \int_{\mathbf{R}^n} G(x-y, t) u_0(y) dy = \int_{\mathbf{R}^n} G(y, t) u_0(x-y) dy \\ &= t^{-n/2} \int_{\mathbf{R}^n} G(y/\sqrt{t}) u_0(x-y) dy. \end{aligned}$$

In terms of $y' = y/\sqrt{t}$, we obtain

$$u(x,t) = \int_{\mathbf{R}^n} G(y) u_0(x - \sqrt{t}y) dy$$

and hence

$$u(x,t) - u_0(x) = \int_{\mathbf{R}^n} G(y)^{(1/p)+(1/p')} \left[u_0(x - \sqrt{t}y) - u_0(x) \right] dy$$

follows, where $p' \in (1, \infty]$ denotes the dual exponent of p. Therefore, Hölder's inequality guarantees that

$$|u(x,t) - u_0(x)| \le \left(\int_{\mathbf{R}^n} G(y) dy \right)^{1/p'} \cdot \left(\int_{\mathbf{R}^n} G(y) \left| u_0(x - \sqrt{t}y) - u_0(x) \right|^p dy \right)^{1/p},$$

which implies

$$\begin{aligned}
\int_{\mathbf{R}^n} |u(x,t) - u_0(x)|^p dx &\le \int_{\mathbf{R}^n} dx \int_{\mathbf{R}^n} dy G(y) \left| u_0(x - \sqrt{t}y) - u_0(x) \right|^p \\
&= \int_{\mathbf{R}^n} G(y) dy \left\{ \int_{\mathbf{R}^n} \left| u_0(x - \sqrt{t}y) - u_0(x) \right|^p dx \right\} \\
&= \int_{|y|<R} G(y) dy \left\{ \int_{\mathbf{R}^n} \left| u_0(x - \sqrt{t}y) - u_0(x) \right|^p dx \right\} \\
&\quad + \int_{|y|\ge R} G(y) dy \left\{ \int_{\mathbf{R}^n} \left| u_0(x - \sqrt{t}y) - u_0(x) \right|^p dx \right\}, \qquad (5.41)
\end{aligned}$$

where $R > 0$.

The second term of the right-hand side of (5.41) is estimated from above by

$$\begin{aligned}
&\int_{|y|\ge R} G(y) dy \cdot 2^p \left\{ \int_{\mathbf{R}^n} \left| u_0(x - \sqrt{t}y) \right|^p dx + \int_{\mathbf{R}^n} |u_0(x)|^p dx \right\} \\
&\quad = 2^{p+1} \|u_0\|_p^p \cdot \int_{|y|\ge R} G(y) dy
\end{aligned}$$

because

$$(a+b)^p \le 2^p (a^p + b^p)$$

holds for $a, b \ge 0$ and $p \ge 1$.

On the other hand, the first term of the right-hand side of (5.41) converges to 0 as $t \downarrow 0$. To see this, we note that

$$\lim_{h\to 0} \|u_0(\cdot + h) - u_0\|_p = 0$$

follows from $u_0 \in L^p(\mathbf{R}^n)$ with $p \in [1, \infty)$, so that any $\varepsilon > 0$ admits $\delta > 0$ satisfying $\|u_0(\cdot + h) - u_0\|_p < \varepsilon$ if $|h| < \delta$. Thus, for $t \in (0, \delta^2/R^2)$ it holds that

$$\int_{\mathbf{R}^n} \left| u_0(x - \sqrt{t}y) - u_0(x) \right|^p dx < \varepsilon^p$$

for any $y \in \mathbf{R}^n$ in $|y| < R$, and this term is estimated from above by

$$\int_{|y|<R} G(y)dy \cdot \varepsilon^p < \varepsilon^p.$$

This means that

$$\lim_{t\downarrow 0} \int_{|y|<R} G(y)dy \left\{ \int_{\mathbf{R}^n} \left| u_0(x - \sqrt{t}y) - u_0(x) \right|^p dx \right\} = 0.$$

We have

$$\limsup_{t\downarrow 0} \int_{\mathbf{R}^n} |u(x,t) - u_0(x)|^p \, dx \le 2^{p+1} \|u_0\|_p^p \cdot \int_{|y|\ge R} G(y)dy$$

and hence (5.40) follows. The proof is complete. □

Exercise 5.12 Prove (5.37) and (5.38).

Exercise 5.13 Prove the smoothing effect of $T_t : L^p(\mathbf{R}^n) \to L^p(\mathbf{R}^n)$.

Exercise 5.14 Confirm that Theorem 5.5 is not valid for $p = \infty$.

Exercise 5.15 Show that the Gaussian kernel $G(x,t)$ converges to $\delta(x)$ in $\mathcal{D}'(\mathbf{R}^n)$ as $t \downarrow 0$.

5.2.6 *Fourier Transformation of Distributions*

If $f = f(x)$ is differentiable infinitely many times in $x \in \mathbf{R}^n$, then it is said to be rapidly decreasing if

$$p_{k,\alpha}(f) = \sup_{x\in\mathbf{R}^n} \left(1 + |x|^2\right)^k |D^\alpha f(x)| < +\infty$$

for any non-negative integer k and the multi-index α. Then the set of such functions, denoted by $\mathcal{S}(\mathbf{R}^n)$, becomes a Fréchet space by $\{p_{k,\alpha}\}$. We have the dense inclusions $\mathcal{D}(\mathbf{R}^n) \subset \mathcal{S}(\mathbf{R}^n) \subset \mathcal{E}(\mathbf{R}^n)$ including the topologies, and hence

$$\mathcal{E}'(\mathbf{R}^n) \subset \mathcal{S}'(\mathbf{R}^n) \subset \mathcal{D}'(\mathbf{R}^n)$$

follows. Each element in $\mathcal{S}'(\mathbf{R}^n)$ is called the *tempered distribution*. This paragraph is devoted to the theory of Fourier transformation of tempered distributions. The reader can skip it first.

If $m(dx)$ is a Borel measure satisfying

$$\int_{\mathbf{R}^n} \left(1+|x|^2\right)^{-k} \|m\|(dx) < +\infty,$$

then it is a tempered distribution, where $\|m\|(dx)$ denotes the *total variation* of $m(dx)$.

The Fourier transformation on $\mathcal{S}(\mathbf{R}^n)$ is defined by

$$(\mathcal{F}f)(\xi) = \hat{f}(\xi) = \int_{\mathbf{R}^n} f(x)e^{-\imath\xi\cdot x}dx$$

with the inverse transformation

$$(\mathcal{F}^{-1}f)(x) = \tilde{f}(x) = \left(\frac{1}{2\pi}\right)^n \int_{\mathbf{R}^n} f(\xi)e^{\imath\xi\cdot x}d\xi.$$

Then, it holds that

$$D_\xi^\alpha(\mathcal{F}f) = \mathcal{F}\left((-x)^\alpha f\right) \qquad \text{and} \qquad \xi^\alpha(\mathcal{F}f) = \mathcal{F}(D_x^\alpha f),$$

where

$$D_x^\alpha = \left(\frac{1}{\imath}\frac{\partial}{\partial x_1}\right)^{\alpha_1} \cdots \left(\frac{1}{\imath}\frac{\partial}{\partial x_n}\right)^{\alpha_n} \qquad \text{and} \qquad x^\alpha = x_1^{\alpha_1}\cdots x_n^{\alpha_n}$$

for $\alpha = (\alpha_1, \cdots, \alpha_n)$.

If $T \in \mathcal{S}'(\mathbf{R}^n)$, then $S(f) = T(\mathcal{F}f)$ determines an element in $\mathcal{S}'(\mathbf{R}^n)$. It is denoted by $S = \mathcal{F}T$ and called the *Fourier transformation* of T. It can be shown that the mapping $\mathcal{F} : \mathcal{S}'(\mathbf{R}^n) \to \mathcal{S}'(\mathbf{R}^n)$ is an isomorphism, $(\mathcal{F}^{-1}T)(f) = T(\mathcal{F}^{-1}f)$ for $f \in \mathcal{S}'(\mathbf{R}^n)$,

$$D_\xi^\alpha \mathcal{F}(T) = \mathcal{F}\left((-x)^\alpha T\right), \qquad \xi^\alpha \mathcal{F}(T) = \mathcal{F}(D_x^\alpha T),$$

and so forth. Also, this definition agrees with the usual Fourier transformation for L^1 and L^2 functions, and it is obvious that

$$\mathcal{F}(\delta) = 1 \qquad \text{and} \qquad \mathcal{F}(1) = (2\pi)^n \delta \tag{5.42}$$

hold.

The *convolution* of $\varphi \in \mathcal{D}(\mathbf{R}^n)$ and $T \in \mathcal{D}'(\mathbf{R}^n)$ is given by

$$(T * \varphi)(x) = T_y\left(\varphi(x-y)\right).$$

Then, it follows that

$$T * \varphi \in C^\infty(\mathbf{R}^n) \tag{5.43}$$

$$\operatorname{supp}(T * \varphi) \subset \operatorname{supp} T + \operatorname{supp} \varphi \tag{5.44}$$

$$D^\alpha(T * \varphi) = T * D^\alpha \varphi = (D^\alpha T) * \varphi. \tag{5.45}$$

The convolution $T * \varphi$ of $\varphi \in \mathcal{E}(\mathbf{R}^n)$ and $T \in \mathcal{E}'(\mathbf{R}^n)$ is defined similarly. Now, we note the following.

Lemma 5.2 *We have $(T * \varphi) * \psi = T * (\varphi * \psi)$ for $\varphi, \psi \in \mathcal{D}(\mathbf{R}^n)$ and $T \in \mathcal{D}'(\mathbf{R}^n)$. The same conclusion holds for $\varphi, \psi \in \mathcal{E}(\mathbf{R}^n)$ and $T \in \mathcal{E}'(\mathbf{R}^n)$.*

Proof. In fact, we have

$$\begin{aligned} ((T * \varphi) * \psi)(x) &= \int_{\mathbf{R}^n} (T * \varphi)(x-y)\psi(y)dy \\ &= \int_{\mathbf{R}^n} T_z\left(\varphi(x-y-z)\right)\psi(y)dy \\ &= \int_{\mathbf{R}^n} T_z\left(\psi(y)\varphi(x-y-z)\right)dy \\ &= T_z\left(\int_{\mathbf{R}^n} \varphi(x-y-z)\psi(y)dy\right) \\ &= T_z\left((\varphi * \psi)(x-z)\right) = \left(T * (\varphi * \psi)\right)(x) \end{aligned}$$

and the proof is complete. □

Here, we describe the following facts without proof. First, if $T \in \mathcal{D}'(\mathbf{R}^n)$, $S \in \mathcal{E}'(\mathbf{R}^n)$, and $\varphi \in \mathcal{D}(\mathbf{R}^n)$, we have $S * \varphi \in \mathcal{D}(\mathbf{R}^n)$ and hence

$T * (S * \varphi)$ is defined. Similarly, $S * (T * \varphi)$ is well-defined, but actually we have

$$T * (S * \varphi) = S * (T * \varphi).$$

This quantity is equal to $K * \varphi$ with the unique $K \in \mathcal{D}'(\mathbf{R}^n)$, which is written as $K = T * S = S * T$.

Next, if $T \in \mathcal{E}'(\mathbf{R}^n)$, then $\mathcal{F}(T)$ is extended to an entire function in $\mathbf{C}^n$ with the property that

$$\mathcal{F}(T)(\zeta) = T_x(e^{-\imath x\cdot\zeta}) \qquad \text{for} \quad \zeta \in \mathbf{C}^n.$$

If $T, S \in \mathcal{E}'(\mathbf{R}^n)$, then

$$\mathcal{F}(T * S) = \mathcal{F}(T) \cdot \mathcal{F}(S)$$

follows. Finally,

$$T * \delta = \delta * T = T$$

for any $T \in \mathcal{D}'(\mathbf{R}^n)$.

The *theorem of Malgrange and Ehlenpreis* assures that any partial differential operator $P(D)$ with constant coefficients admits the *fundamental solution* $E \in \mathcal{D}'(\mathbf{R}^n)$ so that $P(D)E = \delta$ holds in $\mathcal{D}'(\mathbf{R}^n)$. This means that $u = E * f$ solves $P(D)u = f$ for any $f \in \mathcal{D}(\mathbf{R}^n)$.

The following process is called the *regularization*.

Theorem 5.6 *For $\varphi \in C_0^\infty(\mathbf{R}^n)$ satisfying*

$$\varphi \geq 0 \qquad \text{and} \qquad \int_{\mathbf{R}^n} \varphi(x)dx = 1,$$

let $\varphi_\varepsilon = \varepsilon^{-n}\varphi(x/\varepsilon)$ for $\varepsilon > 0$, and take $T * \varphi_\varepsilon \in C^\infty(\mathbf{R}^n)$ for $T \in \mathcal{D}'(\mathbf{R}^n)$. Then, it holds that $T * \varphi_\varepsilon \rightharpoonup T$ is $*$-weakly in $\mathcal{D}'(\mathbf{R}^n)$, which means the pointwise convergence so that

$$(T * \varphi_\varepsilon)(\psi) \to T(\psi) \tag{5.46}$$

holds as $\varepsilon \downarrow 0$ for any $\psi \in \mathcal{D}(\mathbf{R}^n)$.

Proof. If $S \in \mathcal{D}'(\mathbf{R}^n)$ and $\psi \in \mathcal{D}(\mathbf{R}^n)$, then we have

$$S(\psi) = (S * \hat{\psi})(0),$$

where $\hat{\psi}(x) = \psi(-x)$. Then, it holds that

$$\begin{aligned}(T * \varphi_\varepsilon)(\psi) = \left((T * \varphi_\varepsilon) * \hat{\psi}\right)(0) &= \left(T * (\varphi_\varepsilon * \hat{\psi})\right)(0) \\ &= T\left({}^{\vee}(\varphi_\varepsilon * \hat{\psi})\right)\end{aligned}$$

and

$$\varphi_\varepsilon * \hat{\psi} \to \hat{\psi} \quad \text{in} \quad \mathcal{D}(\mathbf{R}^n).$$

Thus, we obtain (5.46) as

$$(T * \varphi_\varepsilon)(\psi) \to T\left({}^{\vee}(\hat{\psi})\right) = T(\psi)$$

and the proof is complete. □

Exercise 5.16 Show that $L^p(\mathbf{R}^n) \subset \mathcal{S}'(\mathbf{R}^n)$ for $p \in [1, \infty]$.

Exercise 5.17 Confirm that (5.42) holds true.

Exercise 5.18 Confirm (5.43), (5.44), and (5.45).

Exercise 5.19 Show that $T * \varphi \in \mathcal{S}'(\mathbf{R}^n) \cap \mathcal{E}(\mathbf{R}^n)$ is well-defined for $T \in \mathcal{S}'(\mathbf{R}^n)$ and $\psi \in \mathcal{S}(\mathbf{R}^n)$. Show also $\mathcal{F}(T * \varphi) = (\mathcal{F}\varphi) \cdot \mathcal{F}T$.

5.3 Potential

5.3.1 *Harmonic Functions*

Function $u(x)$ satisfying $\Delta u = 0$ in a domain $\Omega \subset \mathbf{R}^n$ is said to be *harmonic* there. It is a fundamental problem in mathematical physics to solve

$$\Delta u = 0 \quad \text{in} \quad \Omega, \qquad u = f \quad \text{on} \quad \partial\Omega, \tag{5.47}$$

where Ω is a bounded domain and $f(\xi)$ is a continuous function on its boundary $\partial\Omega$. Actually, we have the following theorem for the Dirichlet problem to harmonic function, (5.47).

Theorem 5.7 *Problem (5.47) admits the solution $u \in C^2(\Omega) \cap C(\overline{\Omega})$ for arbitrary given $f \in C(\partial\Omega)$ if and only if any point on $\partial\Omega$ is regular.*

Here, the boundary point $\xi \in \partial\Omega$ is said to be *regular* if it has a *barrier* $w(x)$. This means that $w(x)$ is continuous on $\overline{\Omega}$, super-harmonic there, positive in $\overline{\Omega} \setminus \{\xi\}$, and is equal to zero at $x = \xi$. Furthermore, superharmonicity of the continuous function $w(x)$ is defined through the mean value property,

$$\fint_{\partial B(x_0,r)} w \geq w(x_0) \qquad (x_0 \in \Omega,\ 0 < r \ll 1)\,. \tag{5.48}$$

It is actually equivalent to $\Delta w \geq 0$ in Ω if $w(x)$ is twice differentiable. Remember that $B(x_0, r)$ denotes the open ball with the center x_0 and the radius r and

$$\fint_\omega = \frac{1}{|\omega|} \int_\omega .$$

If $u \in C^2(\Omega) \cap C(\overline{\Omega})$ satisfies (5.47), then it is called the *classical solution.* A sufficient condition for the regularity of boundary point is the *outer circumscribing ball condition.*

Theorem 5.7 is proven by *Perron's method.* First, sub-harmonicity for continuous function $w(x)$ is defined by the reverse inequality of (5.48). Then, we take

$$\underline{S}(f) = \left\{ v \in C(\overline{\Omega}) \mid v \text{ is sub-harmonic in } \Omega \quad \text{and} \quad v \leq f \text{ on } \partial\Omega \right\}$$

and set

$$u(x) = \sup \left\{ v(x) \mid v \in \underline{S}(f) \right\}.$$

It is proven that $u(x)$ is harmonic in Ω and furthermore, if $\xi \in \partial\Omega$ is a regular point, it holds that

$$\lim_{x \in \Omega \to \xi} u(x) = f(\xi).$$

Rough description of this theory is as follows. First, we have potential and the Kelvin transformation. In use of those, we can represent the solution to (5.47) by the *Poisson integral* in the case that Ω is a ball. By this we get *mean value theorem* and the *Harnack inequality* to the harmonic function, which imply weak and strong *maximum principles* and the *Harnack principle*, respectively. Then, the method of *harmonic lifting* and the notion of barrier settle down the problem.

Here, likely to the *complex function theory,* the integral formula induces every notion but the idea of lifting from sub-solutions is nothing but that of *real analysis.* This beautiful theory has a vast background.

Henceforth, the harmonic function is supposed to be C^2 in the domain Ω in consideration. If Δ is taken in the sense of distribution, this notion is extended to $\mathcal{D}'(\Omega)$. However, *Weyl's lemma* guarantees that any harmonic distribution is regarded as a harmonic function, and furthermore, any harmonic function is *real-analytic*, which means that it is represented as the Taylor series around any point in Ω. Such a property is generally called the *regularity*.

Exercise 5.20 Suppose that the outer circumscribing condition holds at $\xi \in \partial\Omega$, as there is a disc $B \neq \emptyset$ such that $\overline{B} \cap \overline{\Omega} = \{\xi\}$. Then show that

$$w(x) = \begin{cases} R^{2-n} - |x-y|^{2-n} & (n \geq 3) \\ \log|x-y| - \log R & (n = 2) \end{cases}$$

is a barrier at ξ, where $B = B(y, R)$.

5.3.2 *Poisson Integral*

For the moment, we are concentrated on the two-dimensional case and make use of the complex function theory, identifying $\mathbf{R}^2$ with $\mathbf{C}$, the *complex plane*, that is $x = (x_1, x_2) \in \mathbf{R}^2$ is identified with Re $f = u$. In fact, we have the following theorem.

Theorem 5.8 *If $\Omega \subset \mathbf{R}^2$ is simply connected and (real-valued) $u(x)$ is harmonic in Ω, then there exists a holomorphic $f(z)$ in $z \in \Omega$ such that Re $f = u$.*

Here, $f(z)$ is unique up to an additive pure imaginary constant. Even if Ω is not simply connected, such $f(z)$ is taken locally, thus can be an *analytic function* in Ω in this case. In this way, harmonic functions are associated with the complex function theory in the two space dimension.

Here, we wish to confirm that any harmonic function is taken to be real-valued henceforth. Let $u(z)$ be harmonic and continuous in $|z| < R$ and $|z| \leq R$, respectively, and take a holomorphic function $f(z)$ in $|z| < R$ satisfying Re $f = u$. Then, Cauchy's integral formula guarantees that

$$f(z) = \frac{1}{2\pi\imath} \int_{|\zeta|=\rho} \frac{f(\zeta)}{\zeta - z} d\zeta \qquad (|z| < \rho),$$

where $\rho \in (0, R)$. We take the mirror image of $z = re^{\imath\theta}$ for $r \in (0, \rho)$ with respect to the circle $|\zeta| = \rho$, that is, $z^* = \rho^2 e^{\imath\theta}/r$. Then, we have

$$0 = \frac{1}{2\pi\imath}\int_{|\zeta|=\rho} \frac{f(\zeta)}{\zeta - z^*} d\zeta$$

by $|z^*| > \rho$. This gives that

$$\begin{aligned} f(re^{\imath\theta}) &= \frac{1}{2\pi\imath}\int_{|\zeta|=\rho} f(\zeta)\left\{\frac{1}{\zeta - z} - \frac{1}{\zeta - z^*}\right\} d\zeta \\ &= \frac{1}{2\pi\imath}\int_{|\zeta|=\rho} f(\zeta)\left\{\frac{1}{\zeta - re^{\imath\theta}} - \frac{1}{\zeta - \rho^2 e^{\imath\theta}/r}\right\} d\zeta. \end{aligned}$$

Thus, putting $\zeta = \rho e^{\imath\varphi}$, we obtain

$$f(re^{\imath\theta}) = \frac{1}{2\pi}\int_0^{2\pi} f(\rho e^{\imath\theta})\left\{\frac{1}{\zeta - re^{\imath\theta}} - \frac{1}{\zeta - \rho^2 e^{\imath\theta}/r}\right\}\zeta d\varphi$$

by $d\zeta = \imath\rho e^{\imath\varphi} d\varphi$. Here, we have

$$\begin{aligned} &\left\{\frac{1}{\zeta - re^{\imath\theta}} - \frac{1}{\zeta - \rho^2 e^{\imath\theta}/r}\right\}\zeta \\ &\quad = \left\{1 + \frac{re^{\imath\theta}}{\zeta - re^{\imath\theta}}\right\} - \left\{1 + \frac{\rho^2 e^{\imath\theta}/r}{\zeta - \rho^2 e^{\imath\theta}/r}\right\} \\ &\quad = \frac{r}{\rho e^{\imath(\varphi-\theta)} - r} - \frac{\rho^2/r}{\rho e^{\imath(\varphi-\theta)} - \rho^2/r} \\ &\quad = \frac{r}{\rho e^{\imath(\varphi-\theta)} - r} - \frac{\rho}{re^{\imath(\varphi-\theta)} - \rho} \\ &\quad = \frac{r\left(\rho e^{-\imath(\varphi-\theta)} - r\right)}{\rho^2 - 2r\rho\cos(\varphi-\theta) + r^2} - \frac{\rho\left(re^{-\imath(\varphi-\theta)} - \rho\right)}{r^2 - 2r\rho\cos(\varphi-\theta) + \rho^2} \\ &\quad = \frac{\rho^2 - r^2}{\rho^2 - 2\rho r\cos(\varphi-\theta) + r^2} = \frac{\rho^2 - r^2}{\rho^2 - 2\rho r\cos(\theta-\varphi) + r^2}, \end{aligned}$$

and hence

$$f(re^{\imath\theta}) = \frac{1}{2\pi}\int_0^{2\pi} f(\rho e^{\imath\varphi})\frac{\rho^2 - r^2}{\rho^2 - 2\rho r\cos(\theta-\varphi) + r^2} d\varphi$$

follows. Taking real parts of both sides, we get that

$$u(re^{\imath\theta}) = \frac{1}{2\pi}\int_0^{2\pi} u(\rho e^{\imath\varphi})\frac{\rho^2 - r^2}{\rho^2 - 2\rho r\cos(\theta-\varphi) + r^2} d\varphi$$

for $0 < r < \rho < R$. Letting $\rho \uparrow R$, we obtain *Poisson's integral formula.*

Theorem 5.9 *If $n = 2$ and $u(z)$ is harmonic and continuous in $|z| < R$ and $|z| \leq R$, respectively, then it holds that*

$$u(re^{\imath\theta}) = \frac{1}{2\pi} \int_0^{2\pi} u(Re^{\imath\varphi}) \frac{R^2 - r^2}{R^2 - 2Rr\cos(\theta - \varphi) + r^2} d\varphi \qquad (5.49)$$

for $0 \leq r < R$.

Equality (5.49) gives the *mean value theorem,*

$$u(0) = \frac{1}{2\pi} \int_0^{2\pi} u(Re^{\imath\varphi}) d\varphi. \qquad (5.50)$$

In the case of $u(Re^{\imath\theta}) \geq 0$, this gives that

$$\frac{R - |z|}{R + |z|} u(0) \leq u(z) \leq \frac{R + |z|}{R - |z|} u(0)$$

by

$$\frac{R - r}{R + r} \leq \frac{R^2 - r^2}{R^2 - 2Rr\cos(\theta - \varphi) + r^2} \leq \frac{R + r}{R - r}.$$

In particular, *Harnack's inequality*

$$\frac{1}{3} u(0) \leq u(z) \leq 3u(0) \qquad (|z| < R/2) \qquad (5.51)$$

holds true.

We now show that the *strong maximum principle.*

Theorem 5.10 *Any non-constant harmonic function defined in a domain $\Omega \subset \mathbf{R}^2$ cannot attain the maximum or the minimum there.*

Proof. It suffices to show that the non-constant harmonic function $u(z)$ in Ω cannot attain the maximum. Suppose the contrary, that there exists $z_0 \in \Omega$ such that $u(z) \leq u(z_0)$ for any $z \in \Omega$. Then, we apply the mean value theorem that

$$⨍_{\partial B(z_0,R)} u ds \leq m = \max_{\partial B(z_0,R)} u.$$

This implies that $u = m$ on $\partial B(z_0, R)$ and hence near z_0, because $0 < R \ll 1$ is arbitrary. This means that the non-empty set

$$\{z \in \Omega \mid u(z) = m\}$$

is open. Obviously it is closed and hence coincides with Ω from its connectivity. Thus, $u(z)$ is identically equal to the constant m, and the proof is complete. □

The strong maximum principle implies the *weak maximum principle* indicated as follows.

Theorem 5.11 *If $\Omega \subset \mathbf{R}^2$ is a bounded domain, $u(x)$ is harmonic in Ω, and is continuous on $\overline{\Omega}$, then it holds that*

$$\sup_{\Omega} u = \max_{\partial\Omega} u.$$

A general form of Harnack's inequality is given as follows.

Theorem 5.12 *If $\Omega \subset \mathbf{R}^2$ is a domain and $E \subset \Omega$ is a bounded closed set, then there is $K = K(\Omega, E) > 0$ such that any non-negative harmonic function $u(z)$ in Ω admits the inequality*

$$\sup_{E} u \leq K \inf_{E} u. \tag{5.52}$$

Proof. Because E is compact, it is covered by a finite number of discs with the radius $R > 0$, where

$$0 < R < \operatorname{dist}(E, \partial\Omega) = \inf_{x \in E,\ y \in \partial\Omega} \operatorname{dist}(x, y).$$

This means that there is an integer m and $z_1, \cdots, z_m \in E$ such that

$$E \subset \cup_{i=1}^{m} B(z_i, R/2) \subset \Omega.$$

Therefore, inequality (5.51) guarantees that

$$\frac{1}{3} u(z) \leq u(z_i) \leq 3u(z)$$

for $z \in B(z_i, R/2)$ and $i = 1, \cdots, m$. However, any $w_1 \in E$ can come to the same disc to which any $w_2 \in E$ belongs, and hence

$$3^{-m} u(w_1) \leq 3^m u(w_2)$$

follows. This implies

$$3^{-m} \sup_{E} u \leq 3^m \inf_{E} u,$$

or (5.52) with $K = 3^{2m}$. □

Inequality (5.52) implies the following theorem, indicated as the *Harnack principle.*

Theorem 5.13 *Let $\{u_k(z)\}_{k=1}^{\infty}$ be a sequence of harmonic functions in the domain $\Omega \subset \mathbf{R}^2$, monotone non-decreasing at each point, and bounded at some $z_0 \in \Omega$. Then, there is a harmonic function $u(z)$ in Ω such that $u_k(z) \to u(z)$ locally uniformly in $z \in \Omega$, which means that its convergence is uniform on any compact set in Ω.*

Proof. The function $u_{k+1}(z) - u_k(z)$ is non-negative and harmonic in Ω and hence any compact set $E \subset \Omega$ containing z_0 admits $K > 0$ such that

$$0 \leq u_{k+1}(z) - u_k(z) \leq K\{u_{k+1}(z_0) - u_k(z_0)\}$$

holds for $z \in E$ and $k = 1, 2, \cdots$. From the assumption, we have

$$\sum_{k=1}^{\infty} \{u_{k+1}(z_0) - u_k(z_0)\} < +\infty.$$

This implies that

$$\sum_{k=1}^{\infty} \{u_{k+1}(z) - u_k(z)\} < +\infty$$

uniformly in $z \in E$. We get the limiting function

$$u(z) = \lim_{k\to\infty} u_k(z) = u_1(z) + \sum_{k=1}^{\infty} \{u_{k+1}(z) - u_k(z)\}$$

with the convergence uniform in E. Because E is arbitrary, this convergence is locally uniform in Ω. We get $u \in C(\Omega)$.

On the other hand, $u_k(z)$ is harmonic in Ω and hence we have

$$u_k(z_1 + re^{\imath\theta}) = \frac{1}{2\pi}\int_0^{2\pi} u_k(z_1 + re^{\imath\theta})\frac{R^2 - r^2}{R^2 - 2Rr\cos(\theta - \varphi) + r^2}d\varphi$$

for $z_1 \in \Omega$ and $0 \leq r < R \ll 1$. Letting $k \to \infty$, we have

$$u(z_1 + re^{\imath\theta}) = \frac{1}{2\pi}\int_0^{2\pi} u(z_1 + Re^{\imath\varphi})\frac{R^2 - r^2}{R^2 - 2Rr\cos(\theta - \varphi) + r^2}d\varphi.$$

This implies that $u = u(x)$ is continuously differentiable twice and is harmonic in Ω. □

Exercise 5.21 Prove Theorem 5.8 in use of *Cauchy-Riemann's relation* and the path integral.

Exercise 5.22 Derive the uniqueness of the solution to the Dirichlet problem for harmonic function in use of the weak maximum principle.

5.3.3 *Perron Solution*

Real-valued continuous function $u(z)$ defined on the domain $\Omega \subset \mathbf{R}^2$ is said to be *super-harmonic* if

$$u(z) \geq \fint_{\partial B(z,R)} uds$$

holds for any $z \in \Omega$ and $0 < R \ll 1$. Similarly to the case of harmonic function, we can prove that non-constant super-harmonic function cannot attain the infimum in interior. If $-u(x)$ is super-harmonic, we say that $u(x)$ is *sub-harmonic*. Then, non-constant sub-harmonic function cannot attain the interior maximum. Thus, we can say that a continuous function is harmonic if and only if it is sub- and super-harmonic.

Let $\Omega \subset \mathbf{R}^2$ be a bounded domain, and suppose that the Dirichlet problem (5.47) has a solution $u \in C^2(\Omega) \cap C(\overline{\Omega})$, where $f \in C(\partial\Omega)$. In this case, if $v \in C(\overline{\Omega})$ is sub-harmonic in Ω satisfying $v \leq f$ on $\partial\Omega$, then the function $v - u \in C(\overline{\Omega})$ is sub-harmonic in Ω and $v - u \leq 0$ holds on $\partial\Omega$. Therefore, the weak maximum principle guarantees that $v \leq u$ on $\overline{\Omega}$. This means that $u(z)$ attains the maximum of $v(z)$ in $v \in \underline{S}(f)$, where

$$\underline{S}(f) = \left\{ v \in C(\overline{\Omega}) \mid \text{sub-harmonic in } \Omega, \quad v \leq f \text{ on } \partial\Omega \right\}.$$

This observation leads us to the *Perron solution* to (5.47),

$$u(z) = \sup \left\{ v(z) \mid v \in \underline{S}(f) \right\} \tag{5.53}$$

for each $z \in \Omega$.

Henceforth, the *Poisson integral* $\mathcal{P}(f)$ of $f \in C(\partial B)$ is defined by

$$(\mathcal{P}f)(z) = \frac{1}{2\pi} \int_0^{2\pi} f(z_0 + Re^{\imath\varphi}) \frac{R^2 - r^2}{R^2 - 2Rr\cos(\theta - \varphi) + r^2} d\varphi,$$

where $B = B(z_0, R)$ and $z = z_0 + re^{\imath\theta}$ with $0 \leq r < R$. The following theorem is a counterpart of Theorem 5.9, and the proof is left to the reader.

Lemma 5.3 *For $f \in C(\partial B)$, it holds that $\mathcal{P}f \in C(\overline{B}) \cap C^2(B)$ and $\Delta(\mathcal{P}f) = 0$ in B.*

We shall show that $u(z)$ defined by (5.53) is harmonic in Ω. This is proven by showing that each element in $\underline{S}(f)$ can *lift up* as a harmonic function on any disc contained in Ω. The following lemma indicates that $\underline{S}(f)$ is soft enough to make such a process closed.

Lemma 5.4 *Given $B = B(z_0, R)$ satisfying $\overline{B} \subset \Omega$ and $v \in \underline{S}$, we take*

$$v_B(z) = \begin{cases} (\mathcal{P}f)(z) & (z = z_0 + re^{\imath\theta}) \\ v(z) & (z \in \Omega \setminus B) \end{cases} \tag{5.54}$$

for $f = v|_{\partial B}$. Then, it holds that $v_B \in \underline{S}(f)$.

Proof. From Lemma 5.3, we see that v_B defined by (5.54) is continuous on $\overline{\Omega}$. Because $v_B \le f$ on $\partial\Omega$ is obvious, we shall show that v_B is sub-harmonic in Ω. Let us put $w = v_B$ for the moment. We only have to show that

$$w_D(z) \ge w(z) \tag{5.55}$$

holds for any $z \in \Omega$ and for any disc D satisfying $\overline{D} \subset \Omega$, for that purpose.

In fact $v(x)$ is sub-harmonic and hence the weak maximum principle guarantees $w \ge v$ in B, and hence $w \ge v$ holds on $\overline{\Omega}$. This implies that $w_D \ge v_D$ in Ω. On the other hand, we have $v_D \ge v$ in Ω similarly, and hence

$$w_D \ge v_D \ge v \tag{5.56}$$

follows in Ω. If $z \notin D$, then $w_D(z) = w(z)$ holds. If $z \notin B$, then $v(z) = v_B(z) = w(z)$ and hence $w_D(z) \ge w(z)$ follows from (5.56). Therefore, we have only to take the case that $z \in D \cap B$.

In fact, we have $v = v_B = w$ on $\partial B \cap D$ and $w_D = w$ on $\partial D \cap B$, so that $w_D \ge w$ holds on $\partial(D \cap B)$ by (5.56). However, both w_D and $w = v_B$ are harmonic in $D \cap B$, we get that $w_D \ge w$ in $D \cap B$ by the weak maximum principle. Thus, we obtain (5.55) in Ω, and the proof is complete. □

The function v_B defined by (5.54) is called the *harmonic lifting* of v on B. The proof of the following lemma is also left to the reader.

Lemma 5.5 *Show that if the continuous functions v and w are sub-harmonic in Ω, then so is $\max\{v, w\}$.*

Now, we show the following.

Theorem 5.14 *The function $u(z)$ defined by (5.53) is harmonic in Ω.*

Proof. We note first that $-m \in \underline{S}$ holds for a constant $m \gg 1$ satisfying $-m \le f$ on $\partial\Omega$ and hence $\underline{S}(f) \neq \emptyset$ follows. On the other hand, the weak maximum principle guarantees for $v \in \underline{S}(f)$ that

$$v(z) \le \max_{\partial\Omega} v \le \max_{\partial\Omega} f < +\infty$$

and hence

$$u(z) = \sup\{v(z) \mid v \in \underline{S}(f)\} < +\infty$$

holds for each $z \in \Omega$.

We take $B = B(y,R)$ in $\overline{B} \subset \Omega$, and $\{v_k\} \subset \underline{S}(f)$ such that $v_k(y) \nearrow u(y)$ as $k \to \infty$. If we take $\tilde{v}_k$ $(k = 1, 2, \cdots)$ as

$$\tilde{v}_1 = v_1, \qquad \tilde{v}_2 = \max\{v_1, v_2\}, \qquad \cdots,$$

then it holds that $\{\tilde{v}_k\} \subset \underline{S}(f)$ by Lemma 5.5 and also $\tilde{v}_k(y) = v_k(y) \nearrow u(y)$. Thus, we may suppose that $\{v_k\}$ is non-decreasing at any point in Ω.

If $V_k = (v_k)_B$ denotes the harmonic lifting of $v_k(x)$ on $B = B(y,R)$, then $\{V_k\} \subset \underline{S}(f)$ is non-decreasing at any point in Ω. Furthermore, V_k is harmonic in B and the weak maximum principle guarantees $v_k \le V_k$ in Ω. This implies

$$v_k(y) \le V_k(y) \quad \to \quad u(y) < +\infty \tag{5.57}$$

from $v_k(y) \to u(y)$ and (5.53). The Harnack principle, Theorem 5.13 is applicable, and we have a harmonic function $v(z)$ in B such that $V_k \to v$ locally uniformly in B. If $v = u$ is shown to hold in B, then $u(z)$ is harmonic in B. Thus, it is harmonic in Ω because B is arbitrary.

For this purpose, first we note that $V_k \le u$ holds in Ω by (5.53). This implies

$$v \le u \qquad \text{in} \qquad B.$$

On the other hand, relation (5.57) gives that

$$v(y) = u(y).$$

If there is $z \in B$ satisfying $u(z) > v(z)$, then we have $\tilde{u} \in \underline{S}(f)$ such that

$$u(z) > \tilde{u}(z) > v(z).$$

Let $w_k = \max\{v_k, \tilde{u}\}$ for $\{v_k\}$ taken previously. It holds that $\{w_k\} \subset \underline{S}(f)$ and is non-decreasing in Ω. Take, furthermore, that $W_k = (w_k)_B$. Then, similarly to the case of $\{V_k\}$, we have a harmonic function w in B such that $W_k \to w$ locally uniformly in B and $W_k(y) \to u(y) = w(y)$. Here, we have $w_k \geq v_k$ and hence $W_k \geq V_k$ in Ω, which implies that

$$w \geq v \qquad \text{in} \qquad B.$$

On the other hand, we have $w(y) = u(y) = v(y)$ and hence the strong maximum principle guarantees that $w = v$ in B. This implies $w(z) = v(z)$.

On the other hand, we have

$$w_k \geq \tilde{u} \qquad \text{in} \qquad \Omega,$$

and hence $w_k(z) \geq \tilde{u}(z)$ holds. This implies $w(z) \geq \tilde{u}(z) > v(z)$, a contradiction. □

Exercise 5.23 Give the proof of Lemmas 5.3 and 5.5.

5.3.4 *Boundary Regularity*

Remember that $\xi \in \partial\Omega$ is said to be a regular point if there is a barrier $w(z)$, which means that $w \in C(\overline{\Omega})$ is super-harmonic in Ω, $w(\xi) = 0$, and $w(x) > 0$ for $x \in \overline{\Omega} \setminus \{\xi\}$. Now, we shall show the following.

Theorem 5.15 *If f is continuous on $\partial\Omega$ and any point on $\partial\Omega$ is regular, then the Perron solution $u(z)$ defined by (5.53) is continuous on $\overline{\Omega}$ and satisfies $u|_{\partial\Omega} = f$.*

Proof. We shall show that if $\xi \in \partial\Omega$, $\{x\} \subset \Omega$, and $x \to \xi$, then $u(x) \to f(\xi)$ follows. In fact, because f is continuous on $\partial\Omega$, we have $M = \max_{\partial\Omega} |f| < +\infty$, and any $\varepsilon > 0$ admits $\delta > 0$ such that $|x - \xi| < \delta$ for $x \in \partial\Omega$ implies $|f(x) - f(\xi)| < \varepsilon$.

There is a barrier, denoted by $w(x)$, at ξ. Because $w \in C(\overline{\Omega})$ and $w(x) > 0$ for $x \in \overline{\Omega} \setminus \{\xi\}$, we have $k > 0$ such that

$$kw(x) \geq 2M$$

holds for $x \in \partial\Omega$ in $|x-\xi| \geq \delta$. Furthermore, $w(x)$ is super-harmonic in Ω and hence

$$v_1(x) = f(\xi) - \varepsilon - kw(x)$$

is sub-harmonic in Ω and is continuous on $\overline{\Omega}$. If $x \in \partial\Omega$ satisfies $|x-\xi| < \delta$, then

$$\begin{aligned} v_1(x) &= f(x) + (f(\xi) - f(x) - \varepsilon) - kw(x) \\ &\leq f(x) \end{aligned}$$

follows. On the other hand, for $x \in \partial\Omega$ in $|x-\xi| \geq \delta$, it holds that

$$\begin{aligned} v_1(x) &\leq f(\xi) - kw(x) \leq f(\xi) - 2M \\ &\leq -M \leq f(x). \end{aligned}$$

We obtain $v_1 \in \underline{S}(f)$ and hence $u \geq v_1$ holds in Ω.

Now, we take

$$v_2(x) = f(\xi) + \varepsilon + kw(x).$$

It is super-harmonic and continuous in Ω and on $\overline{\Omega}$, respectively. Similarly, we can show that $v_2 \geq f$ on $\partial\Omega$. Any $v \in \underline{S}(f)$ is sub-harmonic in Ω and satisfies $v \leq f$ on $\partial\Omega$, and therefore, it follows that $v \leq v_2$ in Ω from the weak maximum principle. Hence we obtain $u \leq v_2$ holds in Ω because $v \in \underline{S}(f)$ is arbitrary.

Those relations are summarized as

$$|u(x) - f(\xi)| \leq \varepsilon + kw(x)$$

for $x \in \Omega$. Because of $w \in C(\overline{\Omega})$ and $w(\xi) = 0$, it holds that

$$\limsup_{x \in \Omega \to \xi} |u(x) - f(\xi)| \leq \varepsilon.$$

Here, $\varepsilon > 0$ is arbitrary, we have

$$\lim_{x \in \Omega \to \xi} u(x) = f(\xi)$$

and the proof is complete. □

Exercise 5.24 Confirm that if the Dirichlet problem (5.47) has a solution $u \in C^2(\Omega) \cap C(\overline{\Omega})$ for any $f \in C(\partial\Omega)$, then any point on $\partial\Omega$ is regular.

5.3.5 *The Green's Function*

If $\Omega \subset \mathbf{R}^2$ is a bounded domain, the function $G(x, x_0)$ defined for $(x, x_0) \in \overline{\Omega} \times \Omega$ is said to be the *Green's function*, if, as a function of x, it is harmonic in $\Omega \setminus \{x_0\}$,

$$K(x, x_0) = G(x, x_0) + \frac{1}{2\pi} \log |x - x_0|$$

is harmonic in a neighborhood of x_0, and $\lim_{x \in \Omega \to \xi} G(x, x_0) = 0$ for $\xi \in \partial\Omega$. The Green's function exists if any boundary point of $\partial\Omega$ is regular, and is unique if it exists.

If $r > 0$ is small and $|x - x_0| = r$, we have $G(x, x_0) > 0$ by the second requirement to $G(x, x_0)$. Therefore, $G(x, x_0) > 0$ for $x \in \Omega \setminus B(x_0, r)$ from the third requirement and the maximum principle. Because $r > 0$ is arbitrary, we have $G(x, x_0) > 0$ if $x \in \Omega \setminus \{x_0\}$. Therefore, it holds that

$$\frac{\partial G(\xi, x_0)}{\partial \nu_\xi} \leq 0$$

for $\xi \in \partial\Omega$ if the left-hand side exists, where $\nu = \nu_\xi$ denotes the outer unit normal vector.

The following theorem is the generalization of the Poisson's formula (5.49).

Theorem 5.16 *If $\partial\Omega$ is C^2 and $u(x)$ is harmonic and continuous in Ω and on $\overline{\Omega}$, respectively, then it holds that*

$$u(x) = -\int_{\partial\Omega} u(\xi) \frac{\partial G(\xi, x)}{\partial \nu} dS_\xi \qquad \text{for} \qquad x \in \Omega, \tag{5.58}$$

where dS_ξ denotes the line element.

Proof. In use of Green's formula (1.28) in $\Omega \setminus B(x_0, \rho)$ with small $\rho > 0$, we have

$$\int_{\partial B(x_0,\rho)} \frac{\partial G(x, x_0)}{\partial \nu} dS + \int_{\partial\Omega} \frac{\partial G(x, x_0)}{\partial \nu} dS = 0.$$

Now, we have, again by (1.28) that

$$\int_{\partial B(x_0,\rho)} \left(u(\xi) \frac{\partial G(\xi, x)}{\partial \nu} - G(\xi, x) \frac{\partial u(x)}{\partial \nu} \right) dS$$
$$+ \int_{\partial\Omega} \left(u(\xi) \frac{\partial G(\xi, x)}{\partial \nu} - G(\xi, x) \frac{\partial u(\xi)}{\partial \nu} \right) dS = 0.$$

We have for $x \in \partial B(x_0, \rho)$ and $r = \rho$ that

$$G(x, x_0) = \frac{1}{2\pi} \log \frac{1}{r} + K(x, x_0)$$

and

$$\frac{\partial G(x, x_0)}{\partial \nu} = -\frac{\partial G(x, x_0)}{\partial r} = \frac{1}{2\pi r} - \frac{\partial K(x, x_0)}{\partial r},$$

and hence the first term of the left-hand side is treated as

$$\begin{aligned} &\int_{\partial B(x_0,\rho)} \left(u(\xi) \frac{\partial G(\xi, x)}{\partial \nu} - G(\xi, x) \frac{\partial u(\xi)}{\partial \nu} \right) dS \\ &= \int_0^{2\pi} \left[u(x + re^{\imath\theta}) \left(\frac{1}{2\pi r} - \frac{\partial K}{\partial r} \right) \right. \\ &\left. + \left(\frac{1}{2\pi} \log \frac{1}{r} + K \right) \frac{\partial u(x + re^{\imath\theta})}{\partial r} \right]_{r=\rho} \rho d\theta \to u(x) \end{aligned}$$

as $\rho \downarrow 0$. Then, (5.58) follows because $G(\xi, x_0) = 0$ for $\xi \in \partial\Omega$, and the proof is complete. □

The privilege of the use of the complex function theory is the relation between Riemann's mapping and the Green's function. Actually, we have the following, where a domain surrounded by a Jordan curve is called the *Jordan region.*

Theorem 5.17 *Let $w = f(z)$ be a conformal homeomorphism between the domains D and Ω, and $G(w, w_0)$ the Green's function on Ω. Then, $G_D(z, z_0) = G(f(z), f(z_0))$ is that on D.*

Proof. First, if $u(w)$ is harmonic in Ω, then $v(z) = u(f(z))$ is so in D. Thus, we have only to show that $G_D(z, z_0)$ satisfies the second requirement. In fact,

$$G(w, w_0) + \frac{1}{2\pi} \log |w - w_0|$$

is harmonic at $w = w_0 \in \Omega$, and it holds that

$$\begin{aligned} &G_D(z, z_0) + \frac{1}{2\pi} \log |z - z_0| \\ &= G(f(z), f(z_0)) + \frac{1}{2\pi} \log |f(z) - f(z_0)| - \frac{1}{2\pi} \log \left| \frac{f(z) - f(z_0)}{z - z_0} \right|. \end{aligned}$$

The last term of the right-hand side is also holomorphic at $z = z_0$ by $f'(z_0) \neq 0$. The proof is complete. $\square$

Exercise 5.25 Show that the Green's function on the unit disc $D: |x| < 1$ is given by

$$G(x, x_0) = \frac{1}{2\pi} \log \left| \frac{1 - \overline{z}_0 z}{z - z_0} \right|$$

with $z = x_1 + \imath x_2$ and $z_0 = x_{01} + \imath x_{02}$ for $x = (x_1, x_2)$ and $x_0 = (x_{01}, x_{02})$. Then, apply Theorem 5.17 and show that the Green's function on the Jordan region Ω is expressed as

$$G(x, x_0) = \frac{1}{2\pi} \log \left| \frac{1 - \overline{f(z_0)} f(z)}{f(z) - f(z_0)} \right|,$$

where $f : \Omega \to D$ denotes the Riemann mapping, that is, a conformal homeomorphism.

Exercise 5.26 Confirm that if $f : D \to \Omega$ is a conformal homeomorphism and $u(w)$ is harmonic in Ω, then so is $v(z) = u(f(z))$ in D.

5.3.6 *Newton Potential*

We have seen that the theory of two-dimensional harmonic functions are based on the expression of the solution of Dirichlet problem on the disc, the Poisson integral. The function $\log |x|$ is a two-dimensional harmonic function in $x \neq 0$, because it is the real part of the analytic function $\log z$. It depends only on $r = |x|$. Now, we shall seek such harmonic functions in higher dimensions.

Let n be the space dimension. For that purpose, we take the polar coordinate $x = r\omega \in \mathbf{R}^n$ with $r = |x|$ and $\omega \in S^{n-1} = \{x \in \mathbf{R}^n \mid |x| = 1\}$. In this case, it holds that

$$\begin{aligned}
\Delta &= \frac{\partial^2}{\partial x_1^2} + \cdots + \frac{\partial^2}{\partial x_n^2} \\
&= \frac{\partial^2}{\partial r^2} + \frac{n-1}{r}\frac{\partial}{\partial r} + \frac{1}{r^2}\Lambda \\
&= \frac{1}{r^{n-1}}\frac{\partial}{\partial r}\left(r^{n-1}\frac{\partial}{\partial r}\right) + \frac{1}{r^2}\Lambda,
\end{aligned}$$

where Λ denotes the *Laplace-Bertrami operator* on S^{n-1}. In the case that $n = 2$ and $x_1 = r\cos\varphi$, $x_2 = r\sin\varphi$ with $\varphi \in [0, 2\pi)$ are taken, then it holds that $\Lambda = \frac{\partial^2}{\partial\varphi^2}$. If $n = 3$ and $x_1 = r\cos\theta$, $x_2 = r\sin\theta\cos\varphi$, $x_3 = r\sin\theta\sin\varphi$ with $\theta \in [0, \pi]$, $\varphi \in [0, 2\pi)$, then we have

$$\Lambda = \frac{1}{\sin\theta}\frac{\partial}{\partial\theta}\left(\sin\theta\frac{\partial}{\partial\theta}\right) + \frac{1}{\sin^2\theta}\frac{\partial^2}{\partial\varphi^2}.$$

If $\Gamma = \Gamma(|x|)$ is harmonic, then

$$\frac{1}{r^{n-1}}\frac{\partial}{\partial r}\left(r^{n-1}\frac{\partial\Gamma}{\partial r}\right) = 0$$

and it follows that $\Gamma(r) = c_1 r^{-n+2} + c_2$ for $n \geq 3$, where c_1, c_2 are constants. We take

$$\Gamma(x) = \frac{1}{\omega_n(2-n)}|x|^{2-n},$$

where ω_n denotes the area of S^{n-1} so that it is equal to $2\pi^{n/2}/\gamma(n/2)$. Here, γ indicates the Gamma function:

$$\gamma(z) = \int_0^\infty e^{-s}s^{z-1}ds.$$

In the case of $n = 3$, we have the *Newton potential*,

$$\Gamma(x) = -\frac{1}{4\pi|x|}.$$

Generally, $\Delta\Gamma = 0$ for $x \neq 0$. The singularity at $x = 0$ of Γ is important.

Suppose that $u \in C^2(\overline{B})$ is harmonic in B, where

$$B = \{x \in \mathbf{R}^n \mid |x| < 1\}.$$

We take $x \in B$, $0 < \varepsilon \ll 1$, and $\Omega = B \setminus \overline{B(x,\varepsilon)}$. If ν and dS denote the outer normal vector and the area element on $\partial\Omega$, respectively. In use of Green's formula, we obtain

$$\int_\Omega (\Delta u(y)\cdot\Gamma(x-y) - u(y)\Delta_y\Gamma(x-y))\,dy$$
$$= \int_{\partial\Omega}\left(\frac{\partial u}{\partial\nu_y}(y)\Gamma(x-y) - u(y)\frac{\partial\Gamma}{\partial\nu_y}(x-y)\right)dS_y. \quad (5.59)$$

The left-hand side vanishes because $\Delta u(y) = \Delta_y \Gamma(x-y) = 0$ for $y \in \Omega$. The right-hand side is divided into $\int_{\partial B} + \int_{\partial B(x,\varepsilon)}$. Here, we have

$$\begin{aligned}
&\left| \int_{\partial B(x,\varepsilon)} \frac{\partial u}{\partial \nu_y}(y)\Gamma(x-y)dS_y \right| \leq \|\nabla u\|_{L^\infty(B)} \cdot \int_{|y-x|=\varepsilon} |\Gamma(y-x)|\, dS_y \\
&= \|\nabla u\|_{L^\infty(B)} \cdot \frac{1}{\omega_n(n-2)} \cdot \varepsilon^{2-n} \cdot \omega_n \varepsilon^{n-1} \\
&= \|\nabla u\|_{L^\infty(B)} \frac{\varepsilon}{n-2} \to 0.
\end{aligned}$$

On the other hand $\frac{\partial}{\partial \nu_y} = -\frac{\partial}{\partial \rho}$ on $\partial B(x,\varepsilon)$ for $\rho = |y-x|$. Hence we obtain

$$\begin{aligned}
&\int_{\partial B(x,\varepsilon)} u(y) \frac{\partial \Gamma}{\partial \nu_y}(x-y)dS_y \\
&= \int_{\partial B(x,\varepsilon)} u(y) \cdot \frac{1}{\omega_n(2-n)} \cdot (2-n)\, |x-y|^{1-n}\, (-1)dS_y \\
&= -\frac{1}{\omega_n \varepsilon^{n-1}} \int_{\partial B(x,\varepsilon)} u(y)dS_y = -\fint_{\partial B(x,\varepsilon)} u(y)dS_y \to -u(x)
\end{aligned}$$

as $\varepsilon \downarrow 0$. Those relations are summarized as

$$\int_{\partial B} \left\{ \frac{\partial u}{\partial r}(y)\Gamma(x-y) - u(y)\frac{\partial \Gamma}{\partial r_y}(x-y) \right\} dS_y + u(x) = 0 \tag{5.60}$$

for $x \in B$.

The mirror image with respect to ∂B is associated with the *Kelvin transformation* as follows.

Lemma 5.6 *If $y = x/|x|^2$ and $U(y) = |x|^{n-2} u(x)$, then it holds that $\Delta_y U = |x|^{n+2} \Delta_x u$.*

Proof. We take the polar coordinate $x = r\omega$. This implies $y = r^{-1}\omega$ and hence for $\rho = r^{-1}$ that

$$\begin{aligned}
\Delta_y U &= \frac{1}{\rho^{n-1}} \frac{\partial}{\partial \rho}\left(\rho^{n-1}\frac{\partial U}{\partial \rho}\right) + \frac{1}{\rho^2}\Lambda U \\
&= -r^{n-1}\left(\rho^{n-1}\frac{\partial U}{\partial \rho}\right)_r \cdot r^2 + r^2 \cdot \Lambda U \\
&= r^{n-1}\left(r^{-n+1}\left(r^{n-2}u\right)_r \cdot r^2\right)_r \cdot r^2 + r^2 \cdot \Lambda\left(r^{n-2}u\right) \\
&= r^{n+1}\left(r^{-n+3}\left(r^{n-2}u\right)_r\right)_r + r^n \Lambda u \\
&= r^{n+1}\left(ru_r + (n-2)u\right)_r + r^n \Lambda u
\end{aligned}$$

$$\begin{aligned}
&= r^{n+1}\left(ru_{rr}+(n-1)u_r\right)+r^n\Lambda u \\
&= r^{n+2}\left(u_{rr}+\frac{n-1}{r}u_r\right)+r^n\Lambda u \\
&= r^{n+2}\left\{\frac{1}{r^{n-1}}\left(r^{n-1}u_r\right)_r+\frac{1}{r^2}\Lambda u\right\}.
\end{aligned}$$

This means the conclusion. □

Now, we take

$$\Gamma^x(y)=\Gamma\left(|y|\left(x-y/|y|^2\right)\right).$$

For $z=y/|y|^2$, it holds that

$$\begin{aligned}
\Gamma^x(y)=\Gamma\left(\frac{1}{|z|}(x-z)\right)&=\frac{1}{\omega_n(2-n)}|z|^{n-2}|x-z|^{-(n-2)} \\
&=|z|^{n-2}\Gamma(x-z).
\end{aligned}$$

Namely, Γ^x is the Kelvin transformation of $\Gamma(x-\cdot)$. Thus, we obtain from Lemma 5.6 that

$$\Delta_y\Gamma^x=|z|^{n+2}\Delta_z\Gamma(x-z)=0$$

for $z\neq x$. If $x,y\in B$, then $z\notin\overline{B}$ and hence $x\neq z$. Therefore, Γ^x is harmonic on $\overline{B}$. Replacing Γ by Γ^x in (5.59), we obtain

$$\int_{\partial B}\left\{\frac{\partial u}{\partial r}(y)\Gamma^x(y)-u(y)\frac{\partial\Gamma^x}{\partial r_y}(y)\right\}dS_y=0$$

for (5.60). Here, if $y\in\partial B$, then $z=y$ and hence $\Gamma^x(y)=\Gamma(x-y)$ holds. This implies

$$\int_{\partial B}\left\{\frac{\partial u}{\partial r_y}(y)\Gamma(x-y)-u(y)\frac{\partial\Gamma^x}{\partial r_y}(y)\right\}dS_y=0. \tag{5.61}$$

Equalities (5.60) and (5.61) imply

$$u(x)=\int_{\partial B}\left\{\frac{\partial\Gamma}{\partial r_y}(x-y)-\frac{\partial\Gamma^x}{\partial r_y}(y)\right\}u(y)dS_y. \tag{5.62}$$

It is valid for $u\in C^2(B)\cap C(\overline{B})$ with $\Delta u=0$ because then we take $B(0,r)$ for $B=B(0,1)$ and making $r\uparrow 1$ after deriving an analogous equality.

Thus, the Poisson integral is given as follows if $n \geq 3$, where

$$P(x,y) = \frac{1-|x|^2}{\omega_n |y-x|^n}$$

is called the *Poisson kernel.*

Theorem 5.18 *If $u \in C^2(B)$ is harmonic and continuous in B and on $\overline{B}$, respectively, then it holds that*

$$u(x) = \int_{\partial B} P(x,y)u(y)dS_y \qquad (x \in B). \tag{5.63}$$

Proof. In fact we have

$$\begin{aligned} \frac{\partial \Gamma}{\partial x_j}(x) &= -\frac{1}{\omega_n} \cdot \frac{1}{|x|^{n-1}} \cdot \frac{x_j}{|x|} \\ \frac{\partial^2 \Gamma}{\partial x_j \partial x_k}(x) &= -\frac{1}{\omega_n} \left\{ \delta_{jk} - n\frac{x_j x_k}{|x|^2} \right\} |x|^{-n}. \end{aligned} \tag{5.64}$$

Therefore, because of $y \cdot \nabla_y = r\frac{\partial}{\partial r}$ for $r = |y|$ it holds that

$$\begin{aligned} \frac{\partial \Gamma}{\partial r_y}(x-y) &= \sum_{j=1}^{n} \frac{y_j}{|y|} \frac{\partial}{\partial y_j} \Gamma(x-y) \\ &= \sum_{j=1}^{n} \frac{y_j}{|y|} \cdot \frac{1}{\omega_n} \cdot \frac{1}{|x-y|^{n-1}} \cdot \frac{-(x_j - y_j)}{|x-y|} \\ &= \frac{1}{\omega_n} \cdot \frac{1}{|y|} \cdot \frac{(y, y-x)}{|y-x|^n}. \end{aligned} \tag{5.65}$$

On the other hand, for $\rho = |z| = 1/r$ with $z = y/|y|^2$ we have

$$\begin{aligned} \frac{\partial \Gamma^x}{\partial r_y}(y) &= -\left(\rho^{n-2}\Gamma(x-z)\right)_\rho \cdot r^2 \\ &= -(n-2)\Gamma(x-z) - \frac{\partial}{\partial \rho}\Gamma(x-z) \end{aligned}$$

at $\rho = 1$. In use of (5.65), we see that the right-hand side is equal to

$$-(n-2)\Gamma(x-z) - \frac{1}{\omega_n} \cdot \frac{(z, z-x)}{|z-x|^n} = \frac{1}{\omega_n}|x-y|^{2-n} - \frac{1}{\omega_n} \cdot \frac{(y, y-x)}{|y-x|^n}$$

by $\rho = 1$. Thus, we get

$$\begin{aligned}\frac{\partial \Gamma}{\partial r_y}(x-y) - \frac{\partial \Gamma^x}{\partial r_y}(y) &= \frac{2(y, y-x) - |y-x|^2}{\omega_n |y-x|^n} \\ &= \frac{(y+x, y-x)}{\omega_n |y-x|^n} = \frac{|y|^2 - |x|^2}{\omega_n |y-x|^n}\end{aligned}$$

and hence (5.63) follows. □

Similar formula to (5.63) holds on $B = B(x_0, R)$ with $P(x,y)$ replaced by

$$P(x,y) = \frac{R^2 - |x - x_0|^2}{\omega_n |y-x|^n}.$$

We can reproduce the arguments in §§5.3.2, 5.3.3, 5.3.4, such as mean value theorem, maximum principle, Harnack's inequality, Harnack principle, sub- and super-harmonic functions, Perron solution, harmonic lifting, and barrier. Thus, Theorem 5.7 holds true.

Exercise 5.27 Seek all eigenvalues and eigenfunctions of Λ for $n = 2$. Then, give the answer to the same problem for $n = 3$ in use of the Legendre function.

Exercise 5.28 Confirm (5.64).

5.3.7 *Layer Potentials*

Here, we take a different approach to (5.47), the layer potential, supposing that the bounded domain $\Omega \subset \mathbf{R}^3$ has the smooth boundary $\partial\Omega$. Taking

$$\Gamma(x) = \frac{1}{4\pi |x|},$$

we say that

$$v(x) = \int_{\partial\Omega} f(\eta)\Gamma(x-\eta) dS_\eta \tag{5.66}$$

and

$$w(x) = \int_{\partial\Omega} f(\eta) \frac{\partial}{\partial \nu_\eta} \Gamma(x-\eta) dS_\eta \tag{5.67}$$

the *single layer integral* and the *double layer integral* of $f \in C(\partial\Omega)$, respectively, where ν denotes the outer unit normal vector on $\partial\Omega$. We note that both $v(x)$ and $w(x)$ are harmonic in $\mathbf{R}^3 \setminus \partial\Omega$. We shall show that the normal derivative of the single layer integral $\frac{\partial}{\partial \nu_x} v(x)$ and the value itself of the double layer integral $w(x)$ take gaps across $\partial\Omega$, and that those gaps reduce the Neumann and the Dirichlet problems for harmonic functions in Ω to some integral equations on $\partial\Omega$. Actually, the kernels $\Gamma(\cdot - \eta)$ and $\frac{\partial}{\partial \nu_\eta}\Gamma(\cdot - \eta)$ of those integrals are called the *single layer potential* and the *double layer potential*, respectively.

First, we note that the double layer potential has weaker singularities on the boundary. In fact, given $x_0 \in \partial\Omega$, let us take $\eta \to x_0$ in $\eta \in \partial\Omega$. Then, it holds that

$$\frac{x_0 - \eta}{|x_0 - \eta|} \cdot \nu_\eta \quad \to \quad 0$$

and hence

$$\frac{x_0 - \eta}{|x_0 - \eta|} \cdot \nu_\eta = O\left(|\eta - x_0|\right) \qquad (\eta \in \partial\Omega \to x_0 \in \partial\Omega)$$

follows from the smoothness of $\partial\Omega$. This implies

$$\begin{aligned}\frac{\partial}{\partial \nu_\eta}\Gamma(x_0 - \eta) &= \nu_\eta \cdot \nabla_\eta \Gamma(x_0 - \eta) \\ &= -\frac{1}{4\pi |x_0 - \eta|^2} \cdot \frac{x_0 - \eta}{|x_0 - \eta|} \cdot \nu_\eta = O\left(|\eta - x_0|^{-1}\right). \end{aligned} \tag{5.68}$$

On the other hand, in use of the polar coordinate $\eta = x_0 + r\omega$ with $r = |\eta - x_0|$, we have

$$dS_\eta \approx r dr d\omega \tag{5.69}$$

near $\eta = x_0 \in \partial\Omega$, so that the double layer integral $w(x)$ converges even at each $x = x_0 \in \partial\Omega$.

Now, we show

$$\int_{\partial\Omega} \frac{\partial}{\partial \nu_\eta}\Gamma(x - \eta) dS_\eta = \begin{cases} -1 & (x \in \Omega) \\ -\frac{1}{2} & (x \in \partial\Omega) \\ 0 & (x \in \Omega^c). \end{cases} \tag{5.70}$$

In fact, if $x \in \Omega^c$, then

$$\Delta_\eta \Gamma(x - \eta) = 0$$

holds for $\eta \in \overline{\Omega}$. Then, from Green's formula we obtain

$$\int_{\partial\Omega} \frac{\partial}{\partial \nu_\eta} \Gamma(x-\eta) dS_\eta = 0.$$

In the case of $x \in \Omega$, we take $\varepsilon > 0$ sufficiently small and note that

$$\Delta_\eta \Gamma(x-\eta) = 0$$

holds for $\eta \in \overline{\Omega} \setminus B(x,\varepsilon)$. Then, it follows that

$$\begin{aligned} \int_{\partial\Omega} \frac{\partial}{\partial \nu_\eta} \Gamma(x-\eta) dS_\eta &= -\int_{\partial B(x,\varepsilon)} \frac{\partial}{\partial \nu_\eta} \Gamma(x-\eta) dS_\eta \\ &= \int_{\partial B(x,\varepsilon)} \frac{\partial}{\partial r} \Gamma(x-\eta) dS_\eta \end{aligned}$$

for $r = |\eta - x|$. Therefore, similarly to the case of §5.3.5 we have

$$\int_{\partial B(x,\varepsilon)} \frac{\partial}{\partial r} \Gamma(x-\eta) dS_\eta = -\frac{1}{4\pi} \int_{|\omega|=1} \frac{1}{\varepsilon^2} \cdot \varepsilon^2 d\omega = -1, \tag{5.71}$$

and hence

$$\int_{\partial\Omega} \frac{\partial}{\partial \nu_\eta} \Gamma(x-\eta) dS_\eta = -1$$

follows for $x \in \Omega$. Finally, if $x \in \partial\Omega$, we take small $\varepsilon > 0$ and deform $\partial\Omega$ as

$$\partial\Omega_\varepsilon = [\partial\Omega \cap B(x,\varepsilon)^c] \cup [\Omega \cap \partial B(x,\varepsilon)].$$

Because x is on the outside of Ω_ε, we have

$$\int_{\partial\Omega_\varepsilon} \frac{\partial}{\partial \nu_\eta} \Gamma(x-\eta) dS_\eta = 0 \tag{5.72}$$

by Green's formula. On the other hand, it holds that

$$\int_{\Omega} \frac{\partial}{\partial \nu_\eta} \Gamma(x-\eta) dS_\eta = \lim_{\varepsilon \downarrow 0} \int_{\partial\Omega \setminus B(x,\varepsilon)} \frac{\partial}{\partial \nu_\eta} \Gamma(x-\eta) dS_\eta$$

by (5.68) and (5.69), and that

$$\begin{aligned} \int_{\partial\Omega \setminus B(x,\varepsilon)} \frac{\partial}{\partial \nu_\eta} \Gamma(x-\eta) dS_\eta &= -\int_{\partial B(x,\varepsilon) \cap \Omega} \frac{\partial}{\partial \nu_\eta} \Gamma(x-\eta) dS_\eta \\ &= \int_{\partial B(x,\varepsilon) \cap \Omega} \frac{\partial}{\partial r} \Gamma(x-\eta) dS_\eta \end{aligned}$$

by (5.72). Similarly to (5.71), we can derive that

$$\lim_{\varepsilon\downarrow 0}\int_{\partial B(x,\varepsilon)\cap\Omega}\frac{\partial}{\partial r}\Gamma(x-\eta)dS_\eta=-\frac{1}{2},$$

and hence

$$\int_{\partial\Omega}\frac{\partial}{\partial\nu_\eta}\Gamma(x-\eta)dS_\eta=-\frac{1}{2}$$

follows for $x\in\partial\Omega$. In this way, we have proven (5.70).

From the *extension theorem of Tietze*, the continuous function $f(x)$ defined on $\partial\Omega$ has a continuous extension near $\partial\Omega$, denoted by $\tilde{f}(x)$. In use of the regularization process, this $\tilde{f}(x)$ is approximated uniformly by a family of Hölder continuous functions. If $f(x)$ itself is such a function from the beginning and $x_0\in\partial\Omega$, then we can show that the function

$$E(x)=\int_{\partial\Omega}(f(\eta)-f(x_0))\frac{\partial}{\partial\nu_\eta}\Gamma(x-\eta)dS_\eta$$

defined for $x\in\mathbf{R}^3$ is continuous at $x=x_0$.

In fact, we split $E(x)$ as

$$\begin{aligned}E(x)=&\int_{\partial\Omega}(f(\eta)-f(x))\frac{\partial}{\partial\nu_\eta}\Gamma(x-\eta)dS_\eta\\&+(f(x)-f(x_0))\int_{\partial\Omega}\frac{\partial}{\partial\nu_\eta}\Gamma(x-\eta)dS_\eta\end{aligned}$$

for this purpose, and apply (5.70) to the second term. This implies that

$$\left|(f(x)-f(x_0))\int_{\partial\Omega}\frac{\partial}{\partial\nu_\eta}\Gamma(x-\eta)dS_\eta\right|\le|f(x)-f(x_0)|\to 0$$

as $x\to x_0$ in $\mathbf{R}^3$. Furthermore, we have for $x\in\mathbf{R}^3$ and $\eta\in\partial\Omega$ that

$$\left|(f(\eta)-f(x))\frac{\partial}{\partial\nu_\eta}\Gamma(x-\eta)\right|\le M|x-\eta|^{-2+\theta}$$

with the constants $M>0$ and $\theta\in(0,1)$, and it holds that

$$\lim_{x\to x_0}\int_{\partial\Omega}|x-\eta|^{-2+\theta}\,dS_\eta=\int_{\partial\Omega}|x_0-\eta|^{-2+\theta}\,dS_\eta.$$

This gives

$$\lim_{x \to x_0} \int_{\partial\Omega} (f(\eta) - f(x)) \frac{\partial}{\partial \nu_\eta} \Gamma(x - \eta) dS_\eta$$
$$= \int_{\partial\Omega} (f(\eta) - f(x_0)) \frac{\partial}{\partial \nu_\eta} \Gamma(x_0 - \eta) dS_\eta = E(x_0)$$

from the dominated convergence theorem, and therefore,

$$\lim_{x \to x_0} E(x) = E(x_0)$$

holds true.

On the other hand, we have from (5.70) that

$$E(x_0) = w(x_0) + \frac{1}{2} f(x_0)$$

and

$$E(x) = \begin{cases} w(x) + f(x) & (x \in \Omega) \\ w(x) & (x \in \Omega^c). \end{cases}$$

This gives for $x_0 \in \partial\Omega$ that

$$\begin{aligned} w^-(x_0) &= \lim_{x \in \Omega,\ x \to x_0} w(x) = E(x_0) - f(x_0) \\ &= w(x_0) - \frac{1}{2} f(x_0) \end{aligned}$$

and

$$\begin{aligned} w^+(x_0) &= \lim_{x \in \Omega^c,\ x \to x_0} w(x) = E(x_0) \\ &= w(x_0) + \frac{1}{2} f(x_0). \end{aligned}$$

We have proven the following.

Theorem 5.19 *If $f(\eta)$ has a Hölder continuous extension near $\partial\Omega$, then the double layer integral $w(x)$ defined by (5.67) satisfies that*

$$w^\pm(x_0) = w(x_0) \pm \frac{1}{2} f(x_0) \tag{5.73}$$

for $x_0 \in \partial\Omega$, where

$$w^-(x_0) = \lim_{x \in \Omega,\ x \to x_0} w(x)$$

and

$$w^+(x_0) = \lim_{x \in \Omega^c,\ x \to x_0} w(x).$$

The single layer integral

$$v(\eta) = \int_{\partial\Omega} f(\eta)\Gamma(x-\eta)dS_\eta$$

is easier to handle, because $\Gamma(x-\eta) = O\left(|x-\eta|^{-1}\right)$ holds for $x, \eta \in \mathbf{R}^3$. This implies for an open set U containing $\partial\Omega$ that

$$\sup_{x \in U} \int_{\partial\Omega} \Gamma(x-\eta)dS_\eta = M < +\infty. \tag{5.74}$$

Then, we can show that

$$F(x) \equiv \int_{\partial\Omega} (f(\eta) - f(x_0))\,\Gamma(x-\eta)dS_\eta \quad \to \quad F(x_0) \tag{5.75}$$

follows as $x \to x_0$ in $\mathbf{R}^3$ if $x \in \partial\Omega \mapsto f(x)$ is continuous at $x = x_0 \in \partial\Omega$.

In fact, from the assumption any $\varepsilon > 0$ admits $\delta > 0$ such that

$$|f(\eta) - f(x_0)| < \varepsilon$$

holds for $|\eta - x_0| < \delta$ and $\eta \in \partial\Omega$. Here, splitting $F(x)$ as

$$\begin{aligned} F(x) = &\int_{\eta \in \partial\Omega,\ |\eta - x_0| \geq \delta} (f(\eta) - f(x_0))\,\Gamma(x-\eta)dS_\eta \\ &+ \int_{\eta \in \partial\Omega,\ |\eta - x_0| < \delta} (f(\eta) - f(x_0))\,\Gamma(x-\eta)dS_\eta, \end{aligned}$$

we can estimate the second term as

$$\left| \int_{\eta \in \partial\Omega,\ |\eta - x_0| < \delta} (f(\eta) - f(x_0))\,\Gamma(x-\eta)dS_\eta \right| < M\varepsilon$$

for $x \in U$. On the other hand, we have

$$\lim_{x \to x_0} F_\delta(x) = F_\delta(x_0)$$

for

$$F_\delta(x) = \int_{\eta \in \partial\Omega,\ |\eta - x_0| \geq \delta} (f(\eta) - f(x_0))\,\Gamma(x-\eta)dS_\eta.$$

We have

$$|F(x) - F(x_0)| < M\varepsilon + |F_\delta(x) - F_\delta(x_0)|$$

for $x \in U$, and hence it follows that

$$\limsup_{x \to x_0} |F(x) - F(x_0)| \leq M\varepsilon.$$

This implies (5.75).

Similarly to the double layer integral, the function

$$\int_{\partial\Omega} f(\eta) \frac{\partial}{\partial \nu_{x_0}} \Gamma(x_0 - \eta) dS_\eta$$

is well-defined for $x_0 \in \partial\Omega$. Also, the function

$$\frac{\partial v}{\partial \nu_{x_0}}(x) \equiv \nu_{x_0} \cdot \nabla v(x) = \int_{\partial\Omega} f(\eta) \frac{\partial}{\partial \nu_{x_0}} \Gamma(x - \eta) dS_\eta$$

is well-defined for $x \in \mathbf{R}^3 \setminus \partial\Omega$. Therefore, the function

$$D(x) = w(x) - \frac{\partial v}{\partial \nu_{x_0}}(x)$$

is well-defined for $x \in \partial\Omega$, $x \in \Omega$, and $x \in \Omega^c$, where $w(x)$ denotes the double layer integral of $f = f(\eta)$.

In fact, we have

$$D(x) = \int_{\partial\Omega} f(\eta) \left(\frac{\partial}{\partial \nu_\eta} - \frac{\partial}{\partial \nu_{x_0}} \right) \Gamma(x - \eta) dS_\eta$$

with the kernel satisfying

$$\begin{aligned} \left(\frac{\partial}{\partial \nu_\eta} - \frac{\partial}{\partial \nu_{x_0}} \right) \Gamma(x - \eta) &= -\frac{1}{4\pi |x - \eta|^2} \frac{x - \eta}{|x - \eta|} \cdot (\nu_\eta - \nu_{x_0}) \\ &= O\left(|x - \eta|^{-2} \right). \end{aligned}$$

This implies $\lim_{x \to x_0} D(x) = D(x_0)$ as $x \to x_0 \in \partial\Omega$ in $\mathbf{R}^3$ if $x \in \partial\Omega \mapsto f(x)$ is continuous. Hence we get the following from (5.73).

Theorem 5.20 *If $f(\eta)$ has a Hölder continuous extension near $\partial\Omega$, the single layer integral $v(x)$ defined by (5.66) satisfies that*

$$\frac{\partial v^\pm}{\partial \nu_{x_0}}(x_0) = \frac{\partial v}{\partial \nu_{x_0}}(x_0) \pm \frac{1}{2} f(x_0) \tag{5.76}$$

for $x_0 \in \partial\Omega$*, where*

$$\frac{\partial v^-}{\partial \nu_{x_0}}(x_0) = \lim_{x\in\Omega,\ x\to x_0} \frac{\partial v}{\partial \nu_{x_0}}(x)$$

and

$$\frac{\partial v^+}{\partial \nu_{x_0}}(x_0) = \lim_{x\in\Omega^c,\ x\to x_0} \frac{\partial v}{\partial \nu_{x_0}}(x).$$

5.3.8 *Fredholm Theory*

From the above mentioned properties of layer potentials, boundary value problems for harmonic functions are reduced to the integral equation on the boundary. Here, we take the simplest case of

$$\Delta v = 0 \quad \text{in} \quad \Omega, \qquad v = g \quad \text{on} \quad \partial\Omega \tag{5.77}$$

and describe the guideline.

In fact, if the integral equation on $\partial\Omega$ given as

$$\mu(\xi) - 2\int_{\partial\Omega} \mu(\xi)\frac{\partial}{\partial \nu_\eta}\Gamma(\xi - \eta)dS_\eta = g(\xi) \qquad (\xi \in \partial\Omega) \tag{5.78}$$

has the solution $\mu(\xi)$, then

$$v(x) = -2\int_{\partial\Omega} \mu(\eta)\frac{\partial}{\partial \nu_\eta}\Gamma(x - \eta)dS_\eta$$

is harmonic in Ω. Furthermore, from (5.73) it follows that

$$v^-(\xi) = v(\xi) + \mu(\xi) = g(\xi)$$

and $\mu(\xi)$ becomes a solution to (5.78).

If we define the operator $K : C(\partial\Omega) \to C(\partial\Omega)$ by

$$(K\mu)(\xi) = \int_{\partial\Omega} \mu(\eta)\frac{\partial}{\partial \nu_\eta}\Gamma(\xi - \eta)dS_\eta, \tag{5.79}$$

then (5.78) means for $\lambda = 2$ that

$$(I - \lambda K)\,\mu = g. \tag{5.80}$$

In use of (5.68), we can show that K is compact by Ascoli-Arzelá's theorem and hence *Riesz-Schauder's theorem* is applicable, which implies the

Fredholm alternative. Namely, because $\sigma = \frac{1}{2}$ is not an eigenvalue of K, we have the bounded linear operator

$$(I - \lambda K)^{-1} : C(\partial\Omega) \to C(\partial\Omega).$$

Hence, equation (5.80) is uniquely solvable for given $g \in C(\partial\Omega)$.

Exercise 5.29 Confirm that the operator $K : C(\partial\Omega) \to C(\partial\Omega)$ defined by (5.79) is compact.

5.4 Regularity

5.4.1 *Poisson Equation*

Let us confirm that equality (5.59) is valid even if $u \in C^2(\overline{B})$ is not harmonic, where $B = B(0,1)$ and $\Omega = B \setminus \overline{B(x,\varepsilon)}$ with $\in B$ and $0 < \varepsilon \ll 1$. The left-hand side is equal to

$$\int_{B\setminus B(x,\varepsilon)} \Delta u(y) \cdot \Gamma(x-y)dy \tag{5.81}$$

and hence converges to

$$\int_B \Delta u(y) \cdot \Gamma(x-y)dy \tag{5.82}$$

as $\varepsilon \downarrow 0$. On the other hand, the right-hand side accepts the same treatment and hence converges to

$$u(x) + \int_{\partial B} \left\{ \frac{\partial u}{\partial r_y}(y) \cdot \Gamma(x-y) - u(y)\frac{\partial \Gamma}{\partial r_y}(x-y) \right\} dS_y.$$

Thus, we obtain

$$\int_B \Delta u(y) \cdot \Gamma(x-y)dy = u(x) + \int_{\partial B} \left\{ \frac{\partial u}{\partial r_y}(y)\Gamma(x-y) - u(y)\frac{\partial \Gamma}{\partial r_y}(x-y) \right\} dS_y. \tag{5.83}$$

The analogous equality to (5.61) is similar, and is given as

$$\int_B \Delta u(y) \cdot \Gamma^x(y)dy = \int_{\partial B} \left\{ \frac{\partial u}{\partial r_y}(y)\Gamma^x(y) - u(y)\frac{\partial \Gamma^x}{\partial r_y}(y) \right\} dS_y. \tag{5.84}$$

Thus, we obtain

$$u(x) = \int_{\partial B} P(x,y)f(y)dS_y + \int_B G(x,y)g(y)dy \qquad (x \in B) \tag{5.85}$$

with the *Green's function*

$$G(x,y) = \Gamma^x(y) - \Gamma(x-y),$$

if $u \in C^2(\overline{B})$ satisfies

$$-\Delta u = g \quad \text{in} \quad B, \qquad \text{with} \qquad u = f \quad \text{on} \quad \partial B. \tag{5.86}$$

In particular, if $u \in C^2(\overline{B})$ solves the *Poisson equation*

$$-\Delta u = g \quad \text{in} \quad B \qquad \text{with} \qquad u = 0 \quad \text{on} \quad \partial B, \tag{5.87}$$

then it is given as

$$u(x) = \int_B G(x,y)g(y)dy \qquad (x \in B).$$

However, deriving (5.86) from (5.85) is not so simple.

First, because B satisfies the outer circumscribing ball condition at any boundary point, any $f \in C(\partial B)$ admits a unique $u \in C^2(B) \cap C(\overline{B})$ satisfying (5.86) with $g = 0$. By Theorem 5.18, this $u(x)$ is given by

$$u(x) = \int_{\partial B} P(x,y)f(y)dy \qquad (x \in B).$$

Therefore, if $f \in C(\partial B)$ the first term of the right-hand side of (5.85) is in $C^2(B) \cap C(\overline{B})$ and satisfies

$$-\Delta u = 0 \quad \text{in} \quad B \qquad \text{with} \qquad u = f \quad \text{on} \quad \partial B.$$

On the other hand, $g \in C(\overline{B})$ cannot imply the first term of the right-hand side is in $C^2(\overline{B})$. We recall that a similar discrepancy is observed in §§5.1.2 and 5.1.3.

Namely, in §5.1.3, we asked for the reader to confirm that $u(x,t)$ given by (5.18) becomes the classical solution to (5.7), provided that $u_0(x)$ is continuously differentiable on $[0,\pi]$ and satisfies the compatibility condition that $u_0|_{x=0,\pi} = 0$. If

$$X^1 = \left\{ v \in C^1[0,\pi] \mid v|_{x=0,\pi} = 0 \right\}$$

and

$$X^0 = \left\{ v \in C[0,\pi] \mid v|_{x=0,\pi} = 0 \right\},$$

this means that the mapping $u_0 \in X^1 \mapsto u(t) \in X^0$ is well-defined through (5.7) for each $t \geq 0$. However, those spaces X^1 and X^0 are different and it causes a discrepancy in the regularity between the initial value u_0 and the solution $u(\cdot,t)$ with $t \downarrow 0$. It is compensated by the use of Hölder space. Similar situation arises in the elliptic problem as is suggested in Theorem 5.21. Actually, we have $u \in C^{2,\theta}(\Omega)$ if $g \in C^\theta(\Omega)$ in

$$-\Delta u = g \quad \text{in} \quad \Omega, \qquad u = 0 \quad \text{on} \quad \partial\Omega,$$

where $\Omega \subset \mathbf{R}^n$ is a bounded domain with $C^{2,\theta}$ boundary $\partial\Omega$ for $\theta \in (0,1)$.

Exercise 5.30 Confirm that (5.81) converges to (5.82) if $u \in C^2(\overline{B})$.

5.4.2 *Schauder Estimate*

In §5.4.1 we have mentioned that $u \in C^2(\overline{B})$ does not follow from $g \in C(\overline{B})$ in

$$u(x) = \int_B G(x,y)g(y)dy \qquad (x \in B)$$

for the Green's function $G(x,y)$ of the Poisson equation (5.87). To understand the situation, we take the essential part

$$\Gamma(x-y) = \frac{1}{\omega_n(2-n)} |x-y|^{2-n}$$

of $G(x,y)$, assuming $n \geq 3$. Given $g \in L^1(\mathbf{R}^n) \cap L^\infty(\mathbf{R}^n)$, we take

$$u(x) = -\int_{\mathbf{R}^n} \Gamma(x-y)g(y)dy \qquad (x \in \mathbf{R}^n). \tag{5.88}$$

Because of

$$|\Gamma(x-y)g(y)| \leq \begin{cases} \frac{1}{\omega_n(n-2)} |g(y)| & (|y-x| \geq 1) \\ \frac{1}{\omega_n(n-2)} \cdot \frac{\|g\|_\infty}{|y-x|^{n-2}} & (|y-x| < 1) \end{cases}$$

and $dy = r^{n-1}drd\omega$ for $y = r\omega$ with $r = |y|$ and

$$\omega \in S^{n-1} = \{x \in \mathbf{R}^n \mid |x| = 1\},$$

we have

$$\int_{\mathbf{R}^n} |\Gamma(x-y)g(y)|\, dy < +\infty$$

for each $x \in \mathbf{R}^n$. Thus, the measurable function $u(x)$ is well-defined. On the other hand, from the first relation (5.64), we have

$$\left|\frac{\partial \Gamma}{\partial x_j}(x-y)g(y)\right| \leq \frac{1}{\omega_n} \cdot \frac{1}{|y-x|^{n-1}} \cdot |g(y)| \tag{5.89}$$

and hence

$$u_j(x) = -\int_{\mathbf{R}^n} \frac{\partial \Gamma}{\partial x_j}(x-y)g(y)dy \qquad (x \in \mathbf{R}^n) \tag{5.90}$$

is well-defined as (5.88). We shall show that $u(x)$ is differentiable and $\frac{\partial u}{\partial x_j} = u_j$. For this purpose, we develop the regularization argument instead of applying directly the dominated convergence theorem.

We take $\eta \in C^\infty[0,\infty)$ satisfying $0 \leq \eta \leq 1$ and

$$\eta(s) = \begin{cases} 1 & (t \geq 2) \\ 0 & (0 \leq t \leq 1). \end{cases}$$

Then, the function

$$u_\varepsilon(x) = -\int_{\mathbf{R}^n} \Gamma(x-y)\eta\left(\frac{|x-y|}{\varepsilon}\right) g(y)dy$$

is well-defined for each $\varepsilon > 0$. From the following lemma, it follows that $u \in C^1(\mathbf{R}^n)$ and $\frac{\partial u}{\partial x_j} = u_j$.

Lemma 5.7 *The function u_ε is continuously differentiable in $\mathbf{R}^n$ and satisfies that*

$$\|u_\varepsilon - u\|_\infty \to 0 \qquad \text{and} \qquad \left\|\frac{\partial u_\varepsilon}{\partial x_j} - u_j\right\|_\infty \to 0$$

as $\varepsilon \downarrow 0$.

Proof. We have

$$\frac{\partial}{\partial x_j}\left(\Gamma(x-y)\eta\left(\frac{|x-y|}{\varepsilon}\right)g(y)\right)=\frac{\partial\Gamma}{\partial x_j}(x-y)\cdot\eta\left(\frac{|x-y|}{\varepsilon}\right)g(y)$$
$$+\Gamma(x-y)\cdot\frac{1}{\varepsilon}\eta'\left(\frac{|x-y|}{\varepsilon}\right)\cdot\frac{x_j-y_j}{|x-y|}\cdot g(y)$$

with

$$\left|\frac{\partial\Gamma}{\partial x_j}(x-y)\cdot\eta\left(\frac{|x-y|}{\varepsilon}\right)\cdot g(y)\right|$$
$$\leq\frac{1}{\omega_n|x-y|^{n-1}}\cdot\eta\left(\frac{|x-y|}{\varepsilon}\right)\cdot|g(y)|$$
$$\leq\frac{1}{\omega_n\varepsilon^{n-1}}\cdot|g(y)|$$

and

$$\left|\Gamma(x-y)\cdot\frac{1}{\varepsilon}\eta'\left(\frac{|x-y|}{\varepsilon}\right)\cdot g(y)\right|$$
$$\leq\frac{1}{\omega_n(n-2)}\cdot\frac{1}{\varepsilon^{n-1}}\cdot\|\eta'\|_\infty|g(y)|$$

by $0\leq\eta\leq1$ and $\operatorname{supp}\eta\subset\mathbf{R}^n\setminus B(0,1)$. This implies

$$\left|\frac{\partial}{\partial x_j}\left(\Gamma(x-y)\eta\left(\frac{|x-y|}{\varepsilon}g(y)\right)\right)\right|$$
$$\leq\frac{1}{\omega_n\varepsilon^{n-1}}\left\{1+\frac{\|\eta'\|_\infty}{n-2}\right\}|g(y)|\in L^1(\mathbf{R}^n)$$

and the dominated convergence theorem guarantees that

$$\frac{\partial u_\varepsilon}{\partial x_j}(x)=-\int_{\mathbf{R}^n}\frac{\partial}{\partial x_j}\left(\Gamma(x-y)\eta\left(\frac{|x-y|}{\varepsilon}\right)g(y)\right)dy.\tag{5.91}$$

Hence we obtain

$$-\frac{\partial u_\varepsilon}{\partial x_j}(x)+u_j(x)=\int_{\mathbf{R}^n}\frac{\partial\Gamma}{\partial x_j}(x-y)\left\{\eta\left(\frac{|x-y|}{\varepsilon}\right)-1\right\}g(y)dy$$
$$+\frac{1}{\varepsilon}\int_{\mathbf{R}^n}\Gamma(x-y)\eta'\left(\frac{|x-y|}{\varepsilon}\right)\frac{x_j-y_j}{|x-y|}g(y)dy$$
$$=\int_{|y-x|<2\varepsilon}\frac{\partial\Gamma}{\partial x_j}(x-y)\left\{\eta\left(\frac{|x-y|}{\varepsilon}\right)-1\right\}g(y)dy$$

$$+\frac{1}{\varepsilon}\int_{\varepsilon<|y-x|<2\varepsilon}\Gamma(x-y)\eta'\left(\frac{|x-y|}{\varepsilon}\right)\frac{x_j-y_j}{|x-y|}g(y)dy.$$

Therefore, in use of

$$\left|\frac{\partial\Gamma}{\partial x_j}(x-y)\left\{\eta\left(\frac{|x-y|}{\varepsilon}\right)-1\right\}g(y)\right|\leq\frac{1}{\omega_n|x-y|^{n-1}}\|g\|_\infty$$

and

$$\left|\Gamma(x-y)\eta'\left(\frac{|x-y|}{\varepsilon}\right)\frac{x_j-y_j}{|x-y|}g(y)\right|$$
$$\leq\frac{1}{\omega_n(n-2)}\cdot\frac{1}{|x-y|^{n-2}}\|\eta'\|_\infty\cdot|g|_\infty$$

we have

$$\begin{aligned}\sup_{x\in\mathbf{R}^n}&\left|\frac{\partial u_\varepsilon}{\partial x_j}(x)-u_j(x)\right|\\ &\leq\int_{|z|<2\varepsilon}\frac{\|g\|_\infty}{\omega_n}\frac{dz}{|z|^{n-1}}+\frac{1}{\varepsilon}\int_{\varepsilon<|z|<2\varepsilon}\frac{\|\eta'\|_\infty\cdot\|g\|_\infty}{\omega_n(n-2)}\frac{dz}{|z|^{n-2}}\\ &=\frac{\|g\|_\infty}{\omega_n}\cdot\omega_n\int_0^{2\varepsilon}\frac{r^{n-1}}{r^{n-1}}dr+\frac{\|\eta'\|_\infty\cdot\|g\|_\infty}{\omega_n(n-2)}\cdot\omega_n\cdot\frac{1}{\varepsilon}\int_\varepsilon^{2\varepsilon}\frac{r^{n-1}}{r^{n-2}}dr\\ &=O(\varepsilon)\end{aligned}$$

and hence

$$\lim_{\varepsilon\downarrow 0}\left\|\frac{\partial u_\varepsilon}{\partial x_j}-u_j\right\|_\infty=0$$

follows. The proof of

$$\lim_{\varepsilon\downarrow 0}\|u_\varepsilon-u\|_\infty=0 \tag{5.92}$$

is easier and left to the reader. □

From the second equality of (5.64), it follows that

$$\left|\frac{\partial^2\Gamma}{\partial x_j\partial x_k}\right|\approx|x|^{-n}\notin L^1_{loc}(\mathbf{R}^n)$$

and the above argument in use of the cut-off function fails to take the second derivative of $u(x)$. Here, we introduce the notion of *Hölder continuity.* A function $f(x)$ is said to be Hölder continuous with the exponent $\theta \in (0,1)$ in a domain $\Omega \subset \mathbf{R}^n$ if

$$[f]_\theta \equiv \sup_{x,y\in\Omega,\ x\neq y} \frac{|f(x)-f(y)|}{|x-y|^\theta} < +\infty.$$

Then, $C^{m,\theta}(\Omega)$ denotes the set of C^m functions on Ω such that $D^\alpha f$ is Hölder continuous with the exponent θ for $|\alpha| = m$, where $m = 0, 1, \cdots$ and $\theta \in (0,1)$.

The following theorem illustrates the simplest case of the *Schauder regularity*.

Theorem 5.21 *If $g \in C^\theta(\mathbf{R}^n) \cap L^1(\mathbf{R}^n)$ with $n > 2$, then $u(x)$ defined by (5.88) is in $C^{2,\theta}(\mathbf{R}^n)$ and satisfies that*

$$-\Delta u = g \qquad in \quad \mathbf{R}^n, \tag{5.93}$$

where $\theta \in (0,1)$.

Proof. From Lemma 5.7, we see that

$$u(x) = -\int_{\mathbf{R}^n} \Gamma(x-y)g(y)dy$$

is C^1 and satisfies

$$\frac{\partial u}{\partial x_j} = -\int_{\mathbf{R}^n} \frac{\partial}{\partial x_j}\Gamma(x-y)g(y)dy$$

in $x \in \mathbf{R}^n$. Similarly, we can prove that

$$\begin{aligned} u_{jk}(x) = &-\int_{|x-y|<\rho} \frac{\partial^2}{\partial x_j \partial x_k}\Gamma(x-y)\left(g(y)-g(x)\right)dy \\ &+g(x)\int_{|x-y|=\rho} \frac{\partial}{\partial x_k}\Gamma(x-y)\frac{y_j-x_j}{|y-x|}dS_y \\ &-\int_{|y-x|>\rho} \frac{\partial^2}{\partial x_j \partial x_k}\Gamma(x-y)g(y)dy \end{aligned} \tag{5.94}$$

converges for each $\rho > 0$. Now, we shall show that $\frac{\partial^2 u_\varepsilon}{\partial x_j \partial x_k}$ converges uniformly to u_{jk}. This implies

$$\frac{\partial^2 u}{\partial x_j \partial x_k} = u_{jk}$$

and also (5.93) by $\Delta\Gamma(x) = 0$ for $x \in \mathbf{R}^n \setminus \{0\}$ and

$$\int_{|x-y|=\rho} \frac{\partial}{\partial x_k}\Gamma(x-y)\frac{x_j - y_j}{|x-y|}dS_y = \frac{\delta_{jk}}{n}. \tag{5.95}$$

The Hölder continuity of u_{jk} is left to the reader and thus the proof is reduced to

$$\lim_{\varepsilon\downarrow 0}\left\| \frac{\partial^2 u_\varepsilon}{\partial x_j \partial x_k} - u_{jk} \right\|_\infty = 0. \tag{5.96}$$

First, we confirm that $\frac{\partial^2 u_\varepsilon}{\partial x_j \partial x_k}$ exists. In fact, we have

$$\begin{aligned}
&\left| \frac{\partial}{\partial x_j}\left(\frac{\partial \Gamma}{\partial x_k}(x-y)\eta\left(\frac{|x-y|}{\varepsilon}\right) g(y)\right)\right| \\
&\quad \leq \left| \frac{\partial^2 \Gamma}{\partial x_j \partial x_k}(x-y)\right| \left|\eta\left(\frac{|x-y|}{\varepsilon}\right)\right| |g(y)| \\
&\quad + \left| \frac{\partial \Gamma}{\partial x_k}(x-y)\right| \left|\eta'\left(\frac{|x-y|}{\varepsilon}\right)\right| \frac{1}{\varepsilon}\left|\frac{x_j - y_j}{|x-y|}\right| |g(y)| \\
&\quad \leq \begin{cases} \left(\frac{n-1}{\omega_n \varepsilon^n} + \frac{\varepsilon^{-1}}{\omega_n \varepsilon^{n-1}}\|\eta'\|_\infty\right)|g(y)| & (|x-y| \geq \varepsilon) \\ 0 & (|x-y| < \varepsilon). \end{cases}
\end{aligned}$$

Here, the right-hand side is independent of y and is summable in x, so that we obtain from the dominated convergence theorem that

$$\frac{\partial^2 u_\varepsilon}{\partial x_j \partial x_k} = -\int_{\mathbf{R}^n} \frac{\partial}{\partial x_j}\left(\frac{\partial \Gamma}{\partial x_k}(x-y)\eta\left(\frac{|x-y|}{\varepsilon}\right) g(y)\right) dy$$

with its continuity in $x \in \mathbf{R}^n$.

Next, we have

$$\begin{aligned}
&-\frac{\partial^2 u_\varepsilon}{\partial x_j \partial x_k}(x) + u_{jk}(x) \\
&= \int_{|x-y|<\rho} \frac{\partial^2 \Gamma}{\partial x_j \partial x_k}(x-y)\left(\eta\left(\frac{|x-y|}{\varepsilon}\right) - 1\right)(g(y) - g(x))\,dy \\
&+ g(x)\int_{|x-y|<\rho} \frac{\partial^2 \Gamma}{\partial x_j \partial x_k}(x-y)\eta\left(\frac{|x-y|}{\varepsilon}\right)dy \\
&+ g(x)\int_{|x-y|=\rho} \frac{\partial \Gamma}{\partial x_k}(x-y)\frac{y_j - x_j}{|y-x|}dS_y
\end{aligned}$$

$$+\int_{|x-y|>\rho} \frac{\partial^2\Gamma}{\partial x_j \partial x_k}(x-y)\left(\eta\left(\frac{|x-y|}{\varepsilon}\right)-1\right)g(y)dy$$
$$+\int_{\varepsilon<|x-y|<2\varepsilon} \frac{\partial\Gamma}{\partial x_k}(x-y)\eta'\left(\frac{|x-y|}{\varepsilon}\right)\frac{1}{\varepsilon}\cdot\frac{x_j-y_j}{|x-y|}g(y)dy.$$

Making $\varepsilon \downarrow 0$, we have $2\varepsilon < \rho$, and hence the fourth term disappears. For the second term, we apply Green's formula as

$$-\int_{|x-y|<\rho} \frac{\partial}{\partial x_j}\left(\frac{\partial\Gamma}{\partial x_k}(x-y)\eta\left(\frac{|x-y|}{\varepsilon}\right)\right)dy$$
$$=\int_{|x-y|=\rho} \frac{\partial\Gamma}{\partial x_k}(x-y)\eta\left(\frac{|x-y|}{\varepsilon}\right)\frac{y_j-x_j}{|y-x|}dS_y,$$

where

$$\eta\left(\frac{|x-y|}{\varepsilon}\right)=1$$

follows for $|x-y|=\rho$ if $2\varepsilon<\rho$. Thus, we obtain

$$-\frac{\partial^2 u_\varepsilon}{\partial x_j \partial x_k}(x)+u_{jk}(x)$$
$$=\int_{|x-y|<\rho} \frac{\partial^2\Gamma}{\partial x_j \partial x_k}(x-y)\left(\eta\left(\frac{|x-y|}{\varepsilon}\right)-1\right)(g(y)-g(x))\,dy$$
$$-g(x)\int_{|x-y|<\rho} \frac{\partial}{\partial x_j}\left(\frac{\partial\Gamma}{\partial x_k}(x-y)\eta\left(\frac{|x-y|}{\varepsilon}\right)\right)dy$$
$$+g(x)\int_{|x-y|<\rho} \frac{\partial^2\Gamma}{\partial x_j \partial x_k}(x-y)\eta\left(\frac{|x-y|}{\varepsilon}\right)dy$$
$$+\int_{\varepsilon<|x-y|<2\varepsilon} \frac{\partial\Gamma}{\partial x_k}(x-y)\eta'\left(\frac{|x-y|}{\varepsilon}\right)\frac{1}{\varepsilon}\frac{x_j-y_j}{|x-y|}g(y)dy$$
$$=\int_{|x-y|<2\varepsilon} \frac{\partial^2\Gamma}{\partial x_j \partial x_k}(x-y)\left(\eta\left(\frac{|x-y|}{\varepsilon}\right)-1\right)(g(y)-g(x))\,dy$$
$$+\int_{\varepsilon<|x-y|<2\varepsilon} \frac{\partial\Gamma}{\partial x_k}(x-y)\eta'\left(\frac{|x-y|}{\varepsilon}\right)\frac{1}{\varepsilon}\frac{x_j-y_j}{|x-y|}(g(y)-g(x))\,dy$$
$$=I+II.$$

Here, we have

$$|I| \leq \frac{(n-1)}{\omega_n}[g]_\theta \int_{|x-y|<2\varepsilon} |x-y|^{\theta-n}\,dy = O(\varepsilon^\theta)$$

and

$$|II| \leq \frac{1}{\omega_n} \|\eta'\|_\infty [g]_\theta \int_{\varepsilon<|x-y|<2\varepsilon} \frac{1}{\varepsilon} |x-y|^{-n+1+\theta} dy = O(\varepsilon^\theta)$$

as $\varepsilon \downarrow 0$. Hence (5.96) follows and the proof is complete. □

Exercise 5.31 Confirm (5.92).

Exercise 5.32 Confirm that the right-hand side of (5.94) converges.

Exercise 5.33 Show (5.95).

Exercise 5.34 Try to show that u_{jk} is Hölder continuous.

5.4.3 *Dirichlet Principle*

In §5.3, we studied (5.47) by means of the potential function. However, first this problem was approached differently from those methods of Perron or Fredholm. Namely, the *Dirichlet principle* asserts that the solution to (5.47) is realized as the minimizer of

$$J(v) = \frac{1}{2} \int_\Omega |\nabla v|^2 \tag{5.97}$$

under the constraint that

$$v = f \quad \text{on} \quad \partial\Omega. \tag{5.98}$$

It took a long time to justify this result. The first obstruction was to establish the existence of the minimizer. For this purpose, it is actually necessary to prepare a functional space with the complete metric. Now, we can show the following.

Theorem 5.22 *The functional $J(v)$ defined by (5.97) attains the minimum on*

$$E = \left\{ v \in H^1(\Omega) \mid v = f \quad on \quad \partial\Omega \right\}.$$

The minimizer $u \in E$ satisfies

$$\int_\Omega \nabla u \cdot \nabla v = 0 \tag{5.99}$$

for any $v \in E$.

Remember that $H^1(\Omega)$ denotes the set of functions square integrable up to their first derivatives. It forms a Hilbert space with the norm

$$\|v\|_{H^1} = \left(\|v\|_2^2 + \|\nabla v\|_2^2\right)^{1/2}.$$

If Ω has the *restricted cone property*, then the mapping

$$\gamma : v \in C(\overline{\Omega}) \quad \mapsto \quad v|_{\partial\Omega} \in C(\partial\Omega)$$

has a continuous extension from $H^1(\Omega)$ to $H^{1/2}(\partial\Omega)$. This operation is called the *trace* and the constraint (5.98) is taken in this sense for $v \in H^1(\Omega)$.

The proof of Theorem 5.22 is quite similar to the abstract Riesz' representation theorem in §3.2.2 if we apply the *Poincaré inequality* described in the following section. Let us confirm that $H_0^1(\Omega)$ is the closure of $C_0^\infty(\Omega)$ in $H^1(\Omega)$, and that $v \in H^1(\Omega)$ is in $H_0^1(\Omega)$ if and only if its trace to $\partial\Omega$ vanishes.

Lemma 5.8 *Any bounded domain $\Omega \subset \mathbf{R}^n$ admits a constant $C > 0$ satisfying*

$$\|v\|_2 \leq C\,\|\nabla v\|_2 \tag{5.100}$$

for any $v \in H_0^1(\Omega)$.

Proof. Inequality (5.100) is reduced to the case of $v \in C_0^\infty(\Omega)$. In fact, from the definition, any $v \in H_0^1(\Omega)$ admits $\{v_k\} \subset C_0^\infty(\Omega)$ satisfying $\|v_k - v\|_{H^1} \to 0$. This implies

$$\|\nabla v_k\|_2 \to \|\nabla v\|_2 \qquad \text{and} \qquad \|v_k\|_2 \to \|v\|_2$$

and hence inequality (5.100) for v_k implies that for v.

We may suppose that

$$\Omega \subset \{x = (x_1, x_2, \cdots, x_n) \in \mathbf{R}^n \mid 0 < x_1 < \ell\}.$$

The function $v \in C_0^\infty(\Omega)$ is regarded as an element in $C_0^\infty(\mathbf{R}^n)$ by the zero extension. Then, it holds for $x = (x_1, x_2, \cdots, x_n) \in \mathbf{R}^n$ that

$$v(x) = \int_0^{x_1} v_{x_1}(t, x_2, \cdots, x_n)dt.$$

This implies

$$|v(x)|^2 \leq \left\{\int_0^\ell |v_{x_1}(t,x_2,\cdots,x_n)|\,dt\right\}^2$$
$$\leq \ell\cdot\int_0^\ell |v_{x_1}(t,x_2,\cdots,x_n)|^2\,dt$$

and hence

$$\int_{\mathbf{R}^{n-1}} |v(x_1,x_2,\cdots,x_n)|^2\,dx_2\cdots dx_n$$
$$\leq \ell\int_0^\ell dt\int_{\mathbf{R}^{n-1}} |v_{x_1}(t,x_2,\cdots,x_n)|^2\,dx_2\cdots dx_n$$

follows. We obtain

$$\begin{aligned}\|v\|_2^2 &= \int_{\mathbf{R}^n} |v(x)|^2\,dx \\ &= \int_0^\ell dx_1\int_{\mathbf{R}^{n-1}} |v(x_1,x_2,\cdots,x_n)|^2\,dx_2\cdots dx_n \\ &\leq \ell^2\int_0^\ell dt\int_{\mathbf{R}^{n-1}} |v_{x_1}(t.x_2,\cdots,x_n)|^2\,dx_2\cdots dx_n \\ &= \ell^2\,\|v_{x_1}\|_2^2 \leq \ell^2\,\|\nabla v\|_2^2\end{aligned}$$

and hence inequality (5.100) holds for $C = \ell$. □

Exercise 5.35 Prove Theorem 5.22 in the following way. First, take the minimizing sequence $\{v_k\} \subset E$ of J. Then apply (5.100) to $w_k = v_k - \tilde{f}$ and show that it is bounded in $H_0^1(\Omega)$, where $\tilde{f} \in H^1(\Omega)$ in $\tilde{f}\big|_{\partial\Omega} = f$. Take its subsequence that converges weakly there, and apply the lower semi-continuity.

5.4.4 *Moser's Iteration Scheme*

This paragraph is an introduction to the regularity theory applicable to nonlinear problems. We admit the following fact referred to as *Sobolev's imbedding theorem*. Here, as $H_0^1(\Omega)$ is based on $L^2(\Omega)$, $W_0^{1,p}(\Omega)$ is constructed from $L^p(\Omega)$. That is, $W^{1,p}(\Omega)$ is the set of p-th integrable functions up to their first derivatives, and $W_0^{1,p}(\Omega)$ is the closure of $C_0^\infty(\Omega)$ in

$W^{1,p}(\Omega)$ under the norm

$$\|v\|_{W^{1,p}(\Omega)} = \left(\|\nabla v\|_p^p + \|v\|_p^p\right)^{1/p}.$$

Theorem 5.23 *If $\Omega \subset \mathbf{R}^n$ is a bounded domain, $p \in (1,n)$, and $n > 2$, then the embedding $W_0^{1,p}(\Omega) \subset L^{p^*}(\Omega)$ holds for $p^* = np/(n-p)$. More precisely, there exists a constant determined by n and p such that*

$$\|v\|_{\frac{np}{n-p}} \leq C \|\nabla v\|_p \tag{5.101}$$

holds for any $v \in W_0^{1,p}(\Omega)$.

We will concentrate on the inner regularity, so that $f \in L^p_{loc}(\Omega)$ means that $\varphi \cdot f \in L^p(\Omega)$ for any $\varphi \in C_0^\infty(\Omega)$. The sets $H^1_{loc}(\Omega)$ and $W^{1,p}_{loc}(\Omega)$ are defined similarly.

Henceforth, we take the case $n > 2$ only. We say that $v \in H^1_{loc}(\Omega)$ is sub-harmonic in Ω if

$$\int_\Omega \nabla u \cdot \nabla \varphi \leq 0 \tag{5.102}$$

holds for any non-negative $\varphi \in C_0^\infty(\Omega)$. By the regularization and cut-off process, this φ can be taken to be a non-negative function in $H_0^1(\Omega)$. Given such u, we take $\overline{B(x_0, 2R)} \subset \Omega$, where $B(x_0, 2R)$ denotes the open ball with the radius and the center $2R > 0$ and x_0, respectively. We put $x_0 = 0$ and $B(x_0, 2R) = B_{2R}$ for simplicity. For $0 < \rho < r \leq 2R$, we take a non-negative $\eta = \eta(|x|) \in C_0^\infty(B_r)$ satisfying

$$\eta = 1 \quad \text{on} \quad B_\rho \qquad \text{and} \qquad |\nabla \eta| \leq C/(r-\rho). \tag{5.103}$$

Here and henceforth, $C > 0$ denotes a constant independent of the parameter in consideration.

It is known that $v \in H^1_{loc}(\Omega)$ implies $|v| \in H^1_{loc}(\Omega)$ and

$$\nabla |v| = \begin{cases} \nabla v & (v > 0) \\ 0 & (v = 0) \\ -\nabla v & (v < 0). \end{cases} \qquad \text{(a.e.)}$$

Similarly, $u_\pm^t = (u_\pm \wedge t) \vee t^{-1}$ is in $H^1_{loc}(\Omega)$ for $t > 1$ and it holds that

$$\nabla \left((u_\pm^t)^{\alpha+1}\eta^2\right) = (\alpha+1)(u_\pm^t)^\alpha \eta^2 \nabla u_\pm^t + 2(u_\pm^t)^{\alpha+1}\eta\nabla\eta \qquad \text{(a.e.)}$$

for $\alpha \in \mathbf{R}$, where $a \wedge b = \min\{a, b\}$ and $a \vee b = \max\{a, b\}$. From this and $t^{-1} \leq u_{\pm}^t \leq t$, the non-negative function $\varphi = (u_{\pm}^t)^{\alpha+1}\eta^2$ is in $H_0^1(\Omega)$. Plugging it into (5.102), we get that

$$(\alpha+1)\int_\Omega (u_{\pm}^t)^\alpha \eta^2 \nabla u \cdot \nabla u_{\pm}^t \leq 2\int_\Omega |\nabla u|\,(u_{\pm}^t)^{\alpha+1}\,|\nabla \eta|\,\eta.$$

Here, sending $t \to \infty$, we have

$$\int_\Omega (u_{\pm}^t)^\alpha \eta^2 \nabla u \cdot \nabla u_{\pm}^t = \int_{t^{-1}<u_\pm<t} |\nabla u|^2\, u_\pm^\alpha \eta^2 \to \int_\Omega |\nabla u|^2\, u_\pm^\alpha \eta^2,$$

and similarly

$$\int_\Omega |\nabla u|\,(u_{\pm}^t)^{\alpha+1}\,|\nabla \eta|\,\eta \to \int_\Omega |\nabla u|\, u_\pm^{\alpha+1}\,|\nabla \eta|\,\eta$$

by the *monotone convergence theorem*. Therefore, it holds that

$$(\alpha+1)\int_{\Omega_\pm} |\nabla u|^2\,|u|^\alpha\,\eta^2 \leq 2\int_{\Omega_\pm} |\nabla u|\,|u|^{\alpha+1}\,|\nabla \eta|\,\eta$$

for $\Omega_\pm = \{x \in \Omega \mid \pm u(x) \geq 0\}$, and by adding those terms, we get that

$$(\alpha+1)\int_\Omega |\nabla u|^2\,|u|^\alpha\,\eta^2 \leq 2\int_\Omega |\nabla u|\,|u|^{\alpha+1}\,|\nabla \eta|\,\eta. \tag{5.104}$$

Inequality (5.104) coincides with that derived from the formal calculation obtained by putting $\varphi = |u|^\alpha u\eta^2$ in (5.102), and henceforth we omit to write this justification process.

Let $\alpha > -1$ in (5.104). We see that

$$\int_\Omega \left|\nabla\left(|u|^{\frac{\alpha}{2}} u\right)\cdot \eta\right|^2 = \int_\Omega \left|\left(\frac{\alpha}{2}+1\right)|u|^{\frac{\alpha}{2}}\nabla u\right|^2 \eta^2$$

is equal to the $\frac{\left(\frac{\alpha}{2}+1\right)^2}{\alpha+1}$ times the left-hand side, and that

$$\begin{aligned}
&\int_\Omega \left|\nabla\left(|u|^{\frac{\alpha}{2}} u\right)\cdot \eta\right| |u|^{\frac{\alpha+2}{2}}\,|\nabla\eta| \\
&\quad= \int_\Omega \left|\left(\frac{\alpha}{2}+1\right)|u|^{\frac{\alpha}{2}}\nabla u \cdot \eta\right| |u|^{\frac{\alpha+2}{2}}\,|\nabla\eta|
\end{aligned}$$

is equal to the $\frac{\left(\frac{\alpha}{2}+1\right)}{2}$ times the right-hand side. Therefore, we have

$$\int_\Omega \left|\nabla\left(|u|^{\frac{\alpha}{2}} u\right)\cdot \eta\right|^2 \leq C(\alpha)\int_\Omega \left|\nabla\left(|u|^{\frac{\alpha}{2}} u\right)\cdot \eta\right| |u|^{\frac{\alpha+2}{2}}\,|\nabla\eta| \tag{5.105}$$

with a constant $C(\alpha) > 0$ determined by $\alpha > -1$.

In use of

$$ab \leq \frac{a^2 + b^2}{2}$$

valid for $a, b \geq 0$, the right-hand side of (5.105) is estimated from above by

$$\frac{1}{2} \int_\Omega \left| \nabla \left(|u|^{\frac{\alpha}{2}} u \right) \cdot \eta \right|^2 + C'(\alpha) \int_\Omega |u|^{\alpha+2} |\nabla \eta|^2,$$

with another constant $C'(\alpha) > 0$. The first term of this quantity is absorbed by the left-hand side of (5.105), and therefore, it follows that

$$\int_\Omega \left| \nabla \left(|u|^{\frac{\alpha}{2}} u \right) \cdot \eta \right|^2 \leq 2C'(\alpha) \int_\Omega |u|^{\alpha+2} |\nabla \eta|^2.$$

Then we obtain

$$\begin{aligned} \int_\Omega \left| \nabla \left(|u|^{\frac{\alpha}{2}} u\eta \right) \right|^2 &\leq 2 \int_\Omega \left(\left| \nabla \left(|u|^{\frac{\alpha}{2}} u \right) \cdot \eta \right|^2 + |u|^{\alpha+2} |\nabla \eta|^2 \right) \\ &\leq C(\alpha) \int_\Omega |u|^{\alpha+2} |\nabla \eta|^2. \end{aligned} \tag{5.106}$$

Here and henceforth, $C(\alpha) > 0$ denotes a constant determined by $\alpha > -1$, possibly changing from line to line.

We apply Sobolev's inequality (5.101) to the left-hand side of (5.106). Then, we have

$$\left\{ \int_\Omega \left(|u|^{\frac{\alpha}{2}+1} \eta \right)^{\frac{2n}{n-2}} \right\}^{\frac{n-2}{n}} \leq C(\alpha) \int_\Omega |u|^{\alpha+2} |\nabla \eta|^2.$$

Namely, for $\theta = \frac{n}{n-2} > 1$, $\beta = \alpha + 2$, and $\alpha > -1$ it holds that

$$\left\{ \int_\Omega |u|^{\beta\theta} \eta^{2\theta} \right\}^{1/\theta} \leq C(\alpha) \int_\Omega |u|^\beta |\nabla \eta|^2. \tag{5.107}$$

At this moment, it is only required to recognize that $C(\alpha)$ is a rational function of α.

The following fact is referred to as the *local maximum principle.*

Theorem 5.24 *Any $\gamma > 1$ takes $C = C(\gamma) > 0$ that admits*

$$\|u\|_{L^\infty(B_R)} \leq C \left\{ \fint_{B_{2R}} |u|^\gamma \right\}^{1/\gamma} \tag{5.108}$$

for any sub-harmonic function $u \in H^1_{loc}(\Omega)$ *and* $\overline{B(x_0, 2R)} \subset \Omega$, *where* $B_{2R} = B(x_0, 2R)$ *and*

$$\fint_{\Omega} = \frac{1}{|\Omega|} \int_{\Omega}.$$

Here and henceforth, $|G|$ *denotes the volume of the set* G.

Proof. We have from (5.107) and (5.103) that

$$\left\{ \fint_{B_\rho} |u|^{\beta\theta} \right\}^{1/\theta} \leq C(\alpha)(r-\rho)^{-2} \cdot r^n \cdot \rho^{-n/\theta} \fint_{B_r} |u|^{\beta}. \tag{5.109}$$

Remember that $\theta = n/(n-2) > 1$, $\beta = \alpha + 2$, and $\alpha > -1$.

We take $\gamma > 1$ and define β_0, α_0 by $\gamma = \beta_0 = \alpha_0 + 2$, which implies that $\alpha_0 > -1$. Putting

$$\beta_i = \beta_0 \theta^i \qquad \text{and} \qquad R_i = R(1 + 2^{-i}),$$

we apply (5.109) for $\beta = \beta_i$, $r = R_i$, and $\rho = R_{i+1}$, where $i = 0, 1, \cdots$. In fact, it holds that $0 < R_i \leq 2R$ and $\beta_i \uparrow \infty$.

First, we have

$$r - \rho = 2^{-i-1} R \qquad \text{and} \qquad -2 + n - (n/\theta) = 0$$

and hence it follows that

$$\begin{aligned} (r-\rho)^{-2} \cdot r^n \cdot \rho^{-n/\theta} &= 2^{2(i+1)} \cdot (1 + 2^{-i})^n \cdot (1 + 2^{-i-1})^{-n/\theta} \\ &\leq C^{i+1}. \end{aligned}$$

On the other hand, $C(\alpha)$ is a rational function of α and $\alpha_i = \beta_i - 2 \uparrow \infty$, so that we have $m \gg 1$ such that

$$C(\alpha_i) \leq \theta^{(i+1)m}$$

for $i = 0, 1, 2, \cdots$. Therefore, with some $C = C(\gamma) > 1$ it holds that

$$\left\{ \fint_{B_{R_{i+1}}} |u|^{\beta_{i+1}} \right\}^{1/\theta} \leq C^{i+1} \fint_{B_{R_i}} |u|^{\beta_i}$$

for $i = 0, 1, 2, \cdots$. This inequality means

$$\phi_{i+1} \leq C^{(i+1)/\beta_i} \phi_i \qquad \text{for} \qquad \phi_i = \left\{ \fint_{B_{R_i}} |u|^{\beta_i} \right\}^{1/\beta_i},$$

which implies that

$$\phi_i \le C^{\sum_{\ell=0}^{i}(\ell+1)/\beta_\ell}\phi_0 \;\le\; C^{\sum_{\ell=0}^{\infty}(\ell+1)/(\beta_0\theta^\ell)}\phi_0$$
$$= C'\left\{ ⨍_{B_R} |u|^\gamma \right\}^{1/\gamma}.$$

On the other hand, we have $2R > R_i > R$ and $\beta_i \uparrow \infty$, and therefore it follows that

$$\begin{aligned}\phi_i &= \left\{ \frac{1}{|B_{R_i}|}\int_{B_{R_i}} |u|^{\beta_i} \right\}^{1/\beta_i} \\ &\ge \left\{ \frac{1}{|B_{2R}|}\int_{B_R} |u|^{\beta_i} \right\}^{1/\beta_i} \to \|u\|_{L^\infty(B_R)}.\end{aligned}$$

The proof is complete. □

The super-harmonicity of $u \in H^1_{loc}(\Omega)$ is defined similarly by

$$\int_\Omega \nabla u \cdot \nabla \varphi \ge 0 \tag{5.110}$$

for any non-negative function $\varphi \in C_0^\infty(\Omega)$. We can show an analogous fact to the weak minimum principle to non-negative super-harmonic functions by a more delicate argument.

For the moment, $u \in H^1_{loc}(\Omega)$ denotes a non-negative super-harmonic function and $\overline{B_{4R}} \subset \Omega$. Given $0 < \rho < r \le 4R$, we take the non-negative function $\eta = \eta(|x|) \in C_0^\infty(B_r)$ satisfying (5.103). Then, the reverse inequality

$$(\alpha+1)\int_\Omega |\nabla u|^2 u^\alpha \eta^2 \ge -2\int_\Omega |\nabla u|\, u^{\alpha+1}\, |\nabla\eta|\, \eta$$

to (5.104) is obtained by substituting $\varphi = u^{\alpha+1}\eta^2$ into (5.110). In particular, for $\alpha < -1$ it holds that

$$\int_\Omega |\nabla u|^2 u^\alpha \eta^2 \le \frac{2}{-(\alpha+1)}\int_\Omega |\nabla u|\, u^{\alpha+1}\, |\nabla\eta|\, \eta \tag{5.111}$$

and hence

$$\int_\Omega |\nabla u|^2 u^\alpha \eta^2 \le \frac{2}{-(\alpha+1)}\int_\Omega |\nabla u|\, u^{\alpha+1}\, |\nabla\eta|\, \eta$$

follows.

Repeating the previous argument, we obtain an analogous result to (5.107),

$$\left\{\int_\Omega u^{\beta\theta}\eta 2\theta\right\}^{1/\theta} \le C(\alpha)\int_\Omega u^\beta |\nabla\eta|^2,$$

where $\theta = n/(n-2) > 1$, $\beta = \alpha + 2$, and $\alpha < -1$, which implies that

$$\left\{⨍_{B_\rho} u^{\beta\theta}\right\}^{1/\theta} \le C(\alpha)(r-\rho)^{-2} r^n \rho^{-n/\theta} ⨍_{B_r} u^\beta. \tag{5.112}$$

Taking $\gamma < 0$, we define β_0 and α_0 by $\gamma = \beta_0 = \alpha_0 + 2$, and set

$$\beta_i = \beta_0\theta^i \qquad \text{and} \qquad R_i = R(1 + 3\cdot 2^{-i})$$

for $i = 0, 1, 2, \cdots$. Then, (5.112) is applicable to $\beta = \beta_i$, $r = R_i$ and $\rho = R_{i+1}$, as $\alpha_0 < -1$, $\beta_0 < 0$, $\beta_i \downarrow -\infty$, and $0 < R_i \le 4R$ hold. We have

$$\left\{⨍_{B_{R_{i+1}}} u^{\beta_{i+1}}\right\}^{1/\theta} \le C^{i+1} ⨍_{B_{R_i}} u^{\beta_i}$$

for $i = 0, 1, 2, \cdots$, where $C = C(\gamma) > 1$ is a constant.

This means for

$$\phi_i = \left\{⨍_{B_{R_i}} u^{\beta_i}\right\}^{-1/\beta_i} = \left\{⨍_{B_{R_i}} (u^{-1})^{-\beta_i}\right\}^{1/(-\beta_i)}$$

that

$$\phi_{i+1} \le C^{(i+1)/(-\beta_i)}\phi_i$$

and therefore, it holds that

$$\phi_i \le C'\phi_0 = C'\left\{⨍_{B_{4R}} u^\gamma\right\}^{-1/\gamma}.$$

On the other hand, we have

$$\phi_i \to \left\|u^{-1}\right\|_{L^\infty(B_R)} = \left(\operatorname{ess.}\inf_{B_R} u\right)^{-1}$$

and hence the following lemma is obtained with $\sigma = -\gamma > 0$.

Lemma 5.9 *Given $\sigma > 0$, we have $C = C(\sigma) > 0$ such that*

$$C \operatorname{ess.inf}_{B_R} u \geq \left\{ \fint_{4R} u^{-\sigma} \right\}^{-1/\sigma} \tag{5.113}$$

for any non-negative super-harmonic function $u \in H^1_{loc}(\Omega)$ and $\overline{B_{4R}} \subset \Omega$.

In (5.112), β may be negative as far as $\beta < 1$. In particular, for $\beta \in (0,1)$ and $2 \leq \ell < m \leq 4$ we have $C(\beta, \ell, m) > 0$ such that

$$\left\{ \fint_{B_{\ell R}} u^{\beta\theta} \right\}^{1/(\beta\theta)} \leq C(\beta, \ell, m) \left\{ \fint_{B_{mR}} u^{\beta} \right\}^{1/\beta}.$$

Here, the quantity

$$\left\{ \fint_{B_{4R}} u^{\beta} \right\}^{1/\beta}$$

is monotone increasing in $\beta > 0$. We make use of this fact and the above inequality by finitely many times. Then, for given $\gamma \in (0,1)$ and $\sigma > 0$, we have $C = C(n, \gamma, \sigma) > 0$ such that

$$\left\{ \fint_{B_{2R}} u^{\gamma\theta} \right\}^{1/(\gamma\theta)} \leq C \left\{ \fint_{B_{4R}} u^{\sigma} \right\}^{1/\sigma}. \tag{5.114}$$

Here is the key lemma.

Lemma 5.10 *If $\Omega \subset \mathbf{R}^n$ with $n > 2$, then there is $C = C(n) > 0$ and $\sigma_0 = \sigma_0(n) > 0$ such that*

$$\left\{ \fint_{B_{4R}} u^{\sigma} \right\}^{1/\sigma} \cdot \left\{ \fint_{B_{4R}} u^{-\sigma} \right\}^{1/\sigma} \leq C \tag{5.115}$$

is satisfied for any non-negative super-harmonic function $u \in H^1_{loc}(\Omega)$, $\overline{B_{8R}} \subset \Omega$, and $\sigma > \sigma_0$.

Proof. Let us take $\alpha = -2$ in (5.111). In fact, we have

$$\begin{aligned}
\int_\Omega |\nabla \log u|^2 \eta^2 &= \int_\Omega |\nabla u|^2 u^{-2} \eta \leq 2 \int_\Omega |\nabla u|\, u^{-1} |\nabla \eta|\, \eta \\
&= 2 \int_\Omega |\nabla \log u|\, \eta\, |\nabla \eta| \\
&\leq \frac{1}{2} \int_\Omega |\nabla \log u|^2 \eta^2 + 2 \int_\Omega |\nabla \eta|^2,
\end{aligned}$$

and hence

$$\int_\Omega |\nabla \log u|^2 \eta^2 \leq C \int_\Omega |\nabla \eta|^2$$

follows. This implies

$$\int_{B_\rho} |\nabla \log u|^2 \leq C(r-\rho)^{-2} |B_r|$$

by (5.103), and taking $r = 2\rho$, we obtain

$$\fint_{B_\rho} |\nabla \log u|^2 \leq C\rho^{-2} \tag{5.116}$$

for $0 < \rho < R$.

We now make use of *Poincaré-Sobolev's inequality*,

$$\left\{\fint_{B_r} |v - v_{B_r}|^{2^*}\right\}^{1/2^*} \leq Cr \left\{\fint_{B_r} |\nabla v|^2\right\}^{1/2} \tag{5.117}$$

valid for $v \in H^1(B_r)$, where $2^* = 2n/(n-2)$ and

$$v_{B_r} = \fint_{B_r} v.$$

This will be proven in the following paragraph. Admitting it, we can estimate the left-hand side of (5.117) from below by

$$\fint_{B_r} |v - v_{B_r}|.$$

We apply this form to $v = \log u$ and get from (5.116) that

$$\fint_{B_\rho} |\log u - (\log u)_{B_\rho}| \leq C\rho \left\{\fint_{B_\rho} |\nabla \log u|^2\right\}^{1/2} \leq C. \tag{5.118}$$

Now, we get the notion of the function of *bounded mean oscillation* or BMO in short, namely, a measurable function $v = v(x)$ defined on a domain $\Omega \subset \mathbf{R}^n$ is said to be in BMO if

$$\|v\|_{BMO} = \sup \left\{ \fint_B |v - v_B| \mid B : \text{ball}, \ \overline{B} \subset \Omega \right\} < +\infty, \tag{5.119}$$

where

$$v_B = \fint_B v.$$

Thus, we have

$$\|\log u\|_{BMO} \leq C \tag{5.120}$$

by (5.118).

A typical unbounded BMO function is $\log|x|$. In this connection, we have *John-Nirenberg's inequality*, proven in the next paragraph, indicated as follows:

$$|\{x \in B \mid |v(x) - v_B| > t\}| \leq c_1 |B| \exp\left(-c_2 t / \|v\|_{BMO}\right), \tag{5.121}$$

where $c_1 = c_1(n) > 0$ and $c_2 = c_2(n) > 0$ are the constants determined by the dimension n, $t > 0$, and B is a ball with $2B$, the concentric ball with twice radius, satisfying $2B \subset \Omega$.

We apply this inequality to the BMO function $v = \log u$ and $B = B_{4R}$, putting

$$\mu(t) = |\{x \in B \mid |v(x) - v_B| > t\}|.$$

In fact, for $s \in (0, c_2)$, say $s = c_2/2$, we have

$$\begin{aligned}
&\fint_B \exp\left(s\,|v(x) - v_B| / \|v\|_{BMO}\right) \\
&= \frac{1}{|B|} \int_0^\infty \exp\left(st/\|v\|_{BMO}\right) d\left(-\mu(t)\right) \\
&= \frac{1}{|B|} \left[-\exp\left(st/\|v\|_{BMO}\right) \mu(t)\right]_{t=0}^{t=\infty} \\
&\qquad + \frac{s}{\|v\|_{BMO} |B|} \int_0^\infty \exp\left(st/\|v\|_{BMO}\right) \mu(t) dt \qquad (5.122) \\
&\leq 1 + \frac{c_1 s}{\|v\|_{BMO}} \int_0^\infty \exp\left(-(c_2 - s)t/\|v\|_{BMO}\right) dt \\
&= 1 + c_1 s \left[-\frac{1}{c_2 - s} \exp\left(-(c_2 - s)t/\|v\|_{BMO}\right)\right]_{t=0}^{t=\infty} \\
&= 1 + \frac{c_1 s}{c_2 - s} \equiv \beta. \qquad (5.123)
\end{aligned}$$

Therefore, for $\sigma \leq s / \|v\|_{BMO}$ we have

$$\fint_B \exp\left(\pm\sigma\left(v(x) - v_B\right)\right) \leq \beta,$$

which implies that

$$\fint_B e^{\sigma v} \cdot \fint_B e^{-\sigma v} \leq \beta^2.$$

We have, by $u = \log v$ that

$$\left\{\fint_B u^{\sigma}\right\}^{1/\sigma} \cdot \left\{\fint_B u^{-\sigma}\right\}^{1/\sigma} \leq \beta^{2\|\log u\|_{BMO}/s} \leq C.$$

This means (5.115). By (5.120) we can take $\sigma > \sigma_0$ with $\sigma_0 = \sigma_0(n) > 0$ and the proof is complete. □

The *local minimum principle* is now obtained from (5.113), (5.114), and (5.115).

Theorem 5.25 *Any $\gamma \in (0, n/(n-2))$ takes $C = C(n, \gamma) > 0$ that admits*

$$\left\{\fint_{B_{2R}} u^{\gamma}\right\}^{1/\gamma} \leq C \operatorname{ess.}\inf_{B_R} u \tag{5.124}$$

for any non-negative super-harmonic function $u \in H^1_{loc}(\Omega)$ and $\overline{B_{8R}} \subset \Omega$.

Because the local maximum and the minimum principles are commonly valid to $\gamma \in (1, n/(n-2))$, we obtain the *Harnack inequality*.

Theorem 5.26 *If $\Omega \subset \mathbf{R}^n$ is a domain with $n > 2$, then there is a constant $C = C(n) > 0$ determined by n that admits the estimate*

$$\operatorname{ess.}\sup_{B_R} u \leq C \operatorname{ess.}\inf_{B_R} u \tag{5.125}$$

for any non-negative harmonic function $u \in H^1_{loc}(\Omega)$ and any $\overline{B_{8R}} \subset \Omega$.

It is now well recognized that this type of inequality implies the Hölder continuity of the solution.

Theorem 5.27 *If $\Omega \subset \mathbf{R}^n$ is a domain with $n > 2$, then there is $\alpha \in (0, 1)$ such that any compact set $E \subset \Omega$ admits $C > 0$ such that*

$$|u(x) - u(y)| \leq C |x - y|^{\alpha}$$

holds for any harmonic function $u \in H^1_{loc}(\Omega)$ and $x, y \in E$.

Proof. For simplicity, we shall write inf, sup for ess. inf and ess. sup, respectively. Let $u \in H^1_{loc}(\Omega)$ be a non-negative harmonic function, and take $x_0 \in \Omega$, $0 < \rho < \mathrm{dist}(x_0, \partial\Omega)/16$, $0 < r < 16\rho$,

$$M_r = \sup_{B(x_0,r)} u, \qquad \text{and} \qquad m_r = \inf_{B(x_0,r)} u.$$

Here, it may be worth noting that $u \in L^\infty_{loc}(\Omega)$ follows from the local maximum principle.

Because $u(x) - m_{16\rho}$ is non-negative and harmonic in $B(x_0, 8\rho)$, it follows from (5.125) that

$$\begin{aligned}\sup_{B(x_0,\rho)} \{u - m_{16\rho}\} &= M_\rho - m_{16\rho} \\ &\le C \inf_{B(x_0,\rho)} \{u - m_{16\rho}\} = C\,(m_\rho - m_{16\rho}).\end{aligned}$$

The function $M_{16\rho} - u(x)$ has the same property and it follows that

$$M_{16\rho} - m_\rho \le C\,(M_{16\rho} - M_\rho).$$

Adding those two inequalities, we obtain

$$(C+1)\,(M_\rho - m_\rho) \le (C-1)\,(M_{16\rho} - m_{16\rho}),$$

and therefore, for $\omega(\rho) = M_\rho - m_\rho$ and $0 < \theta = (C-1)/(C+1) < 1$ it holds that

$$\omega(\rho) \le \theta\omega(16\rho).$$

Repeating this inequality, we have

$$\omega(\rho \cdot 16^{-i+1}) \le \theta^i \omega(16\rho)$$

for $i = 1, 2, \cdots$.

Taking α by $\theta = 16^{-\alpha}$, we have, for $i = 1, 2, \cdots$ and

$$0 < \rho < \mathrm{dist}(x_0, \partial\Omega)/16$$

that

$$\omega(\rho \cdot 16^{-i+1}) \le (16^{-i})^\alpha \omega(16\rho),$$

which implies, for $0 < r_1 < r_2 < \mathrm{dist}(x_0, \partial\Omega)/16$ that

$$\omega(r_1) \le \theta^{-1} \left(\frac{r_1}{r_2}\right)^\alpha \omega(r_2).$$

Taking $r_2 = \rho$ and setting $r = r_1$, we obtain for $0 < r < \rho$ and $x \in \partial B(x_0, r)$ that

$$|u(x) - u(x_0)| \leq \omega(r) \leq \theta^{-1}\rho^{-\alpha}\omega(\rho) \cdot |x - x_0|^\alpha .$$

In other words, for $C = \theta^{-1}\rho^{-\alpha}\omega(\rho)$, $x \in B(x_0, \rho)$, and

$$0 < \rho < \frac{\operatorname{dist}(x_0, \partial\Omega)}{16}$$

that

$$|u(x) - u(x_0)| \leq C\,|x - x_0|^\alpha .$$

Then, the conclusion follows from the standard covering argument, and the proof is complete. □

5.4.5 *BMO Estimate*

This section is devoted to the proof of (5.117) and (5.121). First, we show (5.117), that is, *Poincaré-Sobolev's inequality.* Actually, it suffices to prove the following.

Theorem 5.28 *There exists $C = C(n,p) > 0$ determined by $n > 2$ and $p \in [1, n)$ that admits the estimate*

$$\left\{ ⨍_{B(x,r)} |v - v_{B(x,r)}|^{p^*} \right\}^{1/p^*} \leq Cr \left\{ ⨍_{B(x,r)} |\nabla v|^p \right\}^{1/p} \tag{5.126}$$

holds true for $v \in W^{1,p}(\mathbf{R}^n)$, where $p^* = np/(n-p)$.

Proof. Let $n > 2$ and $1 \leq p < n$. First, we show that there is a constant $C = C(n,p) > 0$ that admits the estimate

$$\int_{B(x,r)} |v(y) - v(z)|^p \, dy \leq Cr^{n+p-1} \int_{B(x,r)} |\nabla v(y)|^p \, |y - z|^{1-n} \, dy \tag{5.127}$$

for $z \in B(x,r) \subset \mathbf{R}^n$ and $v \in C^1(\overline{B(x,r)})$.

In fact, we have from

$$\begin{aligned} v(y) - v(z) &= \int_0^1 \frac{d}{dt} v(z + t(y-z))dt \\ &= \int_0^1 \nabla v(z + t(y-z))dt \cdot (y - z) \end{aligned}$$

that

$$|v(y) - v(z)|^p \le |y - z|^p \int_0^1 |\nabla v(z + t(y - z))|^p \, dt.$$

Therefore, for $s > 0$ we obtain

$$\int_{B(x,r)\cap\partial B(z,s)} |v(y) - v(z)|^p \, dS_y$$

$$\le s^p \int_0^1 dt \int_{B(x,r)\cap\partial B(z,s)} |\nabla v(z + t(y - z))|^p \, dS_y. \qquad (5.128)$$

Here, in terms of $w = z + t(y - z)$ it holds that

$$dS_y = t^{1-n} dS_w.$$

Also, $y \in B(x,r) \cap \partial B(z,s)$ implies $|w - z| = ts$ and hence the right-hand side of (5.128) is estimated from above by

$$s^p \int_0^1 \frac{dt}{t^{n-1}} \int_{B(x,r)\cap\partial B(z,ts)} |\nabla v(w)|^p \, dS_w$$

$$= s^p \int_0^1 \frac{dt}{t^{n-1}} \int_{B(x,r)\cap\partial B(z,ts)} |\nabla v(w)|^2 \, |w - z|^{1-n} \, dS_w \cdot (ts)^{-1+n}$$

$$= s^{n+p-1} \int_0^1 dt \int_{B(x,r)\cap\partial B(z,ts)} |\nabla v(w)|^p \, |w - z|^{1-n} \, dS_w. \qquad (5.129)$$

Generally, for the measurable function $g = g(w) : \mathbf{R}^n \to [0, \infty]$ and $z \in \mathbf{R}^n$ it holds that

$$\int_{\mathbf{R}^n} g dw = \int_0^\infty d\rho \int_{\partial B(z,\rho)} g dS_w.$$

Therefore, applying the transformation $ts = \rho$, we have

$$\int_0^1 dt \int_{B(x,r)\cap\partial B(z,ts)} g dS_w = s^{-1} \int_0^s d\rho \int_{\partial B(z,\rho)\cap B(x,r)} g dS_w$$

$$= s^{-1} \int_0^\infty \chi_{(0,s)}(\rho) d\rho \int_{\partial B(z,\rho)} \chi_{B(x,r)} g dS_w$$

$$= s^{-1} \int_{B(z,s)} \chi_{B(x,r)} g dw = s^{-1} \int_{B(z,s)\cap B(x,r)} g dw,$$

where χ_F denotes the characteristic function of F. Therefore, the right-hand side of (5.129) is equal to

$$s^{n+p-2}\int_{B(x,r)\cap B(z,s)} |\nabla v(w)|^p \, |w-z|^{1-n}\, dw,$$

and hence it indicates that

$$\int_{B(x,r)\cap\partial B(z,s)} |v(y)-v(z)|^p\, dS_y \leq s^{n+p-2}\int_{B(x,r)\cap B(z,s)} |\nabla v(w)|^p\, |w-z|^{1-n}\, dw.$$

This implies

$$\begin{aligned}
&\int_0^{2r} ds \int_{B(x,r)\cap\partial B(z,s)} |v(y)-v(z)|^p\, dS_y \\
&= \int_{B(x,r)} |v(y)-v(z)|^p\, dy \\
&\leq \int_0^{2r} s^{n+p-2} ds \int_{B(x,r)\cap B(z,s)} |\nabla v(w)|^p\, |w-z|^{1-n}\, dw \\
&\leq \int_0^{2r} s^{n+p-2} ds \int_{B(x,r)} |\nabla v(w)|^p\, |w-z|^{1-n}\, dw \\
&= Cr^{n+p-1}\int_{B(x,r)} |\nabla v(w)|^p\, |w-z|^{1-n}\, dw
\end{aligned}$$

and inequality (5.127) has been proven.

We turn to the proof of (5.126). Actually, we may assume that $v \in C^1(B(x,r))$. A variant of Sobolev's imbedding Theorem 5.23 is indicated as

$$\|v\|_{p^*} \leq C\,\|v\|_{W^{1,p}(\Omega)}$$

with the constant $C = C(n,p,\Omega) > 0$ independent of $v \in W^{1,p}(\Omega)$, if $\Omega \subset \mathbf{R}^n$ is a bounded domain and $\partial\Omega$ is C^1. Actually, it is reduced to (5.101) in use of the *extension operator*.

In use of this to $\Omega = B(0,1)$, we have a constant $C_1 > 0$ that admits the estimate

$$\left\{\int_{B(0,1)} |g(x)|^{p^*}\, dx\right\}^{1/p^*} \leq C_1\left\{\int_{B(0,1)} \left(|\nabla g(x)|^p + |g(x)|^p\right) dx\right\}^{1/p}$$

for any $g \in W^{1,p}(B(0,1))$. Here, we take the scaling transformation $y = rx$ and $f = rg$. Then, we get that

$$\begin{aligned}\left\{\int_{B(0,1)} |g(x)|^{p^*}\,dx\right\}^{1/p^*} &= \left\{\int_{B(0,r)} r^{-p^*}\,|f(y)|^{p^*}\,r^{-n}dy\right\}^{1/p^*}\\ &= Cr^{-1}\left\{⨍_{B(0,r)} |f|^{p^*}\right\}^{1/p^*}\end{aligned}$$

and

$$\begin{aligned}&\left\{\int_{B(0,1)} (|\nabla g(x)|^p + |g(x)|^p)\,dx\right\}^{1/p}\\ &= \left\{\int_{B(0,r)} (r^p\cdot r^{-p}\,|\nabla f(y)|^p + r^{-p}\,|f(y)|^p)\,r^{-n}dy\right\}^{1/p}\\ &= C\left\{⨍_{B(0,r)} (|\nabla f|^p + r^{-p}\,|f|^p)\right\}^{1/p}.\end{aligned}$$

Thus, we get for $f \in W^{1,p}(B(x,r))$ that

$$\left\{⨍_{B(x,r)} |f|^{p^*}\right\}^{1/p^*} \le C\left\{⨍_{B(x,r)} (r^p\,|\nabla f|^p + |f|^p)\right\}^{1/p}. \tag{5.130}$$

On the other hand, from (5.127) we have for $v \in C^1(\overline{B(x,r)})$ that

$$\begin{aligned}⨍_{B(x,r)} |v - v_{B(x,r)}|^p &= ⨍_{B(x,r)} \left|⨍_{B(x,r)} (v(y) - v(z))\,dz\right|^p dy\\ &\le ⨍_{B(x,r)} ⨍_{B(x,r)} |v(y) - v(z)|^p\,dydz\\ &\le C⨍_{B(x,r)} r^{p-1}dz \int_{B(x,r)} |\nabla v(y)|^p\,|y - z|^{1-n}\,dy\\ &= C\int_{B(x,r)} |\nabla v(y)|^p\,dy\cdot r^{p-1}⨍_{B(x,r)} |y - z|^{1-n}\,dz.\end{aligned} \tag{5.131}$$

If $y, z \in B(x,r)$ then $z \in B(y,2r)$, so that the right-hand side of (5.131) is estimated from above by

$$C\int_{B(x,r)} |\nabla v(y)|^p \, dy \cdot r^{p-1} \cdot \frac{1}{r^n} \int_{B(y,2r)} |y-z|^{1-n} \, dz.$$

Thus, we get that

$$⨍_{B(x,r)} \left| v - v_{B(x,r)} \right|^p \le Cr^p ⨍_{B(x,r)} |\nabla v(y)|^p \, dy. \tag{5.132}$$

From inequalities (5.132) and (5.130) with $f = v - v_{B(x,r)}$, we obtain

$$\begin{aligned} &\left\{ ⨍_{B(x,r)} \left| v - v_{B(x,r)} \right|^{p^*} \right\}^{1/p^*} \\ &\le C \left\{ ⨍_{B(x,r)} \left(r^p |\nabla v|^p + \left| v - v_{B(x,r)} \right|^p \right) \right\}^{1/p} \\ &\le C \left\{ ⨍_{B(x,r)} r^p |\nabla v|^p \right\}^{1/p} = Cr \left\{ ⨍_{B(x,r)} |\nabla v|^p \right\}^{1/p}, \end{aligned}$$

and the proof of (5.117) is complete. □

Let us note that inequality (5.121) is equivalent to saying that v is a BMO function. Actually, if

$$\begin{aligned} \mu(t) &= |\{x \in B \mid |v(x) - v_B| > t\}| \\ &\le c_1 |B| \exp(-\kappa t) \end{aligned}$$

holds for a ball B in $\overline{B} \subset \Omega$ and a constant $\kappa > 0$, then we can derive

$$⨍_B e^{\frac{\kappa}{2}|v - v_B|} \le \beta' = \beta'(n)$$

similarly to (5.123). Then, it follows from *Jensen's inequality* that

$$⨍_B |v - v_B| \le \frac{2}{\kappa} ⨍_B e^{\frac{\kappa}{2}|v - v_B|} \le 2\beta'/\kappa.$$

In the original definition of the BMO function, the ball B in condition (5.119) is taken place by the cubic, denoted by Q. We note that a cube can be divided into smaller cubes with the intersection of the Lebesgue measure 0. We shall apply this *dyadic sub-division* by means of the following.

Lemma 5.11 *There exists a constant $c_0(n)$ determined by the dimension n such that if Q is a cubic and B is the minimal ball containing Q such that $\overline{B} \subset \Omega$, then it holds that*

$$\fint_Q |v - v_Q| \le c_0(n) \, \|v\|_{BMO} \,. \tag{5.133}$$

Proof. We have a constant $c_1(n) > 0$ determined by n such that

$$|B| \le c_1(n) \, |Q| \,.$$

On the other hand, we have

$$\frac{1}{|Q|} \int_Q |v - v_Q| \le \frac{1}{|Q|} \int_Q |v - v_B| + |v_B - v_Q|$$

and

$$|v_B - v_Q| = |(v - v_B)_Q| \le \frac{1}{|Q|} \int_Q |v - v_B| \,.$$

Thus, we obtain (5.133) by

$$\begin{aligned} \frac{1}{|Q|} \int_Q |v - v_Q| &\le \frac{2}{|Q|} \int_Q |v - v_B| \\ &\le \frac{2c_1(n)}{|B|} \int_B |v - v_B| \le 2c_1(n) \, \|v\|_{BMO} \,. \end{aligned}$$

The proof is complete. □

Now, we show the *decomposition theorem of Calderón-Zygmund.*

Theorem 5.29 *If $Q_0 \subset \mathbf{R}^n$ is a cube, $v \in L^1(Q_0)$, and*

$$\fint_{Q_0} |v| \le s,$$

then there is a countable family of disjoint sub-cubes denoted by $\{Q_k\}_{k=1}^{\infty}$ such that

$$|v| \le s \qquad a.e. \ in \quad Q_0 \setminus \cup_{k=1}^{\infty} Q_k \tag{5.134}$$

$$|v_{Q_k}| \le 2^n s \qquad (k = 1, 2, \cdots) \tag{5.135}$$

$$\sum_{k=1}^{\infty} |Q_k| \le s^{-1} \int_{Q_0} |v| \,. \tag{5.136}$$

Proof. We take the sub-division of Q_0 uniformly by 2^n sub-cubes, and classify them into two types, that is, the one on which the mean of $|v|$ is greater than or equal to s and that less than s. Let the former and the latter be $\{Q_{1k}\}$ and $\{Q'_{1k}\}$, respectively. As for Q_{1k} we have

$$\begin{aligned} s\,|Q_{1k}| &\leq \int_{Q_{1k}} |v| \leq 2^n\,|Q_{1k}| \fint_{Q_0} |v| \\ &\leq 2^n s\,|Q_{1k}|\,. \end{aligned}$$

Now, we take the sub-division of $\{Q'_{1k}\}$ uniformly by 2^n sub-cubes, and classify them into the ones on which the mean of $|v|$ is greater than or equal to s, denoted by $\{Q_{2k}\}$, and the others, denoted by $\{Q'_{2k}\}$. Because Q_{2k} is contained in some $Q'_{1k'}$, it holds that

$$\begin{aligned} s\,|Q_{2k}| &\leq \int_{Q_{2k}} |v| \leq 2^n\,|Q_{2k}| \fint_{Q'_{1k'}} |v| \\ &\leq 2^n s\,|Q_{2k}|\,. \end{aligned}$$

Continuing this process, we get a family of sub-cubes $\{Q_{mk}\}$. Let us label it as $\{Q_k\}$. Then, it holds that

$$s\,|Q_k| \leq \int_{Q_k} |v| \leq 2^n s\,|Q_k|\,.$$

This means (5.135). It also implies

$$s \sum_{k=1}^{\infty} |Q_k| \leq \int_{\cup_k Q_k} |v| \leq \int_{Q_0} |v|$$

and hence (5.136) follows.

Finally, if $x_0 \notin \cup_k Q_k$, then there is a shrinking family $\{Q'_k\}$ such that $|Q'_{k+1}| = 2^n\,|Q'_k|$, $x_0 \in Q'_k$, and

$$\fint_{Q'_k} |v| < s.$$

Then, the *differentiation theorem of Lebesgue* guarantees that

$$\lim_{k\to\infty} \fint_{Q'_k} |v| = |v(x_0)| \qquad \text{a.e. } x_0 \notin \cup_k Q_k$$

and the proof is complete. □

In view of (5.133), we that (5.121) is almost equivalent to the following.

Lemma 5.12 *There are the constants $c_3(n) > 0$ and $c_4(n) > 0$ determined by n such that if Q is a cubic, B is the minimum ball containing Q in $\overline{B} \subset \Omega$, and $t > 0$, then it holds that*

$$|\{x \in Q \mid |v(x) - v_Q| > t\}| \leq c_3 |Q| \exp\left(-c_4 t / \|v\|_{BMO}\right). \tag{5.137}$$

Proof. Given $v = v(x)$, we take $a > 0$ satisfying

$$v(x) = aw(x) \qquad \text{and} \qquad \|w\|_{BMO} = \frac{1}{c_0(n)}.$$

Then, it follows that

$$\{x \in Q \mid |v(x) - v_Q| > t\} = \{x \in Q \mid |w(x) - w_Q| > t/|a|\}$$

and

$$\|v\|_{BMO} = |a| \, \|w\|_{BMO},$$

and therefore, inequality (5.137) for w implies that for v. Namely, we can assume

$$\|v\|_{BMO} = \frac{1}{c_0(n)} \tag{5.138}$$

from the beginning.

Let Q be a cubic and $B(Q)$ the minimum ball containing Q. We put for $t > 0$ that

$$S_Q(t) = \{x \in Q \mid |v(x) - v_Q| > t\}$$

and

$$F(t) = \inf\left\{C > 0 \mid |S_Q(t)| \leq C \int_Q |v - v_Q|, \ Q : \text{cubic}, \ \overline{B(Q)} \subset \Omega\right\}.$$

It follows from

$$|S_Q(t)| \leq \frac{1}{t} \int_Q |v - v_Q|$$

that

$$F(t) \leq 1/t. \tag{5.139}$$

Let $t \geq 2^n$ and $s \in [1, 2^{-n}t]$. From (5.138) we have

$$\fint_{Q_0} |v - v_Q| \leq 1 \leq s.$$

We apply Theorem 5.29 with Q_0, $v(x)$ replaced by Q, $v(x) - v_Q$, respectively, and get the family of cubics $\{Q_k\}$. Let

$$S_k(t) = S_{Q_k}(t) \qquad \text{and} \qquad S_0(t) = S(t).$$

We have, from $t \geq s$ and (5.138) that

$$S(t) \subset S(s) \subset \cup_{k=1}^{\infty} Q_k$$

except for a set of Lebesgue measure 0. On the other hand, by (5.135) we have

$$|v(x) - v_{Q_k}| = \left|(v(x) - v_Q) - (v(x) - v_Q)_{Q_k}\right| > t - 2^n s$$

for $x \in S(t) \cap Q_k$. Those relations imply

$$|S(t)| \leq \sum_{k=1}^{\infty} |S(t) \cap Q_k| \leq \sum_{k=1}^{\infty} |\{x \in Q_k \mid |v(x) - v_{Q_k}| > t - 2^n s\}| .$$

Here, from the definition of $F(t)$ and (5.138) we have

$$\begin{aligned} &|\{x \in Q_k \mid |v(x) - v_{Q_k}| \geq t - 2^n s\}| \\ &\quad \leq F(t - 2^n s) \int_{Q_k} |v - v_{Q_k}| \leq F(t - 2^n) |Q_k| . \end{aligned}$$

Therefore, it follows from (5.136) that

$$\begin{aligned} |S_Q(t)| = |S(t)| &\leq F(t - 2^n s) \sum_{k=1}^{\infty} |Q_k| \\ &\leq s^{-1} F(t - 2^n s) \int_Q |v - v_Q| , \end{aligned}$$

and because Q is arbitrary, we have

$$F(t) \leq s^{-1} F(t - 2^n s) \qquad (1 \leq s \leq 2^{-n} t,\ t \geq 2^n) . \tag{5.140}$$

Taking $s = e$ in (5.140), we see that if $F(t) \leq A e^{-\alpha t}$ for $\alpha = 1/(2^n e)$, then

$$F(t + 2^n e) \leq \frac{1}{e} A e^{-\alpha t} = A e^{-\alpha(t + 2^n e)}$$

follows. On the other hand, we have

$$F(t) \leq (e-1)e^{\frac{1}{e-1}} \cdot 2^{-n} \cdot e^{-\alpha t} \tag{5.141}$$

for $\frac{2^n e}{e-1} \leq t \leq \frac{2^n e}{e-1} + 2^n e$ by (5.139). Therefore, this inequality (5.141) continues to hold for $t \geq 2^n e/(e-1)$ and hence we obtain

$$|S_0(t)| \leq Ae^{-\alpha t} \fint_{Q_0} |v - v_{Q_0}| \leq Ae^{-\alpha t}|Q|$$

for $t \geq t_0$, where

$$t_0 = \frac{2^n e}{e-1}, \qquad \alpha = \frac{1}{2^n e}, \qquad A = (e-1)e^{\frac{1}{e-1}}.$$

On the other hand, it is obvious that

$$|S_0(t)| \leq |Q_0| \leq e^{\alpha t_0} \cdot e^{-\alpha t}|Q|$$

holds for $0 < t < t_0$, and therefore, (5.137) follows for

$$c_3 = \max\{A, e^{\alpha t_0}\} \qquad \text{and} \qquad c_4 = \alpha.$$

The proof is complete. □

Now, we complete the proof of John-Nirenberg's inequality.

Theorem 5.30 *There are constants $c_1(n) > 0$ and $c_2(n)$ determined by the dimension n such that if $\overline{2B} \subset \Omega$ and $t > 0$, then it holds that (5.121).*

Proof. Let Q be the minimum cubic containing B. Then, from the assumption, we have $\overline{Q} \subset \Omega$. In use of

$$\{x \in B \mid |v(x) - v_B| > t\} \subset \{x \in Q \mid |v(x) - v_Q| > t - |v_B - v_Q|\}$$

and Lemma 5.12, we obtain

$$\begin{aligned} &|\{x \in B \mid |v(x) - v_B| > t\}| \\ &\leq c_3 |Q| \exp\left(-c_4 \max\{0, t - |v_B - v_Q|\} / \|v\|_{BMO}\right) \\ &\leq c_3 |Q| \exp\left(-c_4 t / \|v\|_{BMO}\right) \cdot \exp\left(c_4 |v_B - v_Q| / \|v\|_{BMO}\right). \end{aligned}$$

On the other hand, from Lemma 5.11 we have

$$\begin{aligned} |v_B - v_Q| &= |(v - v_Q)_B| \leq \frac{1}{|B|}\int_B |v - v_Q| \\ &\leq c_1(n) \fint_Q |v - v_Q| \leq c_1(n) c_0(n) \|v\|_{BMO} \end{aligned}$$

and then (5.121) follows. □

Chapter 6

Nonlinear PDE Theory

Although nonlinear partial differential equations have vast varieties, they share several common features and techniques. This chapter is devoted to the non-negative solution to semilinear heat equation $u_t = \Delta u + u^p$ on the whole space $\mathbf{R}^n$. If the nonlinearity is strong as $p > 1 + \frac{2}{n}$, then small initial data admits the solution globally in time. On the contrary, if it is weak as $1 < p < 1 + \frac{2}{n}$, then any non-trivial initial data make the solution to continue to $t = +\infty$ impossible. This phenomenon was noticed by H. Fujita in 1966, and $p_* = 1 + \frac{2}{n}$ is called *Fujita's critical exponent.*

6.1 Method of Perturbation

6.1.1 *Duhamel's Principle*

In §5.2.4, we have derived from

$$u_t = \Delta u \quad (x \in \mathbf{R}^n,\ t > 0) \qquad u|_{t=0} = u_0(x) \quad (x \in \mathbf{R}^n) \tag{6.1}$$

that

$$u(x,t) = \int_{\mathbf{R}^n} G(x-y,t)u_0(y)dy \qquad (x \in \mathbf{R}^n,\ t > 0), \tag{6.2}$$

where

$$G(x,t) = \left(\frac{1}{4\pi t}\right)^{n/2} e^{-|x|^2/4t}$$

denotes the Gaussian kernel. Actually, if $u_0 \in L^p(\mathbf{R}^n)$ with $p \in [1,\infty)$ then $u(x,t)$ defined by (6.2) satisfies (6.1) and

$$\lim_{t\downarrow 0} \|u(\cdot,t) - u_0\|_p = 0. \tag{6.3}$$

We note that the right-hand side of (6.2) converges for more rough data, say

$$|u_0(x)| \le C\exp\left(|x|^\beta\right) \qquad (x \in \mathbf{R}^n),$$

where $C > 0$ and $\beta \in (0,2)$ are constants.

Duhamel's principle asserts that the solution

$$u_t = \Delta u + f(x,t) \quad (x \in \mathbf{R}^n,\ t \in (0,T]) \qquad u|_{t=0} = u_0(x) \quad (x \in \mathbf{R}^n) \tag{6.4}$$

is given by

$$u(x,t) = \int_{\mathbf{R}^n} G(x-y,t)u_0(y)dy + \int_0^t ds \int_{\mathbf{R}^n} G(x-y,t-s)f(y,s)dy. \tag{6.5}$$

This is obtained from the law that

$$\frac{d}{dt}\int_0^t H(t,s)ds = H(t,t) + \int_0^t H_t(t,s)ds. \tag{6.6}$$

In fact, putting

$$v(x,t) = \int_0^t ds \int_{\mathbf{R}^n} G(x-y,t-s)f(y,s)dy, \tag{6.7}$$

we have formally that

$$\begin{aligned} v_t(x,t) &= \int_{\mathbf{R}^n} G(x-y,0)f(y,t)dy \\ &\quad + \int_0^t ds \frac{\partial}{\partial t}\int_{\mathbf{R}^n} G(x-y,t-s)f(y,s)dy \\ &= f(x,t) + \int_0^t ds\Delta \int_{\mathbf{R}^n} G(x-y,t-s)f(y,s)dy \\ &= f(x,t) + \Delta v \end{aligned}$$

because of $G(x,0) = \delta(x)$. The above argument is formal because the behavior as $t \downarrow 0$ of $G(x,t)$ is not obvious.

Exercise 6.1 Give a sufficient condition to $u_0(x)$ for (6.3) to hold. Also give a sufficient condition to $H(t,s)$ for (6.6) to hold.

Exercise 6.2 Show that $v(x,t)$ given by (6.7) satisfies the first relation of (6.4) if $f(\cdot,t) \in C(\mathbf{R}^n) \cap L^\infty(\mathbf{R}^n)$ and

$$\|f(\cdot,t) - f(\cdot,s)\|_\infty \leq C\,|t-s|^\gamma$$

for $t, s \in [0,T]$, where $C > 0$ and $\gamma \in (0,1)$ are constants.

6.1.2 *Semilinear Heat Equation*

Given $p \in (1,\infty)$, we take the problem

$$u_t = \Delta u + u^p \quad (x \in \mathbf{R}^n,\ t > 0) \qquad u|_{t=0} = u_0(x) \quad (x \in \mathbf{R}^n), \tag{6.8}$$

where the solution $u = u(x,t)$ and the initial value $u_0 = u_0(x)$ are supposed to be non-negative. For $T > 0$, we say that the non-negative $u(x,t)$ is the regular solution to (6.8) on $t \in [0,T]$ if u, u_t, u_{x_i}, and $u_{x_i x_j}$ exist and continuous on $\mathbf{R}^n \times [0,T]$ and satisfies (6.8). We say also that $u(x,t) \in \mathcal{E}[0,T]$ if there exist $M > 0$ and $\beta \in (0,2)$ such that

$$|u(x,t)| \leq M \exp\left(|x|^\beta\right) \qquad (x \in \mathbf{R}^n,\ t \in [0,T]).$$

Finally, we say that $u(x,t) \in \mathcal{E}[0,\infty)$ if $u(x,t) \in \mathcal{E}[0,T]$ for any $T > 0$. Then, the following inclusion is obvious:

$$0 \leq u = u(x,t) \in \mathcal{E}[0,T] \quad \Rightarrow \quad u^p \in \mathcal{E}[0,T]. \tag{6.9}$$

Henceforth, we study the regular non-negative solution to (6.8) belonging to $\mathcal{E}[0,T]$. Furthermore, we assume $u_0 \in \mathcal{B}^2(\mathbf{R}^n)$, which means that it is C^2 and any $D^\alpha u_0$ with $|\alpha| \leq 2$ is bounded on $\mathbf{R}^n$. We show that Duhamel's principle is valid even to this class of the solution.

Theorem 6.1 *If $0 \leq u(x,t) \in \mathcal{E}[0,T]$ is a regular solution to (6.8), then it holds that*

$$u(x,t) = u_0(x,t) + \int_0^t ds \int_{\mathbf{R}^n} G(x-y,t-s)u(y,s)^p dy \tag{6.10}$$

for $x \in \mathbf{R}^n$, $t \in [0,T]$, where

$$u_0(x,t) = \int_{\mathbf{R}^n} G(x-y,t)u_0(y)dy.$$

Proof. It suffices to prove (6.10) for $t \in (0,T]$. We take $\rho \in C_0^\infty(\mathbf{R}^n)$ satisfying $0 \le \rho(x) \le 1$ and

$$\rho(x) = \begin{cases} 1 & (|x| \le 1) \\ 0 & (|x| \ge 2), \end{cases}$$

and put that $\rho_N(x) = \rho(x/N)$ and $v_N = \rho_N u$ for $N = 1, 2, \cdots$. Then, in use of

$$\frac{\partial v_N}{\partial t} = \rho_N \frac{\partial u}{\partial t} = \rho_N (\Delta u + u^p)$$

and

$$\Delta v_N = \rho_N \Delta u + 2\nabla \rho_N \cdot \nabla u + (\Delta \rho_N)u,$$

we get that

$$\frac{\partial v_N}{\partial t} = \Delta v_N + \rho_N u^p - 2\nabla \rho_N \cdot \nabla u - (\Delta \rho_N)u$$

in $\mathbf{R}^n \times [0,T]$ with $v_N|_{t=0} = \rho_N u_0$. Because $v_N(x,t)$ is appropriately smooth and has the compact support in x, we can apply Duhamel's principle as

$$v_N(x,t) = V_1 + V_2 - 2V_3 - V_4$$

with

$$V_1 = \int_{\mathbf{R}^n} G(x-y,t)\rho_N \cdot u_0(y)dy$$

$$V_2 = \int_0^t ds \int_{\mathbf{R}^n} G(x-y,t-s)\rho_N \cdot u(y,s)^p dy$$

$$V_3 = \int_0^t ds \int_{\mathbf{R}^n} G(x-y,t-s)\nabla_y \rho_N \cdot \nabla_y u(y,s)dy$$

$$V_4 = \int_0^t ds \int_{\mathbf{R}^n} G(x-y,t-s)\Delta_y \rho_N \cdot u(y,s)dy.$$

First, if $N \to \infty$, then $\rho_N u_0(y) \to u_0(y)$ for any $y \in \mathbf{R}^n$. We can apply the dominated convergence theorem by

$$|G(x-y,t)\rho_N \cdot u_0(y)| \leq \|u_0\|_\infty G(x-y,t) \in L^1_y(\mathbf{R}^n),$$

and obtain

$$V_1 \to u_0(x,t).$$

To treat the second term, we recall the assumption $0 \leq u(x,t) \in \mathcal{E}[0,T]$ and (6.9). There exist $M > 0$ and $\beta \in (0,2)$ such that

$$0 \leq u(y,s)^p \leq M \exp\left(|y|^\beta\right) \qquad (y \in \mathbf{R}^n,\ s \in [0,T]),$$

and hence we obtain

$$\begin{aligned} 0 &\leq \int_0^t ds \int_{\mathbf{R}^n} G(x-y,t-s)\rho_N \cdot u(y,s)^p dy \\ &\leq M \int_0^t ds \int_{\mathbf{R}^n} G(x-y,t-s) \exp\left(|y|^\beta\right) dy \\ &\leq M \int_0^t ds \int_{\mathbf{R}^n} G(x-y,t-s) \\ &\qquad \cdot \exp\left(2^\beta |x-y|^\beta + 2^\beta |x|^\beta\right) dy. \end{aligned} \tag{6.11}$$

For $t > 0$ fixed, we have

$$\begin{aligned} \int_{\mathbf{R}^n} G(x,t) \exp\left(2^\beta |x|^\beta\right) dx &= \left(\frac{1}{4\pi t}\right)^{n/2} \int_{\mathbf{R}^n} e^{-|x|^2/4t + 2^\beta |x|^\beta} dx \\ &= \pi^{-n/2} \int_{\mathbf{R}^n} e^{-|\eta|^2 + (16t)^{\beta/2}|\eta|^\beta} d\eta \\ &\leq \pi^{-n/2} \int_{\mathbf{R}^n} e^{-|\eta|^2 + (16T)^{\beta/2}|\eta|^\beta} d\eta \equiv M_1 < +\infty \end{aligned}$$

for $t \in (0,T]$ by $x = \sqrt{4t}\eta$. Therefore, the right-hand side of (6.11) is estimated from above by

$$M \exp\left(2^\beta |x|^\beta\right) \cdot TM_1.$$

On the other hand, we have

$$\left| V_2 - \int_0^t ds \int_{\mathbf{R}^n} G(x-y,t-s) u(y,s)^p dy \right|$$

$$\begin{aligned}
&\leq \int_0^t ds \int_{|y|\geq N} G(x-y,t-s)u(y,s)^p dy \\
&\leq M \int_0^t ds \int_{|y|\geq N} G(x-y,t-s)\exp\left(|y|^\beta\right) dy. \qquad (6.12)
\end{aligned}$$

In use of an estimate obtained similarly to the right-hand side of (6.11) and the dominated convergence theorem, we see that the right-hand side of (6.12) converges to 0 as $N \to \infty$.

Next, we have $|\Delta\rho_N(y)| \leq CN^{-2}$ with $C = \|\Delta\rho\|_\infty$, and hence

$$|V_4| \leq CN^{-2} \cdot M \cdot \int_0^t ds \int_{\mathbf{R}^n} G(x-y,t-s)\exp\left(|y|^\beta\right) dy \to 0$$

follows. Finally, we have

$$\begin{aligned}
V_3 &= \int_0^t ds \int_{\mathbf{R}^n} G(x-y,t-s)\nabla_y\rho_N \cdot \nabla_y u(y,s)dy \\
&= \int_0^t ds \left\{ - \int_{\mathbf{R}^n} \nabla_y \cdot \left(G(x-y,t-s)\nabla_y\rho_N\right) u(y,s)dy \right\} \\
&= -\tilde{V}_3 - V_4.
\end{aligned}$$

We shall show that

$$\tilde{V}_3 = \int_0^t ds \int_{\mathbf{R}^n} \nabla_x G(x-y,t-s) \cdot \nabla\rho_N u(y,s)dy \to 0.$$

In fact, we have for $C = \|\nabla\rho\|_\infty$ that

$$|\nabla\rho_N(y)| \leq CN^{-1}, \qquad |\nabla_x G(x,t)| = \left(\frac{1}{4\pi t}\right)^{n/2} e^{-|x|^2/4t} \left|\frac{2}{4t}x\right|,$$

and

$$|u(y,s)| \leq M\exp\left(|y|^\beta\right) \leq M\exp\left(2^\beta |x|^\beta\right) \cdot \exp\left(2^\beta |x-y|^\beta\right).$$

This time we have

$$\begin{aligned}
&\int_{\mathbf{R}^n} |\nabla_x G(x-y,t)| \exp\left(|y|^\beta\right) dy \\
&\leq \exp\left(2^\beta |x|^\beta\right) \int_{\mathbf{R}^n} |\nabla_y G(y,t)| \exp\left(2^\beta |y|^\beta\right) dy \\
&= \exp\left(2^\beta |x|^\beta\right) \left(\frac{1}{4\pi t}\right)^{n/2} \int_{\mathbf{R}^n} \exp\left(-|y|^2/4t + 2^\beta |y|^\beta\right) \cdot \frac{|y|}{2t} dy
\end{aligned}$$

$$= \exp\left(2^\beta |x|^\beta\right) \pi^{-n/2} \int_{\mathbf{R}^n} \exp\left(-|\eta|^2 + (16t)^{\beta/2} |\eta|^\beta\right) |\eta|\, d\eta$$
$$\cdot t^{-1/2}$$
$$= \exp\left(2^\beta |x|^\beta\right) M_1 t^{-1/2}$$

and hence

$$\begin{aligned} \left|\tilde{V}_3\right| &\le CN^{-1} \exp\left(2^\beta |x|^\beta\right) M_1 \int_0^t (t-s)^{-1/2} ds \\ &\le CN^{-1} \exp\left(2^\beta |x|^\beta\right) M_1 \cdot 2T^{1/2} \to 0 \end{aligned}$$

as $N \to +\infty$. Thus, the proof is complete. □

Exercise 6.3 Confirm (6.9).

Exercise 6.4 Show that if $0 \le u_0 \in \mathcal{B}^2(\mathbf{R}^n)$, $u_0(x) \not\equiv 0$, and $0 \le u(x,t) \in \mathcal{E}[0,T]$ satisfies (6.10), then $u(x,t) > 0$ holds for $(x,t) \in \mathbf{R}^n \times (0,T]$ and also $\frac{\partial u}{\partial x_j} \in \mathcal{E}[0,T]$ for $j = 1, \cdots, n$.

Exercise 6.5 Show that if $u_0 \in \mathcal{B}^2(\mathbf{R}^n)$ and $u \in C(\mathbf{R}^n \times [0,T]) \cap L^\infty(\mathbf{R}^n \times (0,T))$ solves (6.10), then it is the regular solution to (6.8).

6.1.3 *Global Existence*

The right-hand side of (6.10) is regarded as a nonlinear operator to $u(x,t)$, and it is the fixed point equation with respect to this operator. If the iterative sequence converges, then we get the solution.

Theorem 6.2 *Let $p > 1 + \frac{2}{n}$ and $u_0 \in \mathcal{B}^2(\mathbf{R}^n)$. If there are $0 < \gamma \ll 1$ and $\delta > 0$ such that*

$$0 \le u_0(x) \le \delta G(x,\gamma) \qquad (x \in \mathbf{R}^n), \tag{6.13}$$

then (6.8) admits the regular solution $0 \le u = u(x,t) \in \mathcal{E}[0,\infty)$.

Proof. We say that $u = u(x,t) \in \mathcal{S}[0,\infty)$ if it is continuous on $\mathbf{R}^n \times [0,\infty)$ and satisfies

$$0 \le u(x,t) \le MG(x,t+\gamma) \qquad (x \in \mathbf{R}^n,\ t \ge 0).$$

Letting

$$(\mathcal{F}u)(x,t) = \int_0^t ds \int_{\mathbf{R}^n} G(x-y,t-s)u(y,s)^p dy,$$

we show that the iterative sequence

$$u_{j+1} = u_0 + \mathcal{F}(u_j) \qquad (j = 0, 1, \cdots) \tag{6.14}$$

converges in $\mathcal{S}[0,\infty)$. Actually, it is the Banach space provided with the norm

$$\|v\| = \sup_{x\in\mathbf{R}^n,\ t\geq 0} \frac{|v(x,t)|}{\rho(x,t)},$$

where $\rho(x,t) = G(x, t+\gamma)$.

First, we show that

$$\|u_0(\cdot,t)\| \leq \delta. \tag{6.15}$$

For this purpose, we recall (5.39):

$$\int_{\mathbf{R}^n} G(x-y,t)G(y,s)dy = G(x,t+s) \qquad (x \in \mathbf{R}^n,\ t,s > 0). \tag{6.16}$$

Then, it holds that

$$\begin{aligned} 0 \ \leq \ & u_0(x,t) = \int_{\mathbf{R}^n} G(x-y,t)u_0(y)dy \\ \leq \ & \delta \int_{\mathbf{R}^n} G(x-y,t)G(y,\gamma)dy \\ = \ & \delta G(x,t+\gamma) = \delta\rho(x,t), \end{aligned}$$

which implies (6.15).

We next show that if $p > 1 + \frac{2}{n}$, then it holds that

$$\begin{aligned} \|\mathcal{F}\rho\| \leq C_0(\gamma,p) \equiv & (4\pi)^{-n(p-1)/2} \\ & \cdot \int_0^\infty (s+\gamma)^{-n(p-1)/2} ds < +\infty. \end{aligned} \tag{6.17}$$

In fact, we have

$$(\mathcal{F}\rho)(x,t) = \int_0^t ds \int_{\mathbf{R}^n} G(x-y,t-s)\rho(y,s)^p dy < +\infty$$

is continuous on $\mathbf{R}^n \times [0, \infty)$. Now, we have

$$\begin{aligned}
\rho(y,s)^{p-1} &= G(y, s+\gamma)^{p-1} \\
&= \left\{\frac{1}{4\pi(s+\gamma)}\right\}^{n(p-1)/2} \exp\left(-\frac{(p-1)\,|y|^2}{4(s+\gamma)}\right) \\
&\leq (4\pi(s+\gamma))^{-n(p-1)/2}\,. \qquad (6.18)
\end{aligned}$$

This implies that

$$\begin{aligned}
0 &\leq (\mathcal{F}\rho)(x,t) \\
&\leq \int_0^t ds \int_{\mathbf{R}^n} G(x-y, t-s)\rho(y,s)\,(4\pi(s+\gamma))^{-n(p-1)/2}\,dy \\
&= \int_0^t ds \cdot (4\pi(s+\gamma))^{-n(p-1)/2} \int_{\mathbf{R}^n} G(x-y,t-s)G(y,s+\gamma)dy \\
&= \int_0^t ds\,(4\pi(s+\gamma))^{-n(p-1)/2}\,G(x,t+\gamma) \\
&= \rho(x,t)C_0(\gamma,p)
\end{aligned}$$

by (6.16). Thus, (6.17) follows.

If $u \in \mathcal{S}[0,\infty)$, then

$$0 \leq u(x,t) \leq \|u\|\,\rho(x,t)$$

holds. This implies $0 \leq u(x,t)^p \leq \|u\|^p\,\rho(x,t)^p$, and hence

$$\begin{aligned}
0 &\leq (\mathcal{F}u)(x,t) \leq \|u\|^p\,(\mathcal{F}\rho)(x,t) \\
&\leq C_0\,\|u\|^p\,\rho(x,t)
\end{aligned}$$

holds by (6.17). Thus, we obtain $\mathcal{F} : \mathcal{S}[0,\infty) \to \mathcal{S}[0,\infty)$ with

$$\|\mathcal{F}u\| \leq C_0(\gamma,p)\,\|u\|^p \qquad (6.19)$$

for $p > 1 + \frac{2}{n}$. The iterative sequence $\{u_j\}_{j=0}^{\infty}$ in $\mathcal{S}[0,\infty)$ is well-defined by (6.14).

Now, we show that $u, v \in \mathcal{S}[0,\infty)$ with $\|u\|,\ \|v\| \leq M$ implies

$$\|\mathcal{F}(u) - \mathcal{F}(v)\| \leq C_0(\gamma,p)pM^{p-1}\,\|u-v\|\,. \qquad (6.20)$$

In fact, for $r, s \geq 0$, we have

$$|r^p - s^p| = \left|\int_r^s \left(\frac{d}{d\tau}\tau^p\right) d\tau\right| \leq p\left|\int_r^s \tau^{p-1} d\tau\right|$$

$$\leq \; p \max\left\{r^{p-1}, s^{p-1}\right\}|r-s|,$$

and hence

$$\begin{aligned}
&|u(y,s)^p - v(y,s)^p| \\
&\quad \leq p \max\left\{u(y,s)^{p-1}, v(y,s)^{p-1}\right\}|u(y,s)-v(y,s)| \\
&\quad \leq pM^{p-1}\rho(y,s)^{p-1}|u(y,s)-v(y,s)|
\end{aligned}$$

follows. In use of (6.18), we have

$$\begin{aligned}
&|u(y,s)^p - v(y,s)^p| \\
&\quad \leq pM^{p-1}\left(\frac{1}{4\pi(s+\gamma)}\right)^{n(p-1)/2}\|u-v\|\,\rho(y,s)
\end{aligned}$$

and therefore, it holds that

$$\begin{aligned}
&|(\mathcal{F}u)(x,t) - (\mathcal{F}v)(x,t)| \\
&\quad = \left|\int_0^t ds \int_{\mathbf{R}^n} G(x-y,t-s)\left[u(y,s)^p - v(y,s)^p\right] dy\right| \\
&\quad \leq pM^{p-1}\|u-v\| \int_0^t ds \left(\frac{1}{4\pi(s+\gamma)}\right)^{n(p-1)/2} \\
&\qquad \cdot \int_{\mathbf{R}^n} G(x-y,t-s)G(y,s+\gamma)dy \\
&\quad = pM^{p-1}\|u-v\| \int_0^t ds \left(\frac{1}{4\pi(s+\gamma)}\right)^{n(p-1)/2} G(x,t+\gamma) \\
&\quad \leq pM^{p-1}\|u-v\|\, C_0(\gamma,p)\rho(x,t).
\end{aligned}$$

This means (6.20).

The sequence $\{u_j\}_{j=0}^{\infty} \subset \mathcal{S}[0,\infty)$ defined by (6.14) satisfies

$$\|u_{j+1}\| \leq \delta + C_0(\gamma,p)\|u_j\|^p$$

by (6.15) and (6.19). If $\delta > 0$ is sufficiently small as $C_0(\gamma,p)(2\delta)^p \leq \delta$, we get

$$\|u_j\| \leq 2\delta \qquad (j = 1, 2, \cdots)$$

by an induction. On the other hand, we have from (6.20) that

$$\begin{aligned}
\|u_{j+2} - u_{j+1}\| &= \|\mathcal{F}u_{j+1} - \mathcal{F}u_j\| \\
&\leq \sigma\|u_{j+1} - u_j\|
\end{aligned}$$

for $j = 1, 2, \cdots$, where $\sigma \equiv C_0(\gamma, p)p(2\delta)^{p-1}$. Making $\delta > 0$ so small as $\sigma < 1$, we have

$$\sum_{j=1}^{\infty} \|u_{j+1} - u_j\| < +\infty.$$

This gives the existence of

$$v = \lim_{k\to\infty} \sum_{j=1}^{k} (u_j - u_{j-1}) \in \mathcal{S}[0, \infty)$$

and hence that of $u = \lim_{j\to\infty} u_j$ in $\mathcal{S}[0, \infty)$.

The mapping $\mathcal{F} : \mathcal{S}[0, \infty) \to \mathcal{S}[0, \infty)$ is continuous by (6.20), and sending $j \to \infty$ in (6.14), we have

$$u = u_0 + \mathcal{F}u.$$

This means (6.10) for $(x, t) \in \mathbf{R} \times [0, \infty)$, and the proof is complete. □

Exercise 6.6 Show that $(\mathcal{F}\rho)(x, t)$ is continuous in $(x, t) \in \mathbf{R}^n \times [0, \infty)$ in use of the dominated convergence theorem.

6.1.4 *Blowup*

In this paragraph, we show the following.

Theorem 6.3 *If $1 < p < 1 + \frac{2}{n}$, $0 \le u_0 \in \mathcal{B}^2(\mathbf{R}^n)$, and $u_0 \not\equiv 0$, then there is no regular solution $u = u(x, t)$ to (6.8) in $0 \le u \in \mathcal{E}[0, \infty)$.*

Proof. Suppose the contrary, and let $0 \le u \in \mathcal{E}[0, \infty)$ be the regular solution to (6.8). First, we shall show that

$$u_0(0, t)^{-(p-1)} - u(0, t)^{-(p-1)} \ge (p - 1)t \qquad (t > 0) \tag{6.21}$$

holds for

$$u_0(x, t) = \int_{\mathbf{R}^n} G(x - y, t)u_0(y)dy.$$

In fact, we take $\varepsilon > 0$ and put

$$J_\varepsilon(s) = \int_{\mathbf{R}^n} v_\varepsilon(x, s)u(x, s)dx \tag{6.22}$$

for $s \in [0,t]$, where $v_\varepsilon(x,s) = G(x, t-s+\varepsilon)$. We have $M > 0$ and $\beta \in (0,2)$ such that

$$0 \le u(x,s) \le M \exp\left(|x|^\beta\right) \qquad (x \in \mathbf{R}^n,\ 0 \le s \le t).$$

Similarly to §6.1.1, we can show that $J_\varepsilon(s)$ is finite and continuous in $s \in [0,t]$ from the dominated convergence theorem. Furthermore, for $\rho \in C_0^\infty(\mathbf{R}^n)$ in the proof of Theorem 6.1, we take $\rho_N(x) = \rho(x/N)$ and set

$$J_\varepsilon^N(s) = \int_{\mathbf{R}^n} v_\varepsilon(x,s) \cdot u(x,s) \cdot \rho_N(x) dx.$$

Then, we can show the convergence $J_\varepsilon^N(s) \to J_\varepsilon(s)$ uniformly in $s \in [0,t]$ as $N \to \infty$.

Now, J_ε^N is continuously differentiable as

$$\begin{aligned} \frac{d}{ds} J_\varepsilon^N(s) &= \int_{\mathbf{R}^n} \left(\frac{\partial}{\partial s} v_\varepsilon \cdot u + v_\varepsilon \cdot \frac{\partial}{\partial s} u \right) \rho_N dx \\ &= \int_{\mathbf{R}^n} \{-\Delta v_\varepsilon \cdot u + v_\varepsilon \Delta u\} \rho_N dx + \int_{\mathbf{R}^n} v_\varepsilon \cdot u^p \cdot \rho_N dx \\ &= I_1 + I_2 \end{aligned}$$

by $\frac{\partial v_\varepsilon}{\partial s} = -\Delta v_\varepsilon$. Here we have from $u^p \in \mathcal{E}[0,t]$ that

$$I_2 \to \int_{\mathbf{R}^n} v_\varepsilon u^p dx$$

uniformly in $s \in [0,t]$ as $N \to \infty$. We shall show $I_1 \to 0$ uniformly in $s \in [0,t]$.

In fact, we have

$$\begin{aligned} I_1 &= \int_{\mathbf{R}^n} \nabla v_\varepsilon \cdot \nabla (u \rho_N) dx + \int_{\mathbf{R}^n} v_\varepsilon \cdot \Delta u \cdot \rho_N dx \\ &= -\int_{\mathbf{R}^n} v_\varepsilon \cdot \Delta (u \rho_N) dx + \int_{\mathbf{R}^n} v_\varepsilon \Delta u \cdot \rho_N dx \\ &= -2 \int_{\mathbf{R}^n} v_\varepsilon \nabla u \cdot \nabla \rho_N dx - \int_{\mathbf{R}^n} v_\varepsilon \cdot u \cdot \Delta \rho_N dx. \end{aligned}$$

Similarly to the proof of Theorem 6.1, we have for the second term of the right-hand side that

$$\left| \int_{\mathbf{R}^n} v_\varepsilon \cdot u \cdot \Delta \rho_N dx \right| \le C N^{-2} \int_{\mathbf{R}^n} v_\varepsilon \cdot u dx \le C' N^{-2}$$

with a constant $C' > 0$ independent of $s \in [0,t]$. The first term is treated similar because of $\frac{\partial u}{\partial x_j} \in \mathcal{E}[0,t]$. Then, it follows that

$$\int_{\mathbf{R}^n} v_\varepsilon \cdot |\nabla u|\, dx \leq C \qquad \text{and} \qquad \left| \int_{\mathbf{R}^n} \nabla \rho_N \cdot \nabla u dx \right| \leq CN^{-1}$$

uniformly in $s \in [0,t]$. In this way, $J_\varepsilon(s)$ defined by (6.22) is continuously differentiable in $s \in [0,t]$ and satisfies

$$\frac{d}{ds} J_\varepsilon(s) = \int_{\mathbf{R}^n} v_\varepsilon(x,s) u(x,s)^p dx.$$

Now, we have

$$\int_{\mathbf{R}^n} v_\varepsilon(x,s) dx = \int_{\mathbf{R}^n} G(x, t-s+\varepsilon) dx = 1$$

and from *Jensen's inequality* that

$$\begin{aligned} \frac{dJ_\varepsilon}{ds} &= \int_{\mathbf{R}^n} v_\varepsilon(x,s) u(x,s)^p dx \\ &\geq \left(\int_{\mathbf{R}^n} v_\varepsilon(x,s) u(x,s) dx \right)^p = J_\varepsilon^p \end{aligned}$$

for $s \in [0,t]$. This implies

$$\frac{d}{ds} J_\varepsilon^{-(p-1)} = -(p-1) J_\varepsilon^{-p} \frac{dJ_\varepsilon}{ds} \leq -(p-1)$$

and hence

$$(p-1)t \leq J_\varepsilon^{-(p-1)}(0) - J_\varepsilon^{-(p-1)}(t) \tag{6.23}$$

follows.

Here, we have

$$\begin{aligned} J_\varepsilon(t) &= \int_{\mathbf{R}^n} v_\varepsilon(x,t) u(x,t) dx = \int_{\mathbf{R}^n} G(x,\varepsilon) u(x,t) dx \\ &= \int_{\mathbf{R}^n} G(0-y,\varepsilon) u(y,t) dy \to u(0,t) \end{aligned}$$

as $\varepsilon \downarrow 0$. Similarly we have

$$\begin{aligned} J_\varepsilon(0) &= \int_{\mathbf{R}^n} v_\varepsilon(x,0) u(x,0) dx = \int_{\mathbf{R}^n} G(x,t+\varepsilon) u_0(x) dx \\ &\to u_0(0,t) \end{aligned}$$

as $\varepsilon \downarrow 0$. Inequality (6.21) is a consequence of (6.23).

Now, we have

$$u_0(0,t)^{-(p-1)} \geq (p-1)t \tag{6.24}$$

for $t > 0$ by (6.21). Recall $u_0(x) \not\equiv 0$. Without loss of generality, we assume $u_0(0) > 0$. There exist $\gamma, \delta > 0$ such that $u_0(x) \geq \gamma$ for $|x| < 2\delta$. For $t \geq \delta^2$ and $|x| \leq 2\delta$ we have

$$|x|^2 / 4t \leq \frac{(2\delta)^2}{4\delta^2} = 1.$$

Therefore, we obtain

$$\begin{aligned}
u_0(0,t) &= \int_{\mathbf{R}^n} G(x,t)u_0(x)dx \geq \int_{|x|\leq 2\delta} G(x,t)u_0(x)dx \\
&\geq \gamma \int_{|x|\leq 2\delta} \left(\frac{1}{4\pi t}\right)^{n/2} e^{-|x|^2/4t} dx \\
&\geq \gamma e^{-1} |B(0,2\delta)| \left(\frac{1}{4\pi t}\right)^{n/2} = C_1 t^{-n/2}
\end{aligned}$$

with a constant $C_1 > 0$. Inequality (6.24) implies

$$C_1^{p-1}(p-1)t \leq t^{n(p-1)/2} \qquad (t \geq \delta^2),$$

which is a contradiction by $1 < p < 1 + \frac{2}{n}$. □

6.2 Method of Energy

6.2.1 *Lyapunov Function*

As is described in the previous section, in 1966, H. Fujita showed that the blowup of the solution occurs to (6.8), involving Fujita's critical exponent. On the other hand, in 1969 he studied the asymptotic behavior of the solution to

$$u_t - \Delta u = \lambda e^u \qquad \text{in} \qquad \Omega \times (0,T)$$

with

$$u|_{\partial\Omega} = 0 \qquad \text{and} \qquad u|_{t=0} = u_0(x)$$

in connection with the stationary solution and discovered the *triple law* of him, which is not described here. Actually, later it was noticed that the work is involved by the blowup of the solution in finite time, but at that time it was thought to have no direct connection to the previous work. However, the forward self-similar transformation to the former equation on the whole space makes the domain compact, and that the critical exponent is realized again by the study of stationary solutions to that system. Meanwhile it has become clear also that the backward self-similar transformation provides sharp descriptions of the blowup mechanism in finite time. In this section, we study

$$u_t - \Delta u = u^p \quad \text{in} \quad \Omega \qquad \text{with} \qquad u|_{\partial\Omega} = 0 \ \text{ and } \ u|_{t=0} = u_0(x) \quad (6.25)$$

by the method of energy, and then derive the critical exponent to (6.1) in use of the forward self-similar transformation, where $\Omega \subset \mathbf{R}^n$ is a bounded domain with smooth boundary $\partial\Omega$.

Henceforth, $u = u(x,t)$ denotes the classical solution to (6.25) and the supremum of its existence time is denoted by $T_{\max} \in (0, +\infty]$. Then, the strong maximum principle guarantees that $u(x,t) > 0$ for $(x,t) \in \overline{\Omega} \times (0, T_{\max})$. Actually, the unique existence of such a solution is assured if $u_0 \in C_0(\overline{\Omega})$, by converting (6.25) to the integral equation

$$u(x,t) = \int_\Omega G(x,y;t)u_0(y)dy + \int_0^t dt' \int_\Omega G(x,y;t-t')u(y,t')^p dy$$

and applying the *contraction mapping principle* in $X_T = C\left([0,T], C_0(\overline{\Omega})\right)$ for $T > 0$ sufficiently small. Here and henceforth, $C_0(\overline{\Omega})$ denotes the set of continuous functions on $\overline{\Omega}$ taking the value 0 on $\partial\Omega$, and $G(x,y;t)$ denotes the fundamental solution of the linear part, so that it holds that

$$(\partial_t - \Delta_x)\, G(x,y;t) = 0 \qquad ((x,y,t) \in \overline{\Omega} \times \overline{\Omega} \times (0,\infty))$$

with

$$G|_{x\in\partial\Omega} = 0 \qquad \text{and} \qquad G|_{t=0} = \delta(x-y).$$

In this argument of showing the well-posedness of (6.25) locally in time, it can be assured that the existence time of the solution $T > 0$ is estimated from below by $\|u_0\|_\infty = \max_{x\in\overline{\Omega}} |u_0(x)|$. Because (6.25) is autonomous in time, then $\liminf_{t\to T} \|u(t)\|_\infty < +\infty$ guarantees that $T_{\max} > T$. In other words, we have

$$T = T_{\max} < +\infty \quad \Rightarrow \quad \lim_{t\to T} \|u(t)\|_\infty = +\infty. \tag{6.26}$$

This justifies the terminology of blowup to indicate the case of $T_{\max} < +\infty$ and also the introducing of the *blowup set* by

$$\begin{aligned} \mathcal{S} = \{ x_0 \in \overline{\Omega} \mid & \text{ there exist } \ t_k \to T, \ \text{ and } \ x_k \to x_0 \\ & \text{such that } \ u(x_k, t_k) \to +\infty \}. \end{aligned}$$

Actually, it is a non-empty compact set in this case of $T_{\max} < +\infty$. In 1993, J.J.L. Velázquez showed if $1 < p < \frac{n+2}{(n-2)_+}$ then the *Hausdorff dimension* of $\mathcal{S}$, denoted by $d_{\mathcal{H}}(\mathcal{S})$ is less than or equal to $n-1$ so that $d_{\mathcal{H}}(\mathcal{S}) \leq n-1$. Here, $p_s = \frac{n+2}{(n-2)_+}$ is referred to as the *Sobolev exponent*, satisfying always that $p_s > p_f$, where $p_f = 1 + \frac{2}{n}$ denotes the Fujita exponent.

This exponent is related to Sobolev's imbedding theorem $H_0^1(\Omega) \hookrightarrow L^{p+1}(\Omega)$ with $p \in (1, p_s]$, where the inclusion is compact if $1 < p < p_s$. If $p = p_s$ on the other hand, the imbedding constant

$$S = \inf \left\{ \|\nabla u\|_2^2 \mid \|u\|_{\frac{2n}{n-2}} = 1 \right\} \tag{6.27}$$

is determined by n and is independent of Ω. Henceforth, this constant is written as $S = S_n > 0$. The role of Sobolev's imbedding theorem is taken from the energy,

$$J(u) = \frac{1}{2} \|\nabla u\|_2^2 - \frac{1}{p+1} \|u\|_{p+1}^{p+1},$$

which acts as the *Lyapunov function*, so that if $u = u(\cdot, t)$ is a solution to (6.25), then it holds that

$$\frac{d}{dt} J(u(t)) = (\nabla u, \nabla u_t) - (u^p, u_t) = - \|u_t\|_2^2 . \tag{6.28}$$

On the other hand, for

$$I(u) = \|\nabla u\|_2^2 - \|u\|_{p+1}^{p+1} = 2J(u) - \frac{p-1}{p+1} \|u\|_{p+1}^{p+1}$$

we have

$$\begin{aligned} \frac{1}{2} \frac{d}{dt} \|u(t)\|_2^2 &= (u_t, u) = -I(u(t)) \\ &= -2J(u(t)) + \frac{p-1}{p+1} \|u(t)\|_{p+1}^{p+1} . \end{aligned} \tag{6.29}$$

If $J(u_0) \leq 0$, then $J(u(t)) \leq 0$ for any $t \in [0, T_{\max})$ and therefore, it follows that

$$\begin{aligned} \frac{1}{2}\frac{d}{dt}\|u(t)\|_2^2 &\geq \frac{p-1}{p+1}\|u(t)\|_{p+1}^{p+1} = \frac{p-1}{p+1}|\Omega| \fint_\Omega u(t)^{p+1} \\ &\geq \frac{p-1}{p+1}|\Omega| \left\{ \fint_\Omega u(t)^2 \right\}^{\frac{p+1}{2}} = \frac{p-1}{p+1}|\Omega|^{-\frac{p-1}{2}}\|u(t)\|_2^{p+1} \end{aligned}$$

for $t \in [0, T_{\max})$. Therefore, $T_{\max} = +\infty$ is impossible by $p + 1 > 2$ and $\|u_0\|_2 \neq 0$. This means that $T_{\max} = +\infty$ implies $J(u_0) > 0$ and translating the initial time, we get the following.

Theorem 6.4 *If $T_{\max} = +\infty$ holds in (6.25) with $p > 1$, then it follows that $J(u(t)) > 0$ for any $t \geq 0$.*

R. Ikehata and the second author showed that if $T_{\max} = +\infty$ and $\liminf_{t\to\infty} \|u(t)\|_\infty > 0$ hold in (6.25) with $p \in (1, p_s]$, then it follows that $J(u(t)) \geq d$ for any $t \geq 0$, where $d > 0$ is a constant determined by Ω. If $p = p_s$, then d depends only on n as $d = \frac{1}{n}S_n^{n/2}$. On the other hand, Y. Giga showed the following in 1986.

Theorem 6.5 *In (6.25) with $1 < p < p_s$ it holds that*

$$T = T_{\max} < +\infty \qquad \Rightarrow \qquad \lim_{t\to T} J(u(t)) = -\infty. \tag{6.30}$$

If Ω is convex, then the work by Y. Giga and R.V. Kohn is applicable. In fact, in this case it follows that $\mathcal{S} \subset \Omega$ and for $x_0 \in \mathcal{S}$ we have

$$\lim_{t\to T}(T-t)^{\frac{1}{p-1}}u(x,t) = \left(\frac{1}{p-1}\right)^{\frac{1}{p-1}} \tag{6.31}$$

locally uniformly in $|x - x_0| \leq C(T-t)^{1/2}$ for any $C > 0$, where $T = T_{\max} < +\infty$ and $1 < p < p_s$. Here, $\{(x,t) \mid |x - x_0| \leq C(T-t)^{1/2}\}$ is the standard parabolic region obtained from the backward self-similar transformation, and

$$u_*(t) = \left(\frac{1}{p-1}\right)^{\frac{1}{p-1}} (T-t)^{-\frac{1}{p-1}}$$

indicates the solution to the ODE part,

$$\dot{u}_* = u_*^p \qquad \text{with} \qquad \lim_{t\to T} u_*(t) = +\infty,$$

or the constant self-similar solution. Relation (6.31) indicates that the blowup mechanism in the parabolic region is controlled by the ODE part if the exponent is sub-critical: $p \in (1, p_s)$.

The actual backward self-transformation is indicated as

$$w(y,s) = (T-t)^{\frac{1}{p-1}} u(y,t), \quad y = (x-x_0)/(T-t)^{1/2}, \quad s = -\log(T-t).$$

Then, it holds that

$$w_s - \Delta w + \frac{1}{2} y \cdot \nabla w + \frac{1}{p-1} w = w^p$$

in

$$\bigcup_{s > -\log T} e^{s/2} \left(\Omega - \{x_0\}\right) \times \{s\}.$$

Relation (6.31) means that

$$w(y,s) \quad \to \quad \left(\frac{1}{p-1}\right)^{\frac{1}{p-1}}$$

locally uniformly in $y \in \mathbf{R}^n$ as $s \to +\infty$. Then, the parabolic regularity guarantees that

$$w_s(y,s) \to 0 \qquad \text{and} \qquad \nabla w(y,s) \to 0$$

locally uniformly in $y \in \mathbf{R}^n$ as $s \to +\infty$. Then, we have for $0 \le t < T = T_{\max}$ and $B(0,1) \subset e^A \left(\Omega - \{x_0\}\right)$ that

$$\begin{aligned} K(t) = \int_0^t \|u_t(t')\|_2^2 \, dt' &= \int_{-\log T}^{-\log(T-t)} \exp\left(\left(\frac{2}{p-1} - \frac{n-2}{2}\right) s\right) ds \\ &\quad \cdot \int_{e^{s/2}(\Omega - \{x_0\})} \left| \frac{1}{p-1} w + w_s + \frac{1}{2} y \cdot \nabla w \right|^2 dy \\ &\ge \int_A^{-\log(T-t)} \exp\left(\left(\frac{2}{p-1} - \frac{n-2}{2}\right) s\right) ds \\ &\quad \cdot \int_{|y|<1} \left| \frac{1}{p-1} w + w_s + \frac{1}{2} y \cdot \nabla w \right|^2 dy. \end{aligned}$$

Because $\frac{2}{p-1} - \frac{n-2}{2} > 0$, we have $\lim_{t \to T} K(t) = +\infty$ and hence

$$J\left(u(t)\right) = J(u_0) - K(t) \quad \to \quad -\infty$$

holds as $t \to T$. The proof is complete.

6.2.2 Solution Global in Time

If $T_{\max} = +\infty$ with $\liminf_{t\to\infty} \|u(t)\|_\infty < +\infty$ holds in (6.25), then there are $t_n \to +\infty$ and $C_1 > 0$ satisfying $\|u(t_n)\|_\infty \leq C_1$. Regarding $u(t_n)$ as the initial value to (6.25), then we can apply the unique existence theorem for the classical solution. The proof guarantees the existence of $\tau > 0$ satisfying that for any $\delta \in (0, \tau)$ we have $C_2 > 0$ such that

$$\|u\|_{C^{2+\theta,1+\theta/2}(\overline{\Omega}\times(t_n+\delta,t_n+\tau))} \leq C_2 \tag{6.32}$$

for $n = 1, 2, \cdots$, where $\theta \in (0,1)$.

On the other hand, we have from (6.28) and Theorem 6.4 that

$$\int_0^\infty \|u_t(t)\|_2^2\, dt \leq J(u_0). \tag{6.33}$$

Taking sub-sequences if necessary, we may suppose that $t_n \to +\infty$ satisfies $t_n + \tau < t_{n+1}$. In this case, we get from (6.33) that

$$\lim_{n\to\infty} \int_{t_n+\delta}^{t_n+\tau} \|u_t(t)\|_2^2\, dt = 0.$$

Therefore, there is $t_n' \in (t_n + \delta, t_n + \tau)$ satisfying

$$\|u_t(t_n')\|_2 \quad \to \quad 0.$$

We also have (6.32) and hence $\|u(t_n')\|_{C^{2+\theta}(\overline{\Omega})} \leq C_2$ follows. Then, passing through a subsequence, we obtain $u(t_n') \to u_\infty$ in $C^2(\overline{\Omega})$ with u_∞ satisfying

$$-\Delta u_\infty = u_\infty^p \quad \text{in} \quad \Omega, \qquad u_\infty|_{\partial\Omega} = 0 \quad \text{on} \quad \partial\Omega,$$

This means that u_∞ is a stationary solution to (6.25). Putting

$$E = \left\{u_\infty \in C^2(\overline{\Omega}) \mid \text{classical solutions to (6.25)}\right\}$$

and

$$\omega(u_0) = \left\{u_\infty \in C^2(\overline{\Omega}) \mid \text{there is } \ t_n' \to +\infty \right. \\ \left. \text{such that } \ u(t_n') \to u_\infty \ \text{ in } \ C^2(\overline{\Omega})\right\}, \tag{6.34}$$

we get the following, where $\omega(u_0)$ is called the *omega-limit set* of the *orbit* $\mathcal{O} = \{u(t) \mid t \geq 0\} \subset C_0(\overline{\Omega})$.

Theorem 6.6 *The asymptotic behavior of the classical solution $u = u(\cdot,t)$ to (6.25) is classified into the following.*

(1) The blowup case $T = T_{\max} < +\infty$, where it holds that

$$\lim_{t\to T} \|u(t)\|_\infty = +\infty.$$

(2) $T_{\max} = +\infty$ and $\liminf_{t\to\infty} \|u(t)\|_\infty < +\infty$, where it holds that $\omega(u_0) \subset E$.

(3) $T_{\max} = +\infty$ and $\lim_{t\to\infty} \|u(t)\|_\infty = +\infty$.

The third case of the above theorem is referred to as the *blowup in infinite time*. In 1980, M. Otani showed that this is not the case for $p \in (1, p_s)$ and more strongly, $\limsup_{t\to\infty} \|u(t)\|_\infty < +\infty$ holds whenever $T_{\max} = +\infty$ in this case. To prove this fact, we make use of the following.

Theorem 6.7 *If the solution $u = u(\cdot,t)$ to (6.25) with $p > 1$ satisfies $T_{\max} = +\infty$, then it holds that*

$$\sup_{t\geq 0} \|u(t)\|_2 < +\infty. \tag{6.35}$$

Proof. We have from (6.29) and (6.28) that

$$\begin{aligned}
\frac{1}{2}\frac{d}{dt}\|u(t)\|_2^2 &= -I(u(t)) \\
&= -(p+1)J(u(t)) + \frac{p-1}{2}\|\nabla u(t)\|_2^2 \\
&\geq (p+1)\int_0^t \|u_t(t')\|_2^2\,dt' - (p+1)J(u_0) \\
&\quad + \frac{p-1}{2}\lambda_1 \|u(t)\|_2^2
\end{aligned} \tag{6.36}$$

for $t \geq 0$, where $\lambda_1 > 0$ denotes the first eigenvalue of $-\Delta$ with the zero Dirichlet boundary condition, which assures the *Poincaré inequality*

$$\|\nabla u\|_2^2 \geq \lambda_1 \|u\|_2^2$$

holds for $u \in H_0^1(\Omega)$. Letting

$$h(t) = \frac{p-1}{2}\lambda_1 \|u(t)\|_2^2 - (p+1)J(u_0),$$

we shall show that $h(t) \leq 0$ for any $t \geq 0$. Then, the conclusion (6.35) is obtained.

In fact, if this is not the case, there is $t_0 \geq 0$ such that $h(t_0) > 0$. This implies from (6.36) that

$$\frac{d}{dt}\|u(t)\|_2^2 \geq h(t_0) \qquad \text{and} \qquad h(t) \geq h(t_0)$$

for any $t \geq t_0$ by the continuation in time. Then, it holds that

$$\lim_{t\to\infty} \|u(t)\|_2 = +\infty.$$

Letting $t_0 = 0$ without loss of generality, we now take

$$f(t) = \int_0^t \|u(t')\|_2^2\, dt'$$

and get that

$$\frac{1}{2}f''(t) \geq (p+1)\int_0^t \|u_t(t')\|_2^2\, dt'$$

by (6.36) and $h(t) \geq 0$, which is combined with

$$\begin{aligned} f(t)\cdot\int_0^t \|u_t(t')\|_2^2\, dt' &= \int_0^t \|u(t')\|_2^2\, dt' \cdot \int_0^t \|u_t(t')\|_2^2\, dt' \\ &\geq \left\{\int_0^t \|u(t')\|_2 \cdot \|u_t(t')\|_2\, dt'\right\}^2 \geq \left\{\int_0^t |(u(t'), u_t(t'))|\, dt'\right\}^2 \\ &= \frac{1}{4}\left\{\int_0^t \frac{d}{dt'}\|u(t')\|_2^2\, dt'\right\}^2 = \frac{1}{4}\left(f'(t) - f'(0)\right)^2. \end{aligned}$$

We have for $\varepsilon = \frac{p-1}{2} > 0$ that

$$f(t)\cdot f''(t) \geq (1+\varepsilon)\left(f'(t) - f'(0)\right)^2.$$

Because $f'(t) = \|u(t)\|_2 \to +\infty$, we have

$$f(t)\cdot f''(t) \geq \left(1 + \frac{\varepsilon}{2}\right) f'(t)^2$$

for t sufficiently large, which means that $f(t)^{-\varepsilon/2} \geq 0$ is concave there. However, again $f'(t) \to +\infty$ implies $f(t) \to +\infty$, which is impossible. We get a contradiction and the proof is complete. □

Now, we proceed to the following.

Theorem 6.8 *If $1 < p < p_s$ and $T_{\max} = +\infty$ hold in (6.25), then we have* $\sup_{t \geq 0} \|u(t)\|_\infty < +\infty$.

Proof. We only take the case of $n \geq 3$. First, we combine (6.29) with (6.35). There is $C_3 > 0$ such that

$$\frac{p-1}{2} \|\nabla u(t)\|_2^2 - (p+1) J(u(t)) = \frac{1}{2} \frac{d}{dt} \|u(t)\|_2^2$$
$$= (u_t(t), u(t)) \leq C_3 \|u_t(t)\|_2 .$$

Thus, we obtain

$$\frac{p-1}{2} \|\nabla u(t)\|_2^2 \leq C_3 \|u_t(t)\|_2 + (p+1) J(u_0). \tag{6.37}$$

Let $\Gamma = \{t > 0 \mid \|u_t(t)\|_2 > 1\}$. We have by (6.33) that

$$\int_0^\infty \|u_t(t)\|_2^2 \, dt < +\infty.$$

Hence we obtain

$$\lim_{t \to \infty} |\Gamma \cap [t, \infty)| = 0, \tag{6.38}$$

where $| \cdot |$ denotes the one-dimensional Lebesgue measure. On the other hand, inequality (6.37) implies

$$\|\nabla u(t)\|_2^2 \leq \frac{2C_3}{p-1} + \frac{2(p+1)}{p-1} J(u_0)$$

for $t \notin \Gamma$, so that it follows that

$$\|u(t)\|_{2^*} \leq C_4 \qquad \text{for} \qquad t \notin \Gamma \tag{6.39}$$

with a constant $C_4 > 0$, where $2^* = p_s + 1 = \frac{2n}{n-2}$.

Now, we take the semi-group $\left\{e^{-tA}\right\}_{t \geq 0}$, where A denotes $-\Delta$ with zero Dirichlet boundary condition. Then, it holds that

$$\left\|e^{-tA} v\right\|_{2^*} \leq \|v\|_{2^*}$$

and

$$\left\|e^{-tA} v\right\|_{2^*} \leq C_5 \left\|A e^{-tA} v\right\|_{2^*/(2^*-1)} \leq C_6 t^{-1} \|v\|_{2^*/(2^*-1)},$$

where the relation $W^{2,2^*/(2^*-1)}(\Omega) \hookrightarrow L^{2^*}(\Omega)$ is made use of. Actually, if A_q denotes A regarded as an operator in $L^q(\Omega)$, it holds that $D(A_q) = W^{2,q}(\Omega) \cap W_0^{1,q}(\Omega)$ for $q \in (1,\infty)$. In use of *Riesz-Thorin's interpolation theorem* we get that

$$\left\|e^{-tA}v\right\|_{2^*} \leq C_7 t^{-\theta} \|v\|_{2^*/p} \tag{6.40}$$

with $\theta = (p-1)/(2^*-2) \in (0,1)$.

In use of $\|u^p\|_{2^*/p} = \|u\|_{2^*}^p$, we have

$$\|u(t+\tau)\|_{2^*} \leq \|u(\tau)\|_{2^*} + C_8 t^{1-\theta} \sup_{s\in[\tau,t+\tau]} \|u(s)\|_{2^*}^p$$

for $t \geq 0$. Letting $f_\tau(t) = \sup_{s\in[\tau,t+\tau]} \|u(s)\|_{2^*}$, we have

$$\begin{aligned} \|u(t+\tau)\|_{2^*} &\leq \|u(\tau)\|_{2^*} + C_8 t^{1-\theta} f_\tau(t)^p \\ &\leq f_\tau(0) + C_8 T^{1-\theta} f_\tau(T)^p \end{aligned}$$

for $t \in [0,T]$, and hence

$$f_\tau(T) \leq f_\tau(0) + C_8 T^{1-\theta} f_\tau(T)^p$$

follows. Writing $T = t$, we obtain

$$f_\tau(t) \leq f_\tau(0) + C_8 t^{1-\theta} f_\tau(t)^p \qquad (t \geq 0).$$

In particular, $f_\tau(t_0) = 2f_\tau(0)$ implies that

$$t_0 \geq 2^{p/(1-\theta)} \cdot C_8^{-1/(1-\theta)} \{f_\tau(0)\}^{-(p-1)/(1-\theta)}$$

and there is $\delta > 0$ such that

$$\tau \notin \Gamma \quad \text{and} \quad f_\tau(t_0) = 2f_\tau(0) \qquad \Rightarrow \qquad t_0 \geq \delta$$

by (6.39). Therefore, it holds by $\|u(\tau)\|_{2^*} \leq C_1$ that

$$\|u(s)\|_{2^*} \leq 2C_1 \qquad \text{for} \qquad \tau \leq s \leq \tau + \delta.$$

Now, coming back to (6.38), we get $t_1 > 0$ satisfying $|\Gamma \cap [t_1,\infty)| < \delta/2$. In particular, any $t \geq t_1$ admits that $[t, t+\delta/2] \cap \Gamma^c \neq \emptyset$, and hence we have

$$\|u(t)\|_{2^*} \leq 2C_1 \qquad \text{for} \qquad t \geq t_2,$$

where $t_2 = t_1 + \delta/2$. Then the following lemma assures the conclusion, and the proof is complete. □

Lemma 6.1 *Let $u^k = u^k(x,t)$ be the solution to (6.25) with $p \in (1, p_s)$ globally in time with the initial value $u_0^k = u^k(\cdot, 0)$ satisfying*

$$\left\|u_0^k\right\|_{2^*} \leq \ell < +\infty$$

for $n = 1, 2, \cdots$. Then, any $\tau \in (0,1)$ admits a positive constant depending only on τ and ℓ, denoted by $C(\tau, \ell) > 0$, such that

$$\sup_{t \in (\tau, 1)} \left\|u^k(t)\right\|_\infty \leq C(\tau, \ell) \tag{6.41}$$

for $n = 1, 2, \cdots$.

Proof. We have from

$$u^k(t) = e^{-tA} u_0^k + \int_0^t e^{-(t-s)A} \left(u^k(s)\right)^p ds$$

and

$$\left\|e^{-tA} v\right\|_q \leq \|v\|_q \qquad (q \in (1, \infty))$$

that

$$\left\|A^\gamma u^k(t)\right\|_q \leq C_9 t^{-\gamma} \left\|u_0^k\right\|_q + C_9 \ell$$

for $t \in (0,1)$, where $\gamma \in (0,1)$ and $q = 2^*/p \in (2^*/(2^*-1), 2^*)$. Given $\tau \in (0,1)$, we get from this inequality that

$$\left\|A^\gamma u^k(t)\right\|_{2^*/p} \leq C_{10} \qquad \text{for} \qquad t \in (\tau, 1).$$

We have $D(A_q^\gamma) \subset W^{2\gamma, q}$ and if $\frac{2^*}{p} > \frac{n}{2}$, then Morrey's theorem guarantees that (6.41) holds. If this is not the case, we apply Sobolev's imbedding, in use of

$$W^{2, \frac{2n}{n+2}(1-\alpha)^{-1}}(\Omega) \subset L^{\frac{2n}{n-2}\left(1 - \frac{n+2}{n-2}\alpha\right)^{-1}}(\Omega)$$

valid for $\alpha \in (0, \frac{n-2}{n+2})$. Thus,

$$\left\|(u_0^k)^p\right\|_{\frac{2n}{n+2}(1-\alpha)^{-1}} \leq C_{11}$$

implies

$$\left\|u^k(t)\right\|_{\frac{2n}{n-2}\left(1 - \frac{n+2}{n-2}\beta\right)^{-1}} \leq C_{12}$$

for $t \in (\tau, 1)$, where $\tau \in (0,1)$ and $\beta \in (0, \alpha)$ are arbitrary. Continuing this procedure finitely many times, we eventually obtain for any $q \in (1, \infty)$ and $\tau \in (0,1)$ that

$$\sup_{t \in (\tau,1)} \left\| u^k(t) \right\|_q \leq C_{13}.$$

Then, (6.41) is obtained similarly, and the proof is complete. □

Exercise 6.7 Confirm that

$$\frac{1-\theta}{2^*} + \frac{\theta}{2^*/(2^*-1)} = \frac{p}{2^*}$$

holds for $\theta = (p-1)/(2^*-2)$ and justify (6.40).

6.2.3 *Unbounded Solution*

In 1986, Y. Giga refined the proof of Theorem 6.8 and showed that the upper bound of $\sup_{t \geq 0} \|u(t)\|_\infty$ depends only on $\|u_0\|_\infty$. This, in particular, implies the following theorem, where $C = \left\{ u_0 \in C_0(\overline{\Omega}) \mid u_0 \geq 0 \right\}$.

Theorem 6.9 *The set K defined by*

$$K = \{ u_0 \in C \mid \textit{It holds that } T_{\max} = +\infty \textit{ in (6.25)} \}$$

is closed in C in the case of $p \in (1, p_s)$.

This means that $C \setminus K$ is relatively open, but F. Merle proved that the blowup time $T_{\max} = T_{\max}(u_0)$ is a continuous function of $u_0 \in C \setminus K$. First, we note the following.

Lemma 6.2 *If $J(u(t_0)) < 0$ holds with some $t_0 \in (0, T_{\max})$, then it holds that*

$$T_{\max} - t_0 \leq C \left| J(u(t_0)) \right|^{-\frac{p-1}{p+1}},$$

where $C > 0$ is a constant determined by Ω and $p > 1$.

Proof. We have from (6.29) that

$$\frac{1}{2}\frac{d}{dt} \|u(t)\|_2^2 = -2J(u(t)) + \frac{p-1}{p+1} \|u(t)\|_{p+1}^{p+1}$$

with $\|u(t)\|_2^2 \leq |\Omega|^{\frac{p-1}{p+1}} \|u(t)\|_{p+1}^2$. Therefore, because $J(u(t))$ is a non-increasing function of t, we have

$$\frac{1}{2}\frac{d}{dt}\|u(t)\|_2^2 \geq -2J(u(t_0)) + \frac{p-1}{p+1}|\Omega|^{-\frac{p+1}{2}}\|u(t)\|_2^{p+1} > 0$$

for $t \in [t_0, T_{\max})$. This means

$$\frac{\frac{1}{2}\frac{d}{dt}\|u(t)\|_2^2}{-2J(u(t_0)) + \frac{p-1}{p+1}|\Omega|^{-\frac{p+1}{2}}\left(\|u(t)\|_2^2\right)^{\frac{p+1}{2}}} \geq 1$$

and hence it follows that

$$\frac{1}{2}\int_{\|u(t_0)\|_2^2}^{\|u(T)\|^2} \frac{ds}{-2J(u(t_0)) + \frac{p-1}{p+1}|\Omega|^{-\frac{p+1}{2}} s^{\frac{p+1}{2}}} \geq T - t_0$$

for any $T \in (t_0, T_{\max})$. Then, this implies

$$T_{\max} - t_0 \leq |J(u(t_0))|^{-\frac{p-1}{p+1}} \int_0^\infty \frac{dt}{1 + ct^{\frac{p+1}{2}}}$$

with a constant $c > 0$ determined by Ω and p. □

Now, we show the following.

Theorem 6.10 *If $1 < p < p_s$, then the mapping*

$$u_0 \in C \setminus K \quad \mapsto \quad T_{\max}(u_0) > 0$$

is continuous, where $T_{\max}(u_0)$ denotes the blowup time for the solution $u = u(\cdot, t)$ to (6.25).

Proof. We take $\{u_0^k\}_{k=1}^\infty \subset C \setminus K$ satisfying that $u_0^k \to u_0$ in C. The solution $u = u(\cdot, t)$ to (6.25) with the initial value u_0^k is denoted by $u^k = u^k(\cdot, t)$ for $k = 0, 1, 2, \cdots$, where $u^0 = u$ and $u_0^0 = u_0$. Let $T_k = T_{\max}(u_0^k)$ be its blowup time. Then, it holds that $\limsup_{k\to\infty} \left\|u_0^k\right\|_\infty < +\infty$. Because $T_{\max}$ is estimated from below by $\|u_0\|_\infty$, any $\varepsilon > 0$ admits k_0 satisfying $T_k > T_0 - \varepsilon$ for any $k \geq k_0$. Thus,

$$\liminf_{k\to\infty} T_k \geq T_0$$

follows.

We now show that

$$\limsup_{k\to\infty} T_k \le T_0 \tag{6.42}$$

holds for $p \in (1, p_s)$. In fact, Theorem 6.5 guarantees that

$$\lim_{t\to T_{\max}} J(u(t)) = -\infty$$

and therefore, any $M > 0$ admits $t_0 \in (0, T_0)$ such that

$$-J(u_0(T_0 - t_0)) \ge M + 1.$$

This implies the existence of k_0 satisfying

$$-J\left(u^k(T_0 - t_0)\right) \ge M$$

for any $k \ge k_0$. Then, Lemma 6.2 gives that

$$T_k - (T_0 - t_0) \le C \left| J\left(u^k(T_0 - t_0)\right)\right|^{-\frac{p-1}{p+1}},$$

or

$$T_k \le T_0 - t_0 + CM^{-\frac{p-1}{p+1}} \le T_0 + CM^{-\frac{p-1}{p+1}}.$$

Letting $M \to +\infty$, we obtain

$$\limsup_{k\to\infty} T_k \le T_0$$

and the proof is complete. □

For the moment, we take the general semilinear parabolic equation

$$u_t - \Delta u = f(u) \quad \text{in} \quad \Omega \times (0, T)$$

with

$$u = 0 \quad \text{on} \quad \partial\Omega \times (0, T) \qquad \text{and} \qquad u|_{t=0} = u_0(x) \quad \text{in} \quad \Omega$$

with the nonlinearity $f : \mathbf{R} \to \mathbf{R}$ is C^1 and $u_0 \in C_0(\overline{\Omega})$. Recall that the omega-limit set $\omega(u_0)$ is defined by (6.34) for $T_{\max} = +\infty$. In the case of

$$\sup_{t\ge 0} \|u(t)\|_\infty < +\infty, \tag{6.43}$$

the orbit $\mathcal{O} = \{u(t)\}_{t\geq 0}$ is compact in $C_0(\overline{\Omega})$ and then the general theory guarantees that $\omega(u_0)$ is a non-empty, compact, and connected set contained in E, the set of stationary solutions, and it holds that

$$\lim_{t\to+\infty} \mathrm{dist}\,(u(t), \omega(u_0)) = 0.$$

Each $u \in E$ is associated with the *linearized operator* $-\Delta - f'(u)$ provided with the zero Dirichlet condition, of which first eigenvalue is denoted by $\lambda_1(u)$. Putting

$$E_{\pm} = \{u \in E \mid \pm\lambda_1(u) > 0\} \qquad \text{and} \qquad E_0 = \{u \in E \mid \lambda_1(u) = 0\},$$

P.L. Lions showed in 1984 the following:

(1) E_+ is composed of at most countable set of isolated points.
(2) Any closed subset of E_0 is an ordered C^1 curve.
(3) If C is a connected component of E_-, then $\overline{C} \subset E_-$ holds.

By this it holds that $\omega(u_0) \subset E_-$, E_0, or E_+, exclusively, in the case of (6.43). It also implies the following, where

$$I_{\pm} = \{u_0 \in K \mid \omega(u_0) \subset E_{\pm}\} \qquad \text{and} \qquad I_0 = \{u_0 \in K \mid \omega(u_0) \subset E_0\}$$

for

$$K = \left\{ u_0 \in C_0(\overline{\Omega}) \mid u_0 \geq 0,\ T_{\max} = +\infty,\ \limsup_{t\to+\infty} \|u(t)\|_\infty < +\infty \right\}.$$

(1) $I_+ \cup I_0$ contains an open dense subset of K.
(2) If $u_0 \in I_+ \cup I_0$, then $\omega(u_0)$ is composed of one element.
(3) If $u_0 \in I_-$, then there are $\varepsilon > 0$ and $u_\pm \in E_\pm \cup E_0$ such that $\tilde{u}_0 \in K$, $\tilde{u}_0 \geq u_0$, $\tilde{u}_0 \not\equiv u_0$, and $\|\tilde{u}_0 - u_0\|_\infty < \varepsilon$ imply $\omega(\tilde{u}_0) = \{u_+\}$, where u_+ is the minimal element of

$$\{u \in E \mid u \geq w \ \text{ for any } \ w \in \omega(u_0)\},$$

and that $\tilde{u}_0 \in K$, $\tilde{u}_0 \leq u_0$, $\tilde{u}_0 \not\equiv u_0$, and $\|\tilde{u}_0 - u_0\|_\infty < \varepsilon$ imply $\omega(\tilde{u}_0) = \{u_-\}$, where u_- is the the maximal element of

$$\{u \in E \mid u \leq w \ \text{ for any } \ w \in \omega(u_0)\}.$$

Those results are proven by the *strong maximum principle.*

Concerning (6.25), we can show that any non-trivial stationary solution u is in E_-. In fact, the linearized operator is $-\Delta - pu^{p-1}$ and it holds that

$$\|\nabla u\|_2^2 - p\|u\|_{p+1}^{p+1} = -(p-1)\|\nabla u\|_2^2 < 0.$$

Therefore, $u \in E_-$ follows from the *Rayleigh* principle. Note also that if $1 < p < p_s$, then $K = \{u_0 \in C_0(\overline{\Omega}) \mid u_0 \geq 0,\ T_{\max} = +\infty\}$.

Kaplan's method is the other tool, which is important in the study of (6.25). There, the first eigenvalue $\lambda_1 > 0$ and the eigenfunction $\varphi_1 = \varphi_1(x) > 0$ of $-\Delta$ provided with the zero Dirichlet boundary condition is taken, so that it holds that

$$-\Delta\varphi_1 = \lambda_1\varphi_1,\ \varphi_1 > 0 \quad \text{in} \quad \Omega, \qquad \varphi_1 = 0 \quad \text{on} \quad \partial\Omega.$$

Adopting the normalization

$$\int_\Omega \varphi_1(x)dx = 1,$$

we put

$$j(t) = \int_\Omega u(x,t)\varphi_1(x)dx$$

and apply Jensen's inequality as

$$\begin{aligned}\frac{dj}{dt} + \lambda_1 j &= \int_\Omega u(x,t)^p\varphi_1(x)dx \\ &\geq \left\{\int_\Omega u(x,t)\varphi_1(x)dx\right\}^p = j^p,\end{aligned}$$

where $t \in [0, T_{\max})$. We have for $j_* = \lambda_1^{\frac{1}{p-1}}$ that $s^p - \lambda_1 s > 0$ for $s > j_*$, and if

$$j(0) = \int_\Omega u_0(x)\varphi_1(x)dx > j_*$$

then $\frac{dj}{dt} > 0$, $j(t) > j_*$ holds for $t \in [0, T_{\max})$. Therefore, $T_{\max} = +\infty$ induces a contradiction as

$$\int_{j(0)}^{+\infty} \frac{dj}{j^p - \lambda_1 j} \geq \int_0^{+\infty} dt = +\infty.$$

Thus, $j(0) > j_*$ implies $T_{\max} < +\infty$, and hence $T_{\max} = +\infty$ gives that $j(0) \leq j_*$. Because (6.25) is autonomous in time, we have that

$$T_{\max} = +\infty \qquad \Rightarrow \qquad \int_\Omega u(x,t)\varphi_1(x)dx \leq j_* \quad (t \geq 0). \tag{6.44}$$

In 1984, W.-M. Ni, P. Sack, and J. Tavantzis provided the following argument. Let $\psi \in C_0(\overline{\Omega})$ with $\psi \geq 0$, $\psi \not\equiv 0$ be fixed, and take $u_0 = \mu\psi$ with $\mu > 0$ in (6.25). The solution and its blowup time are denoted by $u_\mu = u_\mu(\cdot, t)$ and $T_{\max}(\mu) > 0$, respectively. Then, it is proven that $0 < \mu \ll 1$ and $\mu \gg 1$ imply $T_{\max}(\mu) = +\infty$ with $\limsup_{t\to+\infty} \|u_\mu(t)\|_\infty = +\infty$ and $T_{\max} < +\infty$, respectively. Putting

$$\mu_* = \sup\left\{\mu > 0 \mid T_{\max}(\mu) = +\infty, \ \limsup_{t\to+\infty} \|u_\mu(t)\|_\infty < +\infty\right\} > 0,$$

we have $T_{\max}(\mu) = +\infty$ for $\mu \in (0, \mu_*)$ by the comparison theorem, and therefore, it holds that

$$\int_\Omega u_\mu(x,t)\varphi_1(x)dx \leq j_* \qquad \text{for any} \qquad t \geq 0.$$

In use of the Hopf lemma and the monotone convergence theorem, we get the limit function

$$u_* = u_*(x,t) = \lim_{\mu\to\mu_*} u_\mu(x,t)$$

converging in $C\left([0,\infty), L^1(\Omega, \delta(x)dx)\right)$, where $\delta(x) = \operatorname{dist}(x, \partial\Omega)$. Then, we can show that this $u_*(x,t)$ is a *weak solution* to (6.25) so that $u_* = u_*(\cdot, t) \in L^p(\Omega, \delta(x)dx)$ for a.e. $t > 0$, that

$$t \in [0,\infty) \quad \mapsto \quad \int_\Omega u_*(x,t)\varphi(x)dx$$

is locally absolutely continuous if $\varphi \in C^2(\overline{\Omega}) \cap C_\delta(\overline{\Omega})$, which means that $\varphi = \varphi(x)$ is C^2 and satisfies the estimate $|\varphi| \leq C\delta$ with a constant $C > 0$ on $\overline{\Omega}$, and that

$$\frac{d}{dt}\int_\Omega u_*(x,t)\varphi(x)dx = \int_\Omega u_*(x,t)\Delta\varphi(x)dx + \int_\Omega u_*(x,t)^p\varphi(x)dx$$

holds for a.e. $t \in [0,\infty)$.

They showed that if $1 < p < 1 + \frac{2}{n}$ and Ω is convex, then it holds that $\limsup_{t\to+\infty} \|u_*(t)\|_\infty < +\infty$ and hence $u_*(x,t)$ is a classical solution.

Remember that $p_f = 1 + \frac{2}{n}$ is the Fujita exponent, in which case it is known that the L^1 boundedness implies that of L^∞. They also showed that if $p \geq \frac{n+2}{n-2}$ and Ω is star-shaped, then it holds that $\limsup_{t\to+\infty} \|u_*(t)\|_\infty = +\infty$. Again, $p_s = \frac{n+2}{(n-2)_+}$ is the Sobolev exponent, and the set of stationary solutions E is empty if Ω is star-shaped by the Pokhozaev identity. If $T_{\max}(\mu_*) < +\infty$ in that case, the solution blows-up in finite time with a post blowup continuation as the weak solution. If $T_{\max}(\mu_*) = +\infty$ on the contrary, then $u_* = u_*(\cdot, t)$ blows-up in infinite time. That problem was studied by V.A. Galaktionov and J.L. Vazquez in 1997 in more details. The solution $u_* = u_*(x,t)$ in this case is called the *unbounded global solution.*

6.2.4 *Stable and Unstable Sets*

To establish the local well-posed theorem for the discontinuous initial value to (6.25) has been tried by several authors. It is confirmed that if $p \leq p_s = \frac{n+2}{(n-2)_+}$ and $u_0 \in H_0^1(\Omega)$, then there is $T > 0$ that admits the unique solution $u = u(\cdot, t)$ in $C\left([0,T), H_0^1(\Omega)\right)$, which is called the *$H^1$-solution* in this monograph. More precisely, it is the solution to an abstract integral equation in $H_0^1(\Omega)$. However, the parabolic regularity guarantees that it becomes smooth for $t > 0$. Moreover, if $u_0 \in H_0^1(\Omega) \cap C_0(\overline{\Omega})$ then this H^1 solution coincides with the classical solution which we have discussed. In particular, the supremum of the existence time as the classical and the H^1 solutions coincides and is denoted by $T_{\max} \in (0, +\infty]$, and (6.26) holds even in this case of $u_0 \in H_0^1(\Omega)$. On the other hand, if $1 < p < p_s$ we can observe that this $T > 0$ and $\sup_{t\in[0,T]} \|\nabla u(t)\|_2$ are estimated from above and from below by $\|\nabla u_0\|_2$, respectively. Thus, it holds that

$$T = T_{\max} < +\infty \qquad \Rightarrow \qquad \lim_{t\to T} \|\nabla u(t)\|_2 = +\infty, \tag{6.45}$$

although this relation (6.45) does not hold for $p = p_s$.

In any case, $H_0^1(\Omega) \hookrightarrow L^{p+1}(\Omega)$ holds for $p \in (1, p_s]$, and

$$J(u) = \frac{1}{2}\|\nabla u\|_2^2 - \frac{1}{p+1}\|u\|_{p+1}^{p+1} \qquad \text{and} \qquad I(u) = \|\nabla u\|_2^2 - \|u\|_{p+1}^{p+1}$$

are well-defined for $u \in X = H_0^1(\Omega)$. The relations

$$\frac{d}{dt} J\left(u(t)\right) = -\|u_t(t)\|_2^2 \qquad \text{and} \qquad \frac{1}{2}\frac{d}{dt}\|u(t)\|_2^2 = -I\left(u(t)\right)$$

continue to hold for the H^1 solution to (6.25). Combined with Poincaré's inequality

$$\|\nabla u\|_2^2 \geq \lambda_1 \|u\|_2^2 \qquad (u \in H_0^1(\Omega)), \tag{6.46}$$

to realize the orbit $\mathcal{O} = \{u(t) \mid 0 \leq t < T_{\max}\}$ in $X = H_0^1(\Omega)$ becomes meaningful.

The *Nehari manifold* and the *potential depth* indicate the set $\mathcal{N} = \{u \in H_0^1(\Omega) \mid I(u) = 0, u \neq 0\}$ and the constant

$$d = \inf\left\{\sup_{s \geq 0} J(su) \mid u \in H_0^1(\Omega), u \neq 0\right\},$$

respectively. Then, it holds that $E \setminus \{0\} \subset \mathcal{N}$, where E denotes the set of stationary solutions to (6.25):

$$E = \{u \in C^2(\overline{\Omega}) \mid -\Delta u = u^p, \; u \geq 0 \text{ in } \Omega, \quad u = 0 \text{ on } \partial\Omega\}.$$

In the case of $1 < p < p_s$ we have $E \neq \emptyset$ and $d = \inf\{J(u) \mid u \in E \setminus \{0\}\}$. This relation is valid even if E is replaced by

$$\hat{E} = \left\{u \in C^2(\overline{\Omega}) \mid -\Delta u = |u|^{p-1} u \text{ in } \Omega, \quad u = 0 \text{ on } \partial\Omega\right\}.$$

If $p = p_s$, then $d = \frac{1}{n} S^{n/2}$ for the Sobolev constant determined by (6.27).

It is not difficult to see that for any $u \in H_0^1(\Omega) \setminus \{0\}$, the mapping $s \in [0, \infty) \mapsto J(su)$ takes the maximum if and only if $su \in \mathcal{N}$. Furthermore,

$$W_* = \{u \in X \mid J(u) < d, \; I(u) > 0\} \cup \{0\}$$

is a bounded neighborhood of 0 in $X = H_0^1(\Omega)$, and is called the *stable set*. On the other hand,

$$V_* = \{u \in X \mid J(u) < d, \; I(u) < 0\}$$

is called the *unstable set*, and it holds that $0 \notin \overline{V_*}$ in X. Then, it holds that $\overline{W_*} \cap \overline{V_*} = E_*$ for

$$E_* = \{u \in \mathcal{N} \mid J(u) = d\} = \{u \in E \mid J(u) = d\},$$

and each element in E_* is called the *minimum energy solution.* If $1 < p < p_s$, then $E_* \neq \emptyset$, while $E_* = \emptyset$ if $p = p_s$ and Ω is star-shaped, because then *Pohozaev's identity* guarantees that $E = \{0\}$. It is also known that $\inf_{\mathcal{N}} J$ is attained by the element in E_*, which is referred to as the *Nehari principle.*

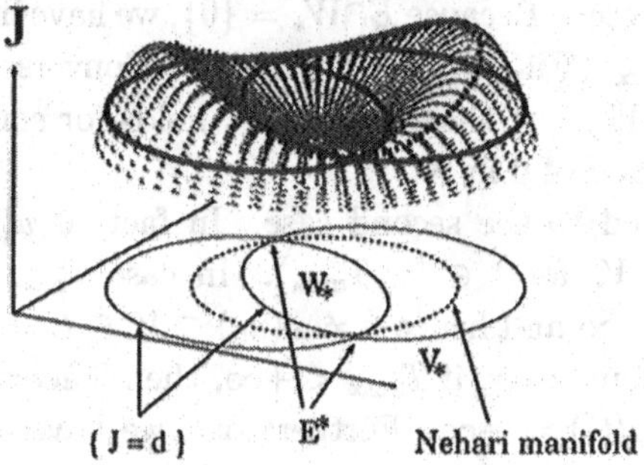

Fig. 6.1

Thus, each W_* and V_* forms the connected component of

$$\{u \in X \mid J(u) < d\}$$

in X and in particular, is a *positively invariant set* of (6.25) as $u_0 \in W_*$ (resp. V_*) implies $u(t) \in W_*$ (resp. V_*) for $t \in [0, T_{\max})$. Now, we can prove the following.

Theorem 6.11 *Let $1 < p < p_s$ and $u = u(\cdot, t)$ be the H^1-solution with the initial value $u_0 \in X = H_0^1(\Omega)$ in $u_0 \geq 0$, and let $T_{\max}$ be the blowup time, that is, the supremum of the existence time of the solution. Then we have the following alternatives.*

(1) *$u(t_0) \in W_*$ for some $t_0 \in [0, T_{\max})$, which is equivalent to $T_{\max} = +\infty$ and $\lim_{t \to +\infty} \|\nabla u(t)\|_2 = 0$.*
(2) *$u(t_0) \in V_*$ for some $t_0 \in [0, T_{\max})$, which is equivalent to $T_{\max} < +\infty$.*
(3) *$T_{\max} = +\infty$ and $u(t) \notin W_* \cup V_*$ for any $t \in [0, T_{\max})$, which is equivalent to $T_{\max} = +\infty$ and $0 \notin \omega(u_0)$.*

In the last case, $\mathcal{O} = \{u(t)\}_{t \geq 0}$ is called the *floating orbit*. Because any orbit global in time is uniformly bounded in this case, then it holds that $\omega(u_0)$ is a compact connected set contained in E_-.

Proof. To prove the first case, we note that $u(t_0) \in W_*$ implies $u(t) \in W_*$ for any $t \in [t_0, T_{\max})$. Because W_* is bounded, we have $\sup_{t \in [t_0, T_{\max})} \|\nabla u(t)\|_2 < +\infty$, and hence $T_{\max} = +\infty$ follows from (6.45). Now, it becomes the classical solution globally in time and it holds that $\sup_{t \geq 1} \|u(t)\|_\infty < +\infty$ and

$\omega(u_0) \subset E$ is connected. Because $E \cap W_* = \{0\}$, we have $\lim_{t\to\infty} \|u(t)\|_{C^2(\overline{\Omega})} = 0$ and hence $\lim_{t\to\infty} \|\nabla u(t)\|_2 = 0$ holds true. Conversely, if $T_{\max} = +\infty$ and $\lim_{t\to\infty} \|\nabla u(t)\|_2 = 0$, then $u(t_0) \in W_*$ holds for some $t_0 \geq 0$ because W_* is a neighborhood of 0 in X.

Now, we proceed to the second case. In fact, if $u(t_0) \in V_*$, then it holds that $u(t) \in V_*$ for $t \in [t_0, T_{\max})$. In case $T_{\max} = +\infty$, we have $\sup_{t\geq 1} \|u(t)\|_\infty < +\infty$ and hence $\emptyset \neq \omega(u_0) \subset V_* \cap E = \emptyset$ follows. This is a contradiction. Conversely, if $T_{\max} < +\infty$, then Theorem 6.5 guarantees that $\lim_{t\to T_{\max}} J(u(t)) = -\infty$. Furthermore, we have $u(t) \notin W_*$ for $t \in [0, T_{\max})$ from the first step. Therefore, it holds that $u(t_0) \in V_*$ for some $t_0 \in [0, T_{\max})$.

The final case is a direct consequence of Theorem 6.8, as $T_{\max} = +\infty$ with $0 \notin \omega(u_0)$ is equivalent to be other than the first and the second cases.

The proof is complete. □

Several parts are open in the case of $p = p_s$. The if part of the first case is obvious, while the only if part is not known. However, we can prove the existence of $\varepsilon_0 > 0$ such that $\|\nabla u(t_0)\|_2 < \varepsilon_0$ for some $t_0 \in [0, T_{\max})$ implies that $T_{\max} = +\infty$ and $\lim_{t\to\infty} \|\nabla u(t)\|_2 = 0$. The only if part of the second case is proven by a different argument, and its if part is true at least for $n \geq 4$. On the other hand, the final part is rather different. From Theorem 6.6, we have the alternatives that $\lim_{t\to\infty} \|u(t)\|_\infty < +\infty$ and $\lim_{t\to\infty} \|u(t)\|_\infty = +\infty$ in this case. Furthermore, if the former occurs then it holds that $\emptyset \neq \omega(u_0) \subset E$. Because $E = \{0\}$ holds if Ω is star-shaped, then we obtain $\lim_{t\to\infty} \|\nabla u(t)\|_2 = \lim_{t\to\infty} \|u(t)\|_\infty = 0$. On the other hand, in the latter case, we have $\lim_{t\to\infty} J(u(t)) = kd$ for some $k = 1, 2, \cdots$. This actually occurs if $n \geq 4$.

6.2.5 *Method of Rescaling*

Let us go back to (6.8). If $u = u(x,t)$ is the solution, then $u_\lambda(x,t) = \lambda^{\frac{2}{p-1}} u(\lambda x, \lambda^2 t)$ is so except for the initial value, where $\lambda > 0$. This transformation is called the *forward self-similar transformation*, and if it is invariant under this transformation the solution u is said to be *self-similar*. This means that $u = u_\lambda$ for any $\lambda > 0$ and hence

$$u(x,t) = t^{-\frac{1}{p-1}} f\left(x/\sqrt{t}\right)$$

holds with some $f = f(y)$ satisfying

$$-\Delta f - \frac{1}{2}x \cdot \nabla f = \frac{1}{p-1}f + f^p, \quad f \geq 0, \ f \not\equiv 0 \quad \text{in} \quad \mathbf{R}^n. \tag{6.47}$$

It is known that if $1 + \frac{2}{n} = p_f < p < p_s = \frac{n+2}{(n-2)_+}$, then some solutions global in time to (6.8) converges to a self-similar solution as $t \to +\infty$. On the other hand, the *backward self-similar transformation* is useful to control the blowup behavior in finite time of the solution as is indicated in §6.2.1.

Concerning the radial symmetry of the self-similar solution, we have the following.

Theorem 6.12 *If*

$$f(x) = o\left(|x|^{-\frac{2}{p-1}}\right) \tag{6.48}$$

holds in (6.47) as $|x| \to +\infty$, then it follows that $f = f(|x|)$.

Thus, from the work on radially symmetric forward self-similar solution, we have the following fact concerning the solution to (6.47) satisfying (6.48).

(1) If $p \geq p_s$, then such a solution does not exist.
(2) If $p_f < p < p_s$, then such a solution exists uniquely.

In fact, for the solution $u = u(r, \alpha)$ to

$$u'' + \left(\frac{n-1}{r} + \frac{r}{2}\right)u' + \frac{1}{p-1}u + |u|^{p-1}u = 0 \quad \text{for} \quad r > 0$$

with

$$u'(0) = 0 \quad \text{and} \quad u(0) = \alpha,$$

we have the following.

(1) The finite value $L(\alpha) = \lim_{r\to\infty} r^{2/(p-1)}u(r,\alpha)$ exists for each $\alpha \in \mathbf{R}$.
(2) If $L(\alpha) = 0$, then it holds that

$$u(r,\alpha) = Ae^{-r^2/4}r^{2/(p-1)-n}\left\{1 + O\left(r^{-2}\right)\right\} \quad \text{as} \quad r \to \infty$$

with some $A \in \mathbf{R} \setminus \{0\}$.
(3) If $p \geq p_s$, then $u(r,\alpha)$ is positive on $[0,\infty)$ and $L(\alpha) > 0$ for each $\alpha > 0$.

(4) If $p_f < p < p_s$, then there is a unique $\alpha_p > 0$ such that $u(r, \alpha_p)$ is positive on $[0, \infty)$ and $L(\alpha_p) = 0$. Moreover, if $\alpha \in (0, \alpha_p)$ it holds that $L(\alpha) > 0$.

Thus, if $p > p_f$, we have the solution $f = f(|x|)$ to (6.47) satisfying $\ell = \lim_{r\to\infty} r^{\frac{2}{p-1}} f(r) > 0$. Then, we have a counterpart of Theorem 6.12, where S^{n-1} denotes the n-dimensional unit sphere.

Theorem 6.13 *If $p > p_f$, then any $A \in C(S^{n-1})$ in $0 \le A(\sigma) \le \ell$ ($\sigma \in S^{n-1}$) admits a solution $f = f(x)$ to (6.47) such that*

$$\lim_{r\to\infty} r^{\frac{2}{p-1}} f(r\sigma) = A(\sigma)$$

uniformly in $\sigma \in S^{n-1}$.

From Theorem 6.3, we see that if $1 < p < p_f$, then there is no solution to (6.47). This is also the case of $p = p_f$.

To describe the relation to the asymptotic profile of the solution to (6.8), we put $K(x) = e^{|x|^2/4}$,

$$L^q(K) = \left\{ v : \mathbf{R}^n \to \mathbf{R} \ \text{ measurable} \mid \int_{\mathbf{R}^n} v(y)^q K(y) dy < +\infty \right\},$$

and

$$H^1(K) = \left\{ v \in L^2(K) \mid \nabla v \in L^2(K)^n \right\},$$

where $q \in [1, \infty)$.

Given $\psi \ge 0$, $\psi \not\equiv 0$ in $H^1(K) \cap L^\infty(\mathbf{R}^n)$, we take $u_0 = \mu\psi$ in (6.1). The classical solution and its blowup time are denoted by $u_\mu = u_\mu(x,t)$ and $T_{\max}(\mu) \in (0, \infty]$, respectively. Then, if $p_f < p < p_s$ we have a unique $\mu_0 > 0$ such that $\mu < \mu_0$ and $\mu > \mu_0$ imply $T_{\max}(\mu) = +\infty$ with $\lim_{t\to\infty} \|u_\mu(t)\|_\infty = 0$ and $T_{\max}(\mu) < +\infty$, respectively. Here, we have $T_{\max}(\mu_0) = +\infty$ and

$$\lim_{t\to\infty} \left\| t^{\frac{1}{p-1}} u_{\mu_0}(t) - f\left(\cdot/\sqrt{t}\right) \right\|_\infty = 0.$$

In 1987, M. Escobedo and O. Kavian proved the following, where

$$Lv = -\frac{1}{K} \nabla \cdot (K \nabla) \qquad \text{and} \qquad \|v\|_{q,K} = \left(\int_{\mathbf{R}^n} |v(y)|^q K(y) dy \right)^{1/q}.$$

(1) $H^1(K) \hookrightarrow L^q(K)$ holds for $1 \le q \le p_s$ and is compact for $1 \le q < p_s$. Furthermore, if $n \ge 3$, then the Poincaré-Sobolev inequality

$$S_n \|v\|_{2^*,K}^2 + \lambda_* \|v\|_{2,K}^2 \le \|\nabla v\|_{2,K}^2 \tag{6.49}$$

holds for $v \in H^1(K)$, where $\lambda_* = \max(1, n/4)$ and S_n is the Sobolev constant:

$$S_n = \inf\left\{\|\nabla v\|_2^2 \mid v \in C_0^\infty(\mathbf{R}^n),\ \|v\|_{2^*} = 1\right\}.$$

(2) The operator L is realized as a self-adjoint operator in $L^2(K)$ with the domain $D(L) = \left\{v \in L^2(K) \mid Lv \in L^2(K)\right\}$. It is positive definite with the compact resolvent and it holds that $D(L^{1/2}) = H^1(K)$. Its eigenvalues are given by $\lambda_k = \frac{n+k-1}{2}$ $(k = 1, 2, \cdots)$ with the multiplicity

$$\begin{pmatrix} n+k-2 \\ n-1 \end{pmatrix}$$

and the eigenfunction $P_{k-1}(D)e^{-|y|^2}$, where $P_{k-1}(\xi)$ denotes the homogeneous polynomial of degree $k-1$.

Applying the *forward self-similar transformation*

$$v(y,s) = (t+1)^{\frac{1}{p-1}} u(x,t), \qquad y = x/(t+1)^{1/2}, \qquad s = \log(t+1)$$

to (6.1), we have

$$v_s + Lv = \frac{1}{p-1} v + v^p \qquad \text{in} \qquad \mathbf{R}^n \times (0, \infty). \tag{6.50}$$

This equation is treated very similarly to (6.25) on the bounded domain, because L has the compact resolvent and the Poincaré-Sobolev inequality (6.49) holds.

Thus, letting

$$J(v) = \frac{1}{2} \|\nabla v\|_{2,K}^2 - \frac{1}{2(p-1)} \|v\|_{2,K}^2 - \frac{1}{p+1} \|v\|_{p+1,K}^{p+1},$$

we have

$$\frac{d}{ds} J(v(s)) \le 0 \qquad (s \ge 0).$$

Now, we apply the Kaplan method in use of $\varphi_1 = e^{-|y|^2/4}$, which satisfies that $L\varphi_1 = \frac{n}{2}\varphi_1$. For $\varepsilon > 0$, we take $\psi = C(\varepsilon)\varphi_1^{1+\varepsilon}$ with $C(\varepsilon) > 0$ satisfying that $\int_{\mathbf{R}^n} \psi K dy = 1$. In fact, we have

$$L\psi \leq \frac{n}{2}(1+\varepsilon)$$

and hence for $u = u(x,t)$ with

$$j(s) = \int_{\mathbf{R}^n} v(s)\psi K dy$$

being well-defined, it holds that

$$\begin{aligned} \frac{dj}{ds} &= \int_{\mathbf{R}^n} v^p \psi K dy + \int_{\mathbf{R}^n} v \cdot \left(\frac{1}{p-1}\psi - L\psi \right) K dy \\ &\geq \left(\int_{\mathbf{R}^n} v\psi K dy \right)^p + \int_{\mathbf{R}^n} \left(\frac{1}{p-1} - \frac{n}{2}(1+\varepsilon) \right) v\psi K dy. \end{aligned}$$

Then, we see that $\frac{1}{p-1} - \frac{n}{2}(1+\varepsilon) > 0$ holds for $0 < \varepsilon \ll 1$ in the case of $1 < p < p_f$. This gives that

$$\frac{dj}{ds} \geq j^p \qquad (s \geq 0)$$

and then non-existence of non-trivial non-negative solution of (6.1) follows in this case.

Chapter 7

System of Chemotaxis

This chapter is devoted to the study of the elliptic-parabolic system of partial differential equations, arising in several areas in mathematical biology and mathematical physics. The first section is the description of the background and the motivation of mathematical study. Then, we shall establish the local wellposedness in the second section.

7.1 Story

7.1.1 *The Keller-Segel System*

System of parabolic partial differential equations is proposed to describe several phenomena in mathematical biology. A typical example is

$$\left.\begin{aligned} u_t &= \nabla\cdot(\nabla u - u\nabla v) \\ 0 &= \Delta v - av + u \end{aligned}\right\} \quad \text{in} \quad \Omega\times(0,T)$$
$$\frac{\partial u}{\partial \nu} = \frac{\partial v}{\partial \nu} = 0 \quad \text{on} \quad \partial\Omega\times(0,T)$$
$$u|_{t=0} = u_0(x) \quad \text{on} \quad \Omega, \tag{7.1}$$

where $\Omega \subset \mathbf{R}^n$ is a bounded domain with smooth boundary $\partial\Omega$, $a > 0$ a constant, and ν the outer unit vector on $\partial\Omega$. It is proposed by T. Nagai in 1995 as a simplified form of the one given by E.F. Keller and L.A. Segel in 1970. Here, $u = u(x,t)$ and $v = v(x,t)$, respectively, stand for the density of cellular slime molds and the concentration of chemical substances secreted by themselves at the position $x \in \Omega$ and the time $t > 0$.

The first equation describes the conservation of mass, where flux of u is given by $\mathcal{F} = -\nabla u + u\nabla v$, as

$$\frac{d}{dt}\int_{\omega} u = -\int_{\partial\omega} \mathcal{F}\cdot\nu$$

holds for any subdomain $\omega \subset \Omega$ with $\overline{\omega} \subset \Omega$. The first term $-\nabla u$ of $\mathcal{F}$ is the vector field with the direction where u decreases mostly, and with the rate equal to its derivative to that direction. The second term $u\nabla v$, on the other hand, indicates that u is carried by the vector field ∇v with the direction where v increases mostly, and with the rate equal to its derivative to that direction. Thus, the effect of diffusion $-\nabla u$ and that of chemotaxis $u\nabla v$ are competing for u to vary. In this context, the boundary condition for u is preferably replaced by the null flux condition,

$$\mathcal{F}\cdot\nu \equiv \frac{\partial u}{\partial\nu} - u\frac{\partial v}{\partial\nu} = 0 \qquad \text{on} \quad \partial\Omega\times(0,T).$$

A general form of this equation is given in §4.1. Each of them is derived from the different principle, and each feature of the solution is also different from the others.

However, a similar system to (7.1) is found in statistical mechanics. There, domain Ω is usually replaced by $\mathbf{R}^n$, and the second equation takes the form

$$v(x,t) = \int_{\mathbf{R}^n} \Gamma(x,y)u(y,t)dy, \tag{7.2}$$

where

$$\Gamma(x,y) = \begin{cases} \frac{1}{2}|x-y| & (n=1) \\ \frac{1}{2\pi}\log\frac{1}{|x-y|} & (n=2) \\ \frac{1}{4\pi|x-y|} & (n=3) \end{cases} \tag{7.3}$$

denotes the (-1) times potential of the gravitational force. It is derived from Langevin and then Fokker-Planck equations, describing the motion of mean field of self-interacting particles. Therefore, while the first equation of (7.1) is concerned with the mass conservation of particles, the second one replaced by (7.2) is the description of the total field of gravitational force made by those particles. This form (7.2) is a natural extension of the second equation of (7.1) to the whole space, as the latter is equivalent to the equality in use of the Green's function for $-\Delta_N + a$, denoted by $G(x,y)$:

$$v(x,t) = \int_\Omega G(x,y)u(y,t)dy. \tag{7.4}$$

In fact, we have

$$G(x,y) = H(x,y) + \begin{cases} \Gamma(x,y) & (y \in \Omega) \\ 2\Gamma(x,y) & (y \in \partial\Omega) \end{cases} \tag{7.5}$$

with $H = H(x,y)$ standing for the regular part.

Other forms of the second equation are also proposed. Taking account of the boundary condition to v, they are totally described as

$$\tau\frac{dv}{dt} + Av = u \quad \text{in} \quad L^2(\Omega), \tag{7.6}$$

where $A > 0$ is a self-adjoint operator $A > 0$ with the compact resolvent. Here, τ is a non-negative constant. As we have seen, if $\tau = 0$ the field created by those particles is physical. In this case we call it the *simplified system*. On the other hand, $\tau > 0$ arises when the field is formed through the chemical material. This case is realistic in some biological media, and (7.1) with the second equation replaced by (7.6) of $\tau > 0$ the *full system*. There the additional initial condition $v|_{t=0} = v_0(x)$ is imposed. In the other case, equation (7.6) is reduced to the ordinary differential equation such as

$$\tau\frac{\partial v}{\partial t} = u.$$

Actually, it follows from the statistical model of cellular automaton as is described in §4.1. There, the effect of transmissive action is restricted to each cell and the field is not formed in the classical sense.

7.1.2 *Blowup Mechanism*

It will be shown that the classical solution to (7.1) exists locally in time if the initial value is smooth, and becomes positive if it is non-negative and not identically zero. Let $T_{\max} > 0$ be the supremum of the existence time of the solution. If $T_{\max} < +\infty$, we say that the solution blows-up in finite time. The blowup mechanism of (7.1) depends sensitively on the space dimension n, and in the case of $n = 2$, spiky patterns are formed as $t \uparrow T_{\max}$.

Henceforth, $\mathcal{M}(\overline{\Omega})$ denotes the set of measures on $\overline{\Omega}$, $\rightharpoonup$ the $*$-weak convergence there, and $\delta_{x_0}(dx)$ the delta function, respectively. Therefore, $\mathcal{M}(\Omega)$ is the dual space of $C(\overline{\Omega})$, the set of continuous functions on $\overline{\Omega}$, and

$$\langle \varphi, \delta_{x_0}(dx) \rangle = \varphi(x_0)$$

for $\varphi \in C(\overline{\Omega})$. Now, if $n = 2$ and $T_{\max} < +\infty$, the solution $u(x,t)$ to (7.1) satisfies

$$u(x,t)dx \quad \rightharpoonup \quad \sum_{x_0 \in \mathcal{S}} m(x_0)\delta_{x_0}(dx) + f(x)dx \tag{7.7}$$

in $\mathcal{M}(\overline{\Omega})$ as $t \uparrow T_{\max}$ with $m(x_0) = m_*(x_0)$ for

$$m_*(x_0) \equiv \begin{cases} 8\pi & (x_0 \in \Omega) \\ 4\pi & (x_0 \in \partial\Omega) \end{cases}$$

and $0 \leq f \in L^1(\Omega) \cap C(\overline{\Omega} \setminus \mathcal{S})$. On the other hand, we have

$$\lim_{t \uparrow T_{\max}} \|u(t)\|_\infty = +\infty$$

and $\mathcal{S}$ coincides with the blowup set of u. That is, $x_0 \in \mathcal{S}$ if and only if there exist $x_k \to x_0$ and $t_k \uparrow T_{\max}$ such that $u(x_k, t_k) \to +\infty$. This means $\mathcal{S} \neq \emptyset$ if $T_{\max} < +\infty$. Here, we have

$$\|u(t)\|_1 = \|u_0\|_1 \tag{7.8}$$

and hence

$$2 \cdot \sharp(\Omega \cap \mathcal{S}) + \sharp(\partial\Omega \cap \mathcal{S}) \leq \|u_0\|_1 / (4\pi) \tag{7.9}$$

follows from (7.7) and (7.8). In particular, $\|u_0\|_1 < 4\pi$ implies $T_{\max} = +\infty$. This fact is related to the conjecture of S. Childress and J.K. Percus in 1981 concerning the threshold in L^1 norm of the initial value for the blowup of the solution.

It was obtained by semi-analysis, derivation of the stationary problem in use of the free energy and numerical study to its bifurcation diagram. On the other hand, relation (7.7) was conjectured by V. Nanjundiah and is referred to as the formation of *chemotactic collapses*, although the terminology is not consistent with that in §4.1. In this context of biology, each collapse

$$m(x_0)\delta_{x_0}(dx)$$

is supposed to express a spore made from the slime molds. Inequality (7.9) indicates that the phenomenon of threshold in $\|u_0\|_1$ concerning the blowup of the solution is a consequence of the formation of collapses in the blowup process. Equality $m(x_0) = m_*(x_0)$ is referred to as the *mass quantization of collapses.* It means that the spore is formed in the normalized mass. It is related to the optimality of the condition $\|u_0\|_1 < 4\pi$ for $T_{\max} = +\infty$ to occur, proven by T. Nagai and the authors. It was observed that the mass quantization holds if the solution is continued after the blowup time, or it blows-up in infinite time. In this connection it was noted that the Fokker-Planck equation admits the weak solution globally in time, provided that the initial value has a finite second moment and is bounded and summable. Fokker-Planck equation is concerned with the case that the distribution of particles is thin, and system (7.1) is regarded as its adiabatic limit. Although post blowup continuation does not hold, mass quantization is valid in (7.1). Here, we emphasize that the mass quantization agrees with the blowup mechanism in the stationary problem, which arises as a nonlinear elliptic eigenvalue problem. We call this story the *nonlinear quantum mechanics.* If the concentration speed is rapid, then the particles are thin near the blowup oint, which makes the blowup mechanism simple. Actually, this case is referred to as the *type II blowup point,* and then the whole blowup mechanism is contained in infinitely small parabolic region in (x,t) space, called the *hyper-parabola.* The family of blowup solutions constructed by M.A. Herrero and J.J.L. Velázquez in 1996 by the method of matched asymptotic expansion is of this type. In the other case, referred to as the *type I blowup point,* the feature of the blowup mechanism is rather different from the previous one. Actually, infinitely wide parabolic region, called the *parabolic envelope,* is necessary to describe the whole blowup mechanism, but the local free energy gets to $+\infty$. It is open whether such a blouwp point actually exists or not.

It is known that the blowup mechanism of the parabolic equation

$$u_t - \Delta u = u^p, \quad u \geq 0 \qquad \text{in} \qquad \Omega \times (0,T)$$

with $u|_{\partial\Omega} = 0$ is controlled by the ordinary differential part $\dot{u} = u^p$ if the nonlinearity is sub-critical as $p \in (1, \frac{n+2}{n-2})$, where $\Omega \subset \mathbf{R}^n$ is a bounded convex domain. Namely, if x_0 is a blowup point, then

$$u(x,t) = (T-t)^{-\frac{1}{p-1}} \left(\frac{1}{p-1} \right)^{\frac{1}{p-1}} \{1 + o(1)\}$$

holds as $t \uparrow T = T_{\max}$ uniformly in $|x - x_0| \leq C(T-t)^{1/2}$. Here, the concentration is so slow that $u(x,t)$ becomes flat in any parabolic region, and total blowup mechanism is not enveloped there in this case. On the other hand, the blowup solution of Herrero and Velázquez to (7.1) has the form

$$u(x,t) = \frac{1}{r(t)^2}\overline{u}\left(\frac{x}{r(t)}\right)\{1+o(1)\} + O\left(\frac{e^{-\sqrt{2}|\log(T-t)|^{1/2}}}{|x|^2}\cdot 1_{\{|x|\geq r(t)\}}\right)$$

as $t \uparrow T = T_{\max}$ uniformly in $|x| \leq C(T-t)^{1/2}$, where

$$r(t) = C(T-t)^{1/2}\cdot e^{-\sqrt{2}/2\,|\log(T-t)|^{1/2}} \cdot |\log(T-t)|^{\frac{1}{4}\log^{-1/2}(T-t)-\frac{1}{4}}(1+o(1))$$

and $\overline{u}(y) = 8\cdot\left(1+|y|^2\right)^{-2}$. We have $0 < r(t) \ll b(T-t)^{1/2}$ for any $b > 0$. This solution creates collapses again, under the backward self-similar transformation $z(y,s) = (T-t)u(x,t)$ for $y = x/(T-t)^{1/2}$ and $s = \log(T-t)$. Thus, super-critical nonlinearity of (7.1) admits type II blowup point with high concentration.

Exercise 7.1 Introduce the stationary problem and the Lyapunov function for (7.1), following the idea to DD model.

7.1.3 *Free Energy*

Parabolic-elliptic system of partial differential equations is found in several areas of applied and theoretical physics. The drift-diffusion model for semiconductor device is written as

$$\left.\begin{aligned} n_t &= \nabla\cdot(\nabla n - n\nabla\varphi) \\ p_t &= \nabla\cdot(\nabla p + p\nabla\varphi) \\ \Delta\phi &= n - p \end{aligned}\right\} \quad \text{in} \quad \Omega\times(0,T)$$

$$\left.\begin{aligned} \frac{\partial n}{\partial\nu} - n\frac{\partial\varphi}{\partial\nu} &= 0 \\ \frac{\partial p}{\partial\nu} + p\frac{\partial p}{\partial\nu} &= 0 \\ \varphi &= 0 \end{aligned}\right\} \quad \text{on} \quad \partial\Omega\times(0,T),$$

where $n = n(x,t)$ and $p = p(x,t)$ are the densities of electron and positron, respectively, and $\varphi = \varphi(x,t)$ is the electric charge field. Particles of the same kind are self-repulsive, while the attractive force acts between different kind of particles, and thus, the system is dissipative totally. Vortex system is given by

$$\left. \begin{array}{l} \omega_t = \nabla \cdot (\nabla \omega - \omega \nabla^{\perp} \psi) \\ -\Delta \psi = \omega \end{array} \right\} \quad \text{in} \quad \mathbf{R}^2 \times (0,T),$$

where

$$\nabla^{\perp} = \begin{pmatrix} -\frac{\partial}{\partial x_2} \\ \frac{\partial}{\partial x_1} \end{pmatrix}$$

for $x = (x_1, x_2)$. It comes from the Navier-Stokes system

$$\left. \begin{array}{r} u_t - \Delta u + u \cdot \nabla u = \nabla p \\ \nabla \cdot u = 0 \end{array} \right\} \quad \text{in} \quad \mathbf{R}^3 \times (0,T),$$

where

$$u = \begin{pmatrix} u_1 \\ u_2 \\ u_3 \end{pmatrix} \quad \text{and} \quad \nabla = \begin{pmatrix} \frac{\partial}{\partial x_1} \\ \frac{\partial}{\partial x_2} \\ \frac{\partial}{\partial x_3} \end{pmatrix}$$

denote the velocity and the gradient operator, respectively. If we take the two-dimensional model with $x = (x_1, x_2, 0)$ and $u_3 = 0$, then we get

$$\nabla \times u = \begin{pmatrix} 0 \\ 0 \\ \omega \end{pmatrix} \quad \text{for} \quad \omega = \omega(x_1, x_2).$$

This system is also dissipative but some underlying chaotic features are observed. Directions of self-interacting forces of those systems, chemotaxis, semi-conductor device, and vortices are different, but some common structures are noticed. Let us recall that the second law of thermodynamics; the mean field of many particles is governed by the free energy, decreasing in time. Its local minimum is an equilibrium state, while transient dynamics are controlled by the critical points, especially, non-local minima.

We note that free energy is given by inner energy minus entropy. If $\rho = \rho(x) \geq 0$ denotes the density of particles, entropy on the domain

$\Omega \subset \mathbf{R}^n$ is given as

$$-\int_\Omega \rho(\log\rho - 1).$$

On the other hand, inner energy is composed of kinetic and potential energies so that it is given as

$$-\frac{1}{2}\int\int_{\Omega\times\Omega}\Gamma(x,y)\rho(x)\rho(y)dxdy + \int_\Omega \rho V,$$

where $-\Gamma(x,y)$ and $V(x)$ denote the potentials of self-interactions and external force, respectively. Note that Newton's third law implies

$$\Gamma(x,y) = \Gamma(y,x).$$

Actually, it is given as (7.3) if the self-interaction is caused by the gravitational force. Thus, physical question is to derive mean field equation of which free energy is given by

$$\mathcal{F}(\rho) = \int_\Omega \rho(\log\rho - 1) - \frac{1}{2}\int\int_{\Omega\times\Omega}\Gamma(x,y)\rho(x)\rho(y)dxdy + \int_\Omega \rho V.$$

It has been known that such a system is realized by introducing friction and fluctuations of particles.

The classical theory of Jeans and Vlasov starts with the Newton equation

$$\frac{dx_i}{dt} = v_i, \qquad m\frac{dv_i}{dt} = \nabla_{x_i}\left\{-mV(x_i) + m^2\sum_{j\neq i}\Gamma(x_j,x_i)\right\} \tag{7.10}$$

for $1 \leq i \leq N$. Letting $N \to \infty$ with $M = mN$ preserved, it asserts the convergence

$$\mu^N(dx,dv,t) = m\sum \delta_{x_i(t)}(dx)\otimes\delta_{v_i(t)}(dv) \rightharpoonup f(x,v,t)dxdv$$

with $f(x,v,t)$ satisfying the kinetic model, referred to as the Jeans-Vlasov equation. In the normal form, it is given as

$$f_t = -\nabla_x\cdot(vf) + \gamma\nabla_v\cdot[f\nabla_x(U+V)]$$
$$U(x,t) = -\int\int\Gamma(x,y)f(y,v,t)dvdt.$$

In the process of $(dv_i)/(dt) \to 0$, the distribution function $f(x,v,t)$ is replaced by the Maxwellian $\omega(x,t)\pi^{-n/2}e^{-v^2/2}$. If $n = 2$, then $\omega(x,t)$ is subject to the vorticity equation derived from the Euler equation, that is,

$$-\Delta\psi = \omega, \qquad \omega_t = -\nabla \cdot \left(\omega\nabla^{\perp}\left(\psi + V\right)\right).$$

The stationary state of this equation, $\omega = \omega(x)$ is associated with the elliptic problem

$$-\Delta\psi = g(\psi + V)$$

with the nonlinearity g unknown. If mass is so concentrated as

$$\omega(x,t) = \sum \delta_{x_j(t)}(dx),$$

then the positions are subject to the Hamiltonian system

$$\frac{dx_i}{dt} = \nabla_{x_i}^{\perp}\mathcal{H}(x_1, x_2, \cdots, x_N) \qquad (i = 1, 2, \cdots, N),$$

where

$$\mathcal{H}(x_1, x_2, \cdots, x_N) = -\sum_i V(x_i) + \sum_{j \neq i} \Gamma(x_i, x_j).$$

If $K(x,y)$ is replaced by $G(x,y)$ in (7.10), then $\frac{1}{2}\sum_i R(x_i)$ is added to the right-hand side, where $R(x)$ is the regular part of $K(x,y)$ so that $R(x) = H(x,x)$ with $H(x,y)$ defined by (7.5). However, the Newton equation is time reversible and this hierarchy of systems is not subject to the second law of thermodynamics, that is, decreasing of the free energy. Actually, this hierarchy is governed by three laws of conservation; mass, momentum, and energy. As a consequence, it has a feature of chaotic motion of particles.

An answer that we know to derive systems provided with free energy is to replace the Newton equation by the Langevin equation. More precisely, this requirement is realized when the particles are subject to the friction and random fluctuations:

$$\begin{aligned} dx_i &= v_i dt \\ m dv_i &= \nabla_{x_i}\left\{-mV(x_i) + m^2\sum_{j\neq i}\Gamma(x_j, x_i)\right\} - \beta v_i dt \\ &\quad +(2\beta kT)^{1/2} dW_t^i. \end{aligned}$$

Here, k, T, and β are Boltzmann constant, temperature, friction coefficient, respectively, and (W_t^i) denotes the white noise. Its kinetic model, referred to as the Fokker-Planck equation is given as

$$f_t = -\nabla_x \cdot (vf) + \nabla_v \cdot [f\nabla_x (U+V)] + \beta kT \nabla_v \cdot (vf + \Delta_v f)$$
$$U(x,t) = -\int\int \Gamma(x,y) f(y,v,t)\,dydv,$$

where

$$\rho(x,t) = \int f(x,v,t)\,dv \qquad \text{and} \qquad \lambda = \int \rho(x,t)\,dx$$

stand for the density and the total mass, respectively. Then, in the adiabatic limit $\beta \to +\infty$, we have

$$\rho_t = \nabla \cdot (\rho \nabla U) + \nabla \cdot (\rho \nabla V) + \Delta \rho.$$

If $V = 0$ and the kernel $\Gamma(x,y)$ is replaced by $G(x,y)$, it is nothing but the simplified system of chemotaxis. Semi-conductor device equation is obtained similarly by taking the opposite sign of the kernel $G(x,y)$. In those systems of chemotaxis and semi-conductor device the interaction acts attractively and repulsively, respectively, and in the Euler equation, particles receive the force perpendicular to the level lines of the field.

As is mentioned, stationary state of the above equation is described by elliptic problem with exponential nonlinearity. Furthermore, the localized densities are subject to the gradient flow. In this way, this hierarchy of equations starts with the free energy as the physical principle. On the other hand, mathematically it is characterized by the quantization of the blowup mechanism as is described in the previous paragraph, and it comes from the quantized structure of the set of stationary solutions. Another important consequence of this observation is the variational structure of the stationary problems derived from the free energy. Actually, it is regarded as the dual variation of the standard one, and remarkably those variational structures are equivalent up to Morse indices.

7.2 Well-posedness

7.2.1 *Summary*

As is described in the previous section, E.F. Keller and L.A. Segel proposed a mathematical model describing the chemotactic aggregation of cellular slime molds which move preferentially toward the area with relatively high concentration of a chemical substance secreted by the amoebae themselves. Then, V. Nanjundiah introduced a simplified form and conjectured the formation of collapses. Later in 1996, M. Mimura and T. Tsujikawa modelled the formation of some kind of bacterium's colony by another system, and studied the asymptotic behavior of the solution. Here, we show the unique existence of the classical solution locally in time.

Namely, we take the following system denoted by (CS):

$$u_t = \nabla \cdot (\nabla u + \chi(u,v)\nabla v) + f(u,v) \quad \text{in} \quad \Omega \times (0,T) \tag{7.11}$$

$$\tau v_t = \Delta v + g(u,v) \quad \text{in} \quad \Omega \times (0,T) \tag{7.12}$$

$$\frac{\partial u}{\partial \nu} = 0 \quad \text{on} \quad \partial\Omega \times (0,T) \tag{7.13}$$

$$\frac{\partial v}{\partial \nu} = 0 \quad \text{on} \quad \partial\Omega \times (0,T) \tag{7.14}$$

$$u|_{t=0} = u_0 \quad \text{on} \quad \Omega \tag{7.15}$$

$$v|_{t=0} = v_0 \quad \text{on} \quad \Omega, \tag{7.16}$$

where Ω is a bounded domain in $\mathbf{R}^n$ $(n = 1, 2, 3)$ with smooth boundary $\partial\Omega$, and χ, f and g are smooth functions of u and v. Furthermore, $\tau > 0$ is a constant and henceforth it is put to 1 for simplicity. We say that (u, v) is a classical solution to (CS) in $\Omega \times (0,T)$, if $u = u(x,t)$ and $v = v(x,t)$ are in $u, v \in C(\overline{\Omega} \times [0,T))$,

$$\frac{\partial u}{\partial x_j}, \ \frac{\partial v}{\partial x_j}, \ \frac{\partial^2 u}{\partial x_j \partial x_i}, \ \frac{\partial^2 v}{\partial x_j \partial x_i}, \ u_t, \ v_t \in C(\Omega \times (0,T))$$

and (CS) is satisfied.

First, we study the linear system

$$\text{(HE)} \begin{cases} w_t = \Delta w + \mathbf{a} \cdot \nabla w + h & \text{in} \quad \Omega \times (0,T) \\ \dfrac{\partial w}{\partial \nu} = 0 & \text{on} \quad \partial\Omega \times (0,T) \\ w|_{t=0} = w_0 & \text{on} \quad \Omega, \end{cases}$$

where $\mathbf{a} = (a_1, \cdots, a_n) : \overline{\Omega} \times (0,T) \to \mathbf{R}^n$, $h : \overline{\Omega} \times (0,T) \to \mathbf{R}$, and $w_0 : \overline{\Omega} \to \mathbf{R}$ are given functions. Henceforth, $\| \cdot \|_p$ and $[\,\cdot\,]_\theta$ denote the L^p and the Hölder norms, respectively, where $p \in [1,\infty]$ and $\theta \in (0,1)$. Then, for $\theta \in (0,1)$ and $m = 1,2,3,\cdots$, we put

$$[w]_{m,\theta} \equiv \sum_{|\alpha|<m} \|D_x^\alpha w\|_\infty + \sum_{|\alpha|=m} [D_x^\alpha w]_\theta .$$

For a function w defined on $Q_T \equiv \Omega \times (0,T)$, we set

$$[w]_{\theta;Q_T} \equiv [w]^{(x)}_{\theta;Q_T} + [w]^{(t)}_{\theta/2;Q_T} ,$$

where

$$[w]^{(x)}_{\theta;Q_T} \equiv \sup \left\{ \frac{|w(x,t) - w(x',t)|}{|x-x'|^\theta} \mid x, x' \in \overline{\Omega},\ x \neq x',\ t \in [0,T] \right\},$$

$$[w]^{(t)}_{\theta/2;Q_T} \equiv \sup \left\{ \frac{|w(x,t) - w(x,t')|}{|t-t'|^{\theta/2}} \mid t, t' \in [0,T],\ t \neq t',\ x \in \overline{\Omega} \right\},$$

and

$$C^\theta(Q_T) \equiv \left\{ w \in C(\Omega) \mid [w]_{\theta;Q_T} < \infty \right\}.$$

Finally, putting

$$[w]_{m,\theta;Q_T} \equiv \sum_{2r+|\alpha|<m} \|\partial_t^r D_x^\alpha w\|_\infty + \sum_{2r+|\alpha|=m} [\partial_t^r D_x^\alpha w]_{\theta;Q_T} ,$$

we define

$$C^{m,\theta}(Q_T) \equiv \left\{ w \in C(Q_T) \mid [w]_{m,\theta;Q_T} < \infty \right\}.$$

The following theorem is contained in Theorem 5.2 of chapter IV of the monograph written by O.A. Ladyženskaja, V.A. Solonnikov, and N. N. Ural'ceva.

Theorem 7.1 *Suppose that*

$$(a_1, \cdots, a_n) \in C^{m,\theta}(Q_T)^n \qquad \textit{and} \qquad h \in C^{m,\theta}(Q_T)$$

for $m = 1,2,3,\cdots$ *and* $\theta \in (0,1)$, *and also that* $w_0 \in C^{m+2,\theta}(Q_T)$ *satisfies the compatibility condition up to the order* $[(m+1)/2]$. *Then, there exists a unique classical solution* $w \in C^{m+2,\theta}(Q_T)$ *to (HE). Moreover, it holds that*

$$[w]_{m+2,\theta;Q_T} \leq C\left([h]_{m,\theta;Q_T} + [w_0]_{m+2,\theta}\right)$$

with a constant $C > 0$ *determined by* $\sup_{1\leq j\leq n}[a_j]_{m,\theta,Q_T}$.

We say that w_0 satisfies the compatibility condition with the order $k = 0$ if $w_0^{(0)} = w_0$ satisfies

$$\frac{\partial}{\partial \nu} w_0^{(k)} = 0 \qquad \text{on} \quad \partial\Omega \tag{7.17}$$

for $k = 0$. The compatibility condition with the order $k = 1$ indicates equality (7.17) with $k = 1$, where $w_0^{(1)} = \Delta w_0 + \mathbf{a}(\cdot, 0) \cdot \nabla w(\cdot, 0) + h(\cdot, 0)$. The other cases are defined similarly.

This section is devoted to the proof of the following.

Theorem 7.2 *If* $\Omega \subset \mathbf{R}^n$ *is a convex domain with smooth boundary* $\partial\Omega$, *and* u_0 *and* v_0 *are in* $C^{4,\theta}(\Omega)$ *for* $\theta \in (0,1)$ *and satisfy the compatibility condition up to the order* 1 *for (7.11) with (7.13), (7.15), and for (7.12) with (7.14) and (7.16), respectively, then, there exists a unique classical solution* (u, v) *to (CS) in* Q_T *for some* $T > 0$. *Moreover,* u *and* v *are in* $C^{2,\theta}(Q_T)$.

To prove the above theorem, we make use of the following system denoted by (IS), where $p > n + 2$ is an even integer, $\theta \in (0, 1 - (n+2)/p)$, $U, V \in C^{\theta}(Q_T)$ with $(U, V)|_{t=0} = (u_0, v_0)$, and $T \in (0, 1]$:

$$\begin{aligned} u_t = \Delta u + \chi_u(U, v)\nabla v \cdot \nabla u + \chi_v(U, v)\,|\nabla v|^2 & \\ +\chi(U, v)\Delta v + f(U, V) \qquad \text{in} \quad \Omega \times (0, T), & \qquad (7.18) \end{aligned}$$

$$v_t = \Delta v + g(U, V) \qquad \text{in} \quad \Omega \times (0, T), \tag{7.19}$$

$$\frac{\partial u}{\partial \nu} = 0 \qquad \text{on} \quad \partial\Omega \times (0, T) \tag{7.20}$$

$$\frac{\partial v}{\partial \nu} = 0 \qquad \text{on} \quad \partial\Omega \times (0, T), \tag{7.21}$$

$$u|_{t=0} = u_0 \qquad \text{on} \quad \Omega \tag{7.22}$$

$$v|_{t=0} = v_0 \qquad \text{on} \quad \Omega. \tag{7.23}$$

In the following, first we show the uniqueness and existence of the solution (u, v) to (IS) for each smooth function U and V with $U(\cdot, 0) = u_0$ and $V(\cdot, t) = v_0$ in use of Theorem 7.1 together with some estimates on (u, v).

Thus, we can define the mapping $\mathcal{F} : C^\theta(Q_T)^2 \to C^\theta(Q_T)^2$ by $(u,v) = \mathcal{F}(U,V)$, and then we show that this $\mathcal{F}$ has a fix point, which gives the unique classical solution to (CS).

7.2.2 *The Linearized System*

Studying (IS), first we note that each $(U,V) \in C^\theta(Q_T)^2$ with $(U,V)|_{t=0} = (u_0, v_0)$ admits the unique classical solution $(u,v) \in C^{2,\theta}(Q_T)^2$. In fact, in this case $g(U,V) \in C^\theta(Q_T)$ follows from the assumption, and therefore, Theorem 7.1 guarantees the unique existence of the classical solution $v \in C^{2,\theta}(Q_T)$ to (7.19) with (7.21) and (7.23). Then, for this v, it holds that

$$\chi_u(U,v)\nabla v \in C^\theta(Q_T)^n$$
$$\chi_v(U,v)\left|\nabla v\right|^2 + \chi(U,v)\Delta v + f(U,V) \in C^\theta(Q_T),$$

and again by Theorem 7.1, we have the unique existence of the classical solution $u \in C^{2,\theta}(Q_T)$ to (7.18) with (7.20) and (7.22). Henceforth, we say $(U,V) \in C^\theta(Q_T)^2$ if $U,\ V \in C^\theta(Q_T)$ and $(U,V)|_{t=0} = (u_0, v_0)$. Thus, putting $(u,v) = \mathcal{F}((U,V);T)$, we have the mapping $\mathcal{F} = \mathcal{F}(\cdot;T)$ defined on $C^\theta(Q_T)^2$.

The space $C^{2,\theta}(Q_T)^2$ is defined similarly. In particular, for $(U,V) \in C^{2,\theta}(Q_T)^2$ it holds that $\mathcal{F}(U,V;T) \in C^{2,\theta}(Q_T)^2$. Therefore, $\mathcal{F}$ is regarded as a mapping on $C^{2,\theta}(Q_T)^2$.

To prescribe its range in more details, we make use of the following.

Lemma 7.1 *If* $\mathbf{f},\ \mathbf{g} \in C^1(\overline{\Omega})^n$, *and*

$$\mathbf{g}\cdot\nu = 0 \qquad on \qquad \partial\Omega, \tag{7.24}$$

then it holds that

$$\int_\Omega (\nabla\cdot\mathbf{f})(\nabla\cdot\mathbf{g})\,dx$$
$$= -\sum_{i,j=1}^n \int_{\partial\Omega} \nu_i \frac{\partial f_i}{\partial x_j} g_j\, d\sigma + \sum_{i,j=1}^n \int_\Omega \frac{\partial f_i}{\partial x_j}\frac{\partial g_j}{\partial x_i}\, dx, \tag{7.25}$$

where $\mathbf{f} = (f_1, f_2, \cdots, f_n)$ *and* $\mathbf{g} = (g_1, g_2, \cdots, g_n)$.

Proof. If $\mathbf{f}, \mathbf{g} \in C^2(\overline{\Omega})^n$, we have

$$\nabla \cdot [(\nabla \cdot \mathbf{f})\mathbf{g}] = (\nabla \cdot \mathbf{f})(\nabla \cdot \mathbf{g}) + \sum_{i=1}^{n} g_j \frac{\partial}{\partial x_j}(\nabla \cdot \mathbf{f})$$
$$= (\nabla \cdot \mathbf{f})(\nabla \cdot \mathbf{g}) + \sum_{i=1}^{n} \nabla \cdot \left(g_j \frac{\partial \mathbf{f}}{\partial x_j} \right) - \sum_{i,j=1}^{n} \frac{\partial g_j}{\partial x_i} \frac{\partial f_i}{\partial x_j}$$

and hence

$$\int_{\partial\Omega} (\nabla \cdot \mathbf{f})(\nu \cdot \mathbf{g}) d\sigma = \int_{\Omega} (\nabla \cdot \mathbf{f})(\nabla \cdot \mathbf{g}) dx$$
$$+ \sum_{i,j=1}^{n} \int_{\partial\Omega} g_j \nu_i \frac{\partial f_i}{\partial x_j} d\sigma - \sum_{i,j=1}^{n} \int_{\Omega} \frac{\partial g_j}{\partial x_i} \frac{\partial f_i}{\partial x_j} dx$$

follows from Green's formula. This equality is extended to $\mathbf{f}, \mathbf{g} \in C^1(\overline{\Omega})^n$ and therefore, (7.25) holds in the case of (7.24). □

Lemma 7.2 *If $w \in C^2(\overline{\Omega})$,*

$$\frac{\partial w}{\partial \nu} = 0 \qquad on \qquad \partial\Omega, \tag{7.26}$$

and $\Omega \subset \mathbf{R}^n$ is a convex domain, then it holds that

$$\frac{\partial}{\partial \nu} |\nabla w|^2 \leq 0 \qquad on \qquad \partial\Omega.$$

Proof. In the case where $n = 1$, it holds that

$$\frac{\partial}{\partial \nu} |\nabla w|^2 = \pm 2 \frac{dw}{dx} \frac{d^2 f}{dx^2} = 0 \quad \text{on} \quad \partial\Omega.$$

In the case where $n = 2$, we take place of Ω by $\Omega \times \mathbf{R}$. Thus, the lemma is reduced to the case of $n = 3$.

Given $x_0 \in \partial\Omega$, we can assume taking principal directions parallel to x_1 and x_2 coordinates and $\nu = (0, 0, 1)$. In this case we have

$$\frac{\partial \nu_i}{\partial x_j} = \frac{\delta_{ij}}{R_i} \ (i, j = 1, 2) \qquad \text{and} \qquad \frac{\partial \nu_3}{\partial x_i} = 0 \ (i = 1, 2, 3)$$

at $x = x_0$ by Exercise 1.25. On the other hand, it holds that

$$\frac{1}{R_j} \geq 0 \qquad (i = 1, 2)$$

because Ω is convex. Thus, we obtain

$$\begin{aligned}\frac{\partial}{\partial\nu}|\nabla w|^2 &= \nu\cdot\nabla|\nabla w|^2 = 2\sum_{i,j=1}^{3}\nu_i\frac{\partial^2 w}{\partial x_i\partial x_j}\frac{\partial w}{\partial x_j}\\ &= 2\sum_{i,j=1}^{3}\frac{\partial}{\partial x_j}\left(\nu_i\frac{\partial w}{\partial x_i}\right)\frac{\partial w}{\partial x_j} - 2\sum_{i,j=1}^{3}\frac{\partial \nu_i}{\partial x_j}\frac{\partial w}{\partial x_i}\frac{\partial w}{\partial x_j}\\ &\leq 2\nabla\frac{\partial w}{\partial\nu}\cdot\nabla w.\end{aligned}$$

Here, we have

$$\frac{\partial}{\partial x_j}\frac{\partial w}{\partial\nu} = 0 \ (j=1,2) \qquad \text{and} \qquad \frac{\partial w}{\partial x_3} = \frac{\partial w}{\partial\nu} = 0$$

at $x = x_0$ by (7.26), and the proof is complete. □

In (IS), $p > n+2$ is an even integer, $\theta \in (0, 1-(n+2)/p)$, $U, V \in C^\theta(Q_T)$, and $T \in (0,1]$. Now, we take

$$\begin{aligned}&\|(u_0, v_0)\|_{W^{2,p}(\Omega)\times W^{3,p}(\Omega)}\\ &\qquad \equiv \left(\|\Delta u_0\|_p^p + \|u_0\|_p^p + \|\nabla\Delta v_0\|_p^p + \|v_0\|_p^p\right)^{1/p},\\ &M_0^p \equiv \|(u_0, v_0)\|^p_{W^{2,p}(\Omega)\times W^{3,p}(\Omega)},\\ &M = 4(M_0+1),\end{aligned}$$

and

$$\begin{aligned}\mathcal{O}(M,T) = \{(u,v) \in C^{2,\theta}(Q_T)^2 \cap C([0,T]; W^{2,p}(\Omega)\times W^{3,p}(\Omega)) \mid\\ \|(u,v)\|_{C([0,T];W^{2,p}(\Omega)\times W^{3,p}(\Omega))} \leq M\\ \text{and} \quad \frac{\partial u}{\partial\nu} = \frac{\partial v}{\partial\nu} = 0 \quad \text{on} \quad \partial\Omega\times[0,T]\},\end{aligned}$$

where

$$\|(u,v)\|_{C([0,T];W^{2,p}(\Omega)\times W^{3,p}(\Omega))} \equiv \max_{t\in[0,T]}\left(\|u(t)\|_{W^{2,p}(\Omega)} + \|v(t)\|_{W^{3,p}(\Omega)}\right).$$

Then, we can show the following.

Lemma 7.3 *If $\Omega \subset \mathbf{R}^n$ is convex and $(U,V) \in \mathcal{O}(M,T)$, then it holds that*

$$\max_{0\leq t\leq T}\left(\|\nabla\Delta v(t)\|_p^p + \|v(t)\|_p^p\right)$$

$$\leq \left(\|\nabla \Delta v_0\|_p^p + \|\nabla v_0\|_p^p + C_1 T \right) e^{C_1 T} \leq C_2, \tag{7.27}$$

where (u, v) denotes the solution to (IS).

Here and henceforce, $C_i > 0$ $(i = 1, 2, 3, 4)$ denote the positive constants depending on p, M_0, Ω, and g.

Proof. Since $p > n + 2$ is an even integer, we put $p = 2m + 2$. Let $h \in (0, T)$ and $s \in (0, T - h)$. Integrating (7.19) over $[s, s + h]$, we have

$$v(x, s + h) - v(x, s) = \int_s^{s+h} \Delta v(x, \tau) d\tau + \int_s^{s+h} g(U(x, \tau), V(x, \tau)) d\tau. \tag{7.28}$$

We set

$$\mathbf{F}(x, s, h) = \left(\sum_{i=0}^{m} |\nabla \Delta v(x, s + h)|^{2i} |\nabla \Delta v(x, s)|^{2(m-i)} \right)$$
$$\cdot (\nabla \Delta v(x, s + h) + \nabla \Delta v(x, s))$$

and $\mathbf{F} = (F_1, \cdots, F_n)$. Operating $-\Delta$ to (7.28), multiplying $\nabla \cdot \mathbf{F}(x, s, h)$, and integrating over Ω, we have

$$\begin{aligned} &- \int_\Omega (\Delta v(x, s + h) - \Delta v(x, s)) \nabla \cdot \mathbf{F}(x, s, h) dx \\ &= - \int_s^{s+h} \int_\Omega \Delta^2 v(x, \tau) \nabla \cdot \mathbf{F}(x, \tau, h) dx d\tau \\ &\quad - \int_s^{s+h} \int_\Omega \Delta g(U(x, \tau), V(x, \tau)) \nabla \cdot \mathbf{F}(x, \tau, h) dx d\tau \\ &= -I - II. \end{aligned} \tag{7.29}$$

We have

$$\frac{\partial}{\partial \nu} v_t = \frac{\partial U}{\partial \nu} = \frac{\partial V}{\partial \nu} = 0 \qquad \text{on} \quad \partial\Omega \times [0, T]$$

and (7.20), and hence it follows that

$$\frac{\partial}{\partial \nu} \Delta v = 0 \qquad \text{on} \quad \partial\Omega \times (0, T). \tag{7.30}$$

Furthermore, it holds that

$$(\mathbf{a} - \mathbf{b}) \cdot (\mathbf{a} + \mathbf{b}) \sum_{i=0}^{m} |\mathbf{a}|^{2i} |\mathbf{b}|^{2(m-i)} = |\mathbf{a}|^p - |\mathbf{b}|^p$$

for $\mathbf{a}, \mathbf{b} \in \mathbf{R}^n$, and the left-hand side of (7.29) is equal to

$$\int_\Omega |\nabla \Delta v(x, s+h)|^p \, dx - \int_\Omega |\nabla \Delta v(x, s)|^p \, dx. \tag{7.31}$$

We divide this term by h.

Applying Lemma 7.1, first we have

$$\begin{aligned}\frac{I}{h} = &-\frac{1}{h}\int_s^{s+h}\sum_{i,j=1}^n \int_{\partial\Omega} \nu_i \left\{\frac{\partial}{\partial x_j}\frac{\partial \Delta v(x,\tau)}{\partial x_i}\right\} \cdot F_j(x,\tau,h) d\sigma d\tau \\ &+\frac{1}{h}\int_s^{s+h}\sum_{i,j=1}^n \int_\Omega \frac{\partial^2 \Delta v(x,\tau)}{\partial x_j \partial x_i}\frac{\partial}{\partial x_i} F_j(x,\tau,h) dx d\tau.\end{aligned}$$

This implies that

$$\begin{aligned}\lim_{h\to 0}\frac{I}{h} &= -\int_{\partial\Omega}\frac{\partial}{\partial \nu}|\nabla \Delta v(x,s)|^p \, d\sigma \\ &\quad +p\sum_{i,j=1}^n \int_\Omega \frac{\partial^2 \Delta v(x,s)}{\partial x_i x_j}\frac{\partial}{\partial x_i}\left\{\frac{\partial \Delta v(x,s)}{\partial x_j}|\nabla \Delta v(x,s)|^{2m}\right\} \\ &= -\frac{p}{2}\int_{\partial\Omega}|\nabla \Delta v(x,s)|^{p-2}\frac{\partial}{\partial \nu}|\nabla \Delta v(x,s)|^2 \, d\sigma \\ &\quad +p\sum_{i,j=1}^n \int_\Omega |\nabla \Delta v(x,s)|^{p-2}\left(\frac{\partial^2 \Delta v(x,s)}{\partial x_i \partial x_j}\right)^2 dx \\ &\quad +p\sum_{i,j=1}^n \int_\Omega \frac{\partial^2 \Delta v}{\partial x_i \partial x_j}\frac{\partial \Delta v}{\partial x_j}\frac{\partial}{\partial x_i}|\nabla \Delta v(x,s)|^{p-2} dx \\ &= -\frac{p}{2}\int_{\partial\Omega}|\nabla \Delta v(x,s)|^{p-2}\frac{\partial}{\partial \nu}|\nabla \Delta v(x,s)|^2 \, d\sigma \\ &\quad +p\sum_{i,j=1}^n \int_\Omega |\nabla \Delta v(x,s)|^{p-2}\left(\frac{\partial^2 \Delta v(x,s)}{\partial x_i \partial x_j}\right)^2 dx \\ &\quad +\frac{p(p-2)}{4}\int_\Omega |\nabla \Delta v(x,s)|^{p-4}\left|\nabla |\nabla \Delta v(x,s)|^2\right|^2 dx.\end{aligned}$$

Therefore, by Lemma 7.2 we obtain

$$\lim_{h\to 0}\frac{I}{h} \geq p\sum_{i,j=1}^n \int_\Omega |\nabla \Delta v(x,s)|^{p-2}\left(\frac{\partial^2 \Delta v(x,s)}{\partial x_i \partial x_j}\right)^2 dx$$

$$+\frac{p(p-2)}{4}\int_\Omega |\nabla\Delta v(x,s)|^{p-4}\left|\nabla|\nabla\Delta v(x,s)|^2\right|^2 dx. \quad (7.32)$$

We also have

$$\lim_{h\to 0}\frac{II}{h} = p\int_\Omega \Delta g(U(x,s),V(x,s))\nabla\cdot\left(\nabla\Delta v(x,s)\,|\nabla\Delta v(x,s)|^{p-2}\right)dx.$$

Combining this with

$$\begin{aligned}
&\nabla\cdot\left(|\nabla\Delta v(x,s)|^{p-2}\,\nabla\Delta v(x,s)\right)\\
&= |\nabla\Delta v(x,s)|^{p-2}\,\Delta^2 v(x,s)\\
&\quad+\frac{p-2}{2}|\nabla\Delta v(x,s)|^{p-4}\,\nabla\Delta v(x,s)\cdot\nabla|\nabla\Delta v(x,s)|^2
\end{aligned}$$

implies that

$$\begin{aligned}
\lim_{h\to 0}\frac{|II|}{h} &\le p\sum_{i=1}^{n}\left(\frac{1}{2}\int_\Omega |\nabla\Delta v(x,s)|^{p-2}\left(\frac{\partial^2\Delta v(x,s)}{\partial x_i^2}\right)^2 dx\right.\\
&\quad\left.+\frac{p-2}{2p}\int_\Omega |\nabla\Delta v(x,s)|^p\,dx+\frac{1}{p}\int_\Omega |\Delta g(U(x,s),V(x,s))|^p\,dx\right)\\
&\quad+\frac{p(p-2)}{2}\left(\frac{1}{4}\int_\Omega |\nabla\Delta v(x,s)|^{p-4}\left|\nabla|\nabla\Delta v(x,s)|^2\right|^2 dx\right.\\
&\quad\left.+\frac{p-2}{2p}\int_\Omega |\nabla\Delta v(x,s)|^p\,dx+\frac{2^{p/2}}{p}\int_\Omega |\Delta g(U(x,s),V(x,s))|^p\,dx\right).
\end{aligned} \quad (7.33)$$

We have

$$\begin{aligned}
&\frac{1}{h}\int_t^{t+h}\int_\Omega |\nabla\Delta v(x,s)|^p dxds-\frac{1}{h}\int_0^h\int_\Omega |\nabla\Delta v(x,s)|^p dxds\\
&= -\int_0^t\frac{1}{h}\int_s^{s+h}\int_\Omega \Delta^2 v(x,\tau)\nabla\cdot\mathbf{F}(x,\tau,h)dxd\tau\\
&\quad-\int_0^t\frac{1}{h}\int_s^{s+h}\int_\Omega \Delta g(U(x,\tau),V(x,\tau))\nabla\cdot\mathbf{F}(x,\tau,h)dxd\tau
\end{aligned}$$

by (7.31). Sending $h\to 0$, we get that

$$\int_\Omega |\nabla\Delta v(x,t)|^p\,dx-\int_\Omega |\nabla\Delta v_0(x)|^p\,dx$$

$$
\begin{aligned}
&\le -\frac{p}{2}\sum_{i,j=1}^{n}\int_0^t\int_\Omega |v(x,s)|^{p-2}\left(\frac{\partial^2\Delta v(x,s)}{\partial x_i\partial x_j}\right)^2 dxds\\
&-\frac{p(p-2)}{8}\int_0^t\int_\Omega |\nabla\Delta v(x,s)|^{p-4}\left|\nabla|\nabla\Delta v(x,s)|^2\right|^2 dxds\\
&+\left(\frac{p-2}{2}n+\frac{(p-2)^2}{4}\right)\int_0^t\int_\Omega |\nabla\Delta v(x,s)|^p dxds\\
&+\left(n+\frac{p-2}{p}2^{p/2}\right)\int_0^t\int_\Omega |\Delta g(U(x,s),V(x,s))|^p dxds
\end{aligned}
$$

by (7.32) and (7.33).

Henceforce, K_i $(i=1,2,\cdots,7)$ denote the positive constants depending only on Ω and p. Because of $p>n+2$, we have by Sobolev's imbedding theorem that

$$
\|\nabla w\|_\infty + \|w\|_\infty \le K_1\left(\|w\|_p^p + \|\Delta w\|_p^p\right)^{1/p} \tag{7.34}
$$

$$
\|\Delta w\|_\infty + \|\nabla w\|_\infty + \|w\|_\infty \le K_2\left(\|w\|_p^p + \|\nabla\Delta w\|_p^p\right)^{1/p}. \tag{7.35}
$$

Therefore, it holds that

$$
\begin{aligned}
&|\Delta g(U(x,s),V(x,s))|\\
&\le |g_u(U(x,s),V(x,s))|\cdot|\Delta U(x,s)|\\
&\quad +|g_{uu}(U(x,s),V(x,s))|\cdot|\nabla U(x,s)|^2\\
&\quad +2|g_{uv}(U(x,s),V(x,s))|\cdot|\nabla U(x,s)|\cdot|\nabla V(x,s)|\\
&\quad +|g_{vv}(U(x,s),V(x,s))|\cdot|\nabla V(x,s)|^2\\
&\quad +|g_v(U(x,s),V(x,s))|\cdot|\Delta V(x,s)|\\
&\le C_3(K_1M+K_2M+|\Delta U(x,s)|).
\end{aligned}
$$

We obtain

$$
\begin{aligned}
\|\nabla\Delta v(t)\|_p^p &\le \|\nabla\Delta v_0\|_p^p + K_3\int_0^t \|\nabla\Delta v(s)\|_p^p ds\\
&\quad +K_4\int_0^t \|\Delta g(U(s),V(s))\|_p^p ds\\
&\le \|\nabla\Delta v_0\|_p^p + TK_4C_3^p\left[(K_1+K_2)M|\Omega|^{1/p}+M\right]^p\\
&\quad +K_3\int_0^t \|\nabla\Delta v(s)\|_p^p ds.
\end{aligned}
$$

By this and Gronwall's inequality, we have that

$$\|\nabla\Delta v(t)\|_p^p \le \left\{\|\nabla\Delta v_0\|_p^p + TK_4C_3^p\left[(K_1+K_2)\,|\Omega|^{1/p}+1\right]^p M^p\right\} e^{K_3T}. \tag{7.36}$$

Next, multiplying $|v|^{p-2}\,v$ to (7.19), we have

$$\begin{aligned}
&\frac{1}{p}\frac{d}{dt}\int_\Omega |v|^p\,dx + (p-1)\int_\Omega |v|^{p-2}\,|\nabla v|^2\,dx \\
&\le \int_\Omega g(U,V)\,|v|^{p-1}\,dx \\
&\le \frac{p-1}{p}\int_\Omega |v|^p\,dx + \frac{1}{p}\int_\Omega |g(U,V)|^p\,dx.
\end{aligned}$$

Combining this with

$$\begin{aligned}
&|g(V(x,t),V(x,t))| \\
&\le \sup\{|g(U,V)| \mid |U|\le K_1M, \quad |V|\le K_2M\} \\
&= C_4,
\end{aligned}$$

we have

$$\frac{d}{dt}\,\|v(t)\|_p^p \le (p-1)\,\|v(t)\|_p^p + C_4^p\,|\Omega|.$$

This is equivalent to

$$\frac{d}{dt}\left(\|v(t)\|_p^p+1\right) \le (p+C_4^p\,|\Omega|-1)\left(\|v(t)\|_p^p+1\right)$$

and hence

$$\|v(t)\|_p^p \le \left(\|v_0\|_p^p+1\right)\exp\left((p+C_4^p\,|\Omega|-1)T\right) \tag{7.37}$$

follows. Inequality (7.27) follows from (7.36) and (7.37) and the proof is complete. □

7.2.3 *Properties of $\mathcal{F}$*

We make use of the contraction mapping principle to show that $\mathcal{F} = \mathcal{F}(\cdot, T)$ has a fixed point if $T > 0$ is small. In this paragraph, we suppose that the domain $\Omega \subset \mathbf{R}^n$ is convex. We put

$$\begin{aligned}
E_1 &= [-K_1M, K_1M] \times [-K_2C_2^{1/p}, K_2C_2^{1/p}], \\
E_2 &= [-K_1M, K_1M] \times [-K_2M, K_2M],
\end{aligned}$$

and

$$\max\left(\|\chi\|_{C^2(E_1)}, \|f\|_{C^1(E_2)}\right) = C_5. \tag{7.38}$$

Here and henceforce, $C_i > 0$ $(i = 5, 6, 7, \cdots, 22)$ denotes the constants determined by p, M_0, Ω, g, χ, f, and K_j $(j = 1, 2, 3, \cdots, 7)$. We emphasize that those constants are independent of $T \in (0, 1]$.

First, we show the following.

Lemma 7.4 *There exists $T_0 \in (0, 1]$ such that*

$$\mathcal{F}\mathcal{O}(M, T) \subset \mathcal{O}(M, T)$$

for any $T \in (0, T_0]$.

Proof. Given $(U, V) \in \mathcal{O}(M, T)$, we take the solution (u, v) to (IS). First, by Lemma 7.3 and (7.35), we have

$$\|v(t)\|_\infty + \|\nabla v(t)\|_\infty + \|\Delta v(t)\|_\infty \le K_2 C_2^{1/p}. \tag{7.39}$$

Next, multiplying $|u|^{p-2} u$ to (7.18) and integrating over Ω, we have that

$$\begin{aligned}
&\frac{1}{p}\frac{d}{dt}\int_\Omega |u|^p\,dx = \int_\Omega u_t\,|u|^{p-2}\,udx \\
&= \int_\Omega \Delta u \cdot |u|^{p-2}\,udx + \int_\Omega \chi_u(U, v)(\nabla v \cdot \nabla u)\,|u|^{p-2}\,udx \\
&+ \int_\Omega \left(\chi_v(U, v)|\nabla v|^2 + \chi(U, v)\Delta v\right)|u|^{p-2}udx + \int_\Omega f(U, V)|u|^{p-2}udx.
\end{aligned}$$

Combining this with (7.39) and (7.38) implies that

$$\begin{aligned}
&\frac{1}{p}\frac{d}{dt}\|u(t)\|_p^p \\
&\le -(p-1)\int_\Omega |u|^{p-2}\,|\nabla u|^2\,dx + K_2 C_2^{1/p} C_5 \int_\Omega |\nabla u| \cdot |u|^{p-1}\,dx \\
&\quad + \left(K_2^2 C_2^{2/p} + K_2 C_2^{1/p}\right) C_5 \int_\Omega |u|^{p-1}\,dx + C_5 \int_\Omega |u|^{p-1}\,dx \\
&\le -(p-1)\int_\Omega |u|^{p-2}\,|\nabla u|^2\,dx + \frac{1}{2}\int_\Omega |u|^{p-2}\,|\nabla u|^2\,dx \\
&\quad + \frac{1}{2}K_2^2 C_2^{2/p} C_5^2 \int_\Omega |u|^p\,dx + \frac{1}{p}\left(K_2^2 C_2^{2/p} + K_2 C_2^{1/p}\right)^p C_5^p\,|\Omega| \\
&\quad + \frac{p-1}{p}\int_\Omega |u|^p\,dx + \frac{1}{p}C_5^p\,|\Omega| + \frac{p-1}{p}\int_\Omega |u|^p\,dx
\end{aligned}$$

$$\leq C_6 + C_6 \int_\Omega |u|^p \, dx. \tag{7.40}$$

Next, we operate Δ to (7.18), multiply $|\Delta u|^{p-2} \Delta u$, and integrate it over Ω. Then, it follows that

$$\begin{aligned}
&\frac{1}{p}\frac{d}{dt}\int_\Omega |\Delta u|^p \, dx + (p-1)\int_\Omega |\Delta v|^{p-2} |\nabla \Delta v|^2 \, dx = -(p-1) \\
&\cdot \int_\Omega \left[\nabla \left(\chi_u(U,v)(\nabla v \cdot \nabla u) + \chi_v(U,v) |\nabla v|^2\right) \cdot \nabla \Delta u\right] |\Delta u|^{p-2} \, dx \\
&\qquad -(p-1)\int_\Omega [\nabla (\chi(U,v)\Delta v + f(U,V)) \cdot \nabla \Delta u] \, |\Delta u|^{p-2} \, dx \\
&\leq \frac{p-1}{2}\int_\Omega |\Delta u|^{p-2} |\nabla \Delta u|^2 \, dx + \frac{p-1}{2}\int_\Omega |\Delta u|^{p-2} \, III^2 dx
\end{aligned} \tag{7.41}$$

with

$$III = \left|\nabla \left\{\chi_u(U,v)(\nabla v \cdot \nabla u) + \chi_v(U,v) |\nabla v|^2 + \chi(U,v)\Delta v + f(U,V)\right\}\right|.$$

Here, we have

$$\begin{aligned}
III &\leq |\chi_{uu}(U,v)| \cdot |\nabla U| \cdot |\nabla v| \cdot |\nabla u| + |\chi_{uv}(U,v)| \cdot |\nabla v|^2 \cdot |\nabla u| \\
&+ \sqrt{n} \, |\chi_u(U,v)| \\
&\cdot \left\{\left(\sum_{i,j=1}^n \left|\frac{\partial^2 v}{\partial x_i \partial x_j}\right|^2\right)^{1/2} |\nabla u| + |\nabla v| \left(\sum_{i,j=1}^n \left|\frac{\partial^2 u}{\partial x_i \partial x_j}\right|^2\right)^{1/2}\right\} \\
&+ |\chi_{uv}(U,v)| \cdot |\nabla U| \cdot |\nabla v|^2 + |\chi_{vv}(U,v)| \cdot |\nabla v|^3 \\
&+ 2\sqrt{n} \, |\chi_v(U,v)| \left(\sum_{i,j=1}^n \left|\frac{\partial^2 v}{\partial x_i \partial x_j}\right|^2\right)^{1/2} |\nabla v| + |\chi_u(U,v)| \cdot |\nabla U| \cdot |\Delta v| \\
&+ |\chi_v(U,v)| \cdot |\nabla v| \cdot |\Delta v| + |\chi(U,v)| \cdot |\nabla \Delta v| + |f_u(U,V)| \cdot |\nabla U| \\
&\quad + |f_v(U,V)| \cdot |\nabla V| .
\end{aligned}$$

By this inequality, (7.39), (7.34), and (7.38), it holds that

$$\begin{aligned}
III &\leq C_5 \left(K_1 M K_2 C_2^{1/p} + K_2^2 C_2^{2/p}\right) |\nabla u| + \sqrt{n} C_5 \\
&\cdot \left\{\left(\sum_{i,j=1}^n \left|\frac{\partial^2 v}{\partial x_i \partial x_j}\right|^2\right)^{1/2} |\nabla u| + K_2 C_2^{1/p} \left(\sum_{i,j=1}^n \left|\frac{\partial^2 u}{\partial x_i \partial x_j}\right|^2\right)^{1/2}\right\}
\end{aligned}$$

$$+C_5K_1MK_2^2C_2^{2/p}+C_5K_2^3C_2^{3/p}$$
$$+2\sqrt{n}C_5\left(\sum_{i,j=1}^{n}\left|\frac{\partial^2 v}{\partial x_i\partial x_j}\right|^2\right)^{1/2}K_2C_2^{1/p}+C_5K_1MK_2C_2^{1/p}$$
$$+C_5K_2^2C_2^{2/p}+C_5\left|\nabla\Delta v\right|+C_5K_1M+C_5K_2M. \tag{7.42}$$

We have

$$\int_\Omega\left(\sum_{i,j=1}^{n}\left|\frac{\partial^2 f}{\partial x_i\partial x_j}\right|^2\right)^{p/2}dx\le K_5\left(\|\Delta f\|_p^p+\|f\|_p^p\right)$$

for $f\in W^{2,p}(\Omega)$, and therefore, it follows that

$$\int_\Omega|\Delta u|^{p-2}\,III^2dx\le\frac{p-2}{p}\int_\Omega|\Delta u|^p\,dx$$
$$+\frac{2}{p}C_7\int_\Omega\left(1+|\nabla u|^p+|\Delta u|^p+|\Delta v|^p+|\nabla\Delta v|^p\right)dx.$$

By this, Lemma 7.3, and

$$\begin{aligned}\|\nabla w\|_p^p&\le K_6\left(\|w\|_p^p+\|\Delta w\|_p^p\right)\\ \|\Delta w\|_p^p&\le K_7\left(\|w\|_p^p+\|\nabla\Delta w\|_p^p\right),\end{aligned}$$

it holds that

$$\int_\Omega|\Delta u|^{p-2}\,III^2dx\le\frac{p-2}{p}\int_\Omega|\Delta u|^p\,dx$$
$$+\frac{2}{p}C_7\left\{|\Omega|+(K_6+1)\int_\Omega\left(|\Delta u|^p+|u|^p\right)dx\right.$$
$$\left.+(K_7+1)\int_\Omega\left(|\nabla\Delta v|^p+|v|^p\right)dx\right\}$$
$$\le C_8\left(1+\|u\|_p^p+\|\Delta u\|_p^p\right).$$

Therefore, from (7.41) we have

$$\frac{d}{dt}\|\Delta u\|_p^p\le pC_8\left(1+\|u\|_p^p+\|\Delta u\|_p^p\right).$$

Then, (7.40) implies that

$$\frac{d}{dt}\left(\|\Delta u\|_p^p + \|u\|_p^p + 1\right) \le C_9\left(1 + \|\Delta u\|_p^p + \|u\|_p^p\right).$$

We obtain

$$\|\Delta u\|_p^p + \|u\|_p^p \le \left(\|\Delta u_0\|_p^p + \|u_0\|_p^p + 1\right) e^{C_9 T}.$$

Then by Lemma 7.3, we get that

$$\begin{aligned}
&\|(u,v)\|_{C([0,T];W^{2,p}(\Omega)\times W^{3,p}(\Omega))} \\
&\le \left(\|(u_0,v_0)\|_{W^{2,p}(\Omega)\times W^{3,p}(\Omega)} + TC_{10} + 2\right) e^{C_{10}T} \\
&= (M_0 + TC_{10} + 2)\, e^{C_{10}T}.
\end{aligned}$$

Because of $M = 4(M_0+1)$, we have $(u,v) \in \mathcal{O}(M,T)$ for sufficiently small $T > 0$, and the proof is complete. □

Now, we show the following.

Lemma 7.5 *There exists $T_1 \in (0, T_0]$ such that*

$$\begin{aligned}
&\|\mathcal{F}(U_1,V_1) - \mathcal{F}(U_2,V_2)\|_{C([0,T];L^p(\Omega)^2)} \\
&\le \frac{1}{2}\|(U_1,V_1) - (U_2,V_2)\|_{C([0,T];L^p(\Omega)^2)}
\end{aligned}$$

is satisfied for $(U_i,V_i) \in \mathcal{O}(M,T)$ $(i = 1,2)$ and $T \in (0,T_1]$, where

$$\|(U,V)\|_{C([0,T];L^p(\Omega)^2)} \equiv \max_{t\in[0,T]}\left(\|U(t)\|_p + \|V(t)\|_p\right).$$

Proof. We put $(u_i,v_i) = \mathcal{F}(U_i,V_i)$ for $i = 1,2$. By Lemma 7.4, it holds that $(u_i,v_i) \in \mathcal{O}(M,T)$ $(i = 1,2)$. Also, we have

$$\begin{aligned}
&(v_2 - v_1)_t = \Delta(v_2 - v_1) + g(U_2,V_2) - g(U_1,V_1) \\
&(u_2 - u_1)_t = \Delta(u_2 - u_1) + (\chi_u(U_2,v_2)(\nabla v_2\cdot\nabla u_2) \\
&\quad -\chi_u(U_1,v_1)(\nabla v_1\cdot\nabla u_1)) \\
&\quad + \left(\chi_v(U_2,v_2)\,|\nabla v_2|^2 - \chi_v(U_1,v_1)\,|\nabla v_1|^2\right) \\
&\quad + (\chi(U_2,v_2)\Delta v_2 - \chi(U_1,v_1)\Delta v_1) \\
&\quad +(f(U_2,V_2) - f(U_1,V_1)).
\end{aligned} \tag{7.43}$$

By Lemma 7.2, for w with $\partial w/\partial\nu = 0$ on $\partial\Omega$ it holds that

$$\begin{aligned}
&\int_\Omega \Delta\nabla w\cdot|\nabla w|^{p-2}\nabla w dx \\
&\quad = \frac{1}{p}\int_{\partial\Omega}\frac{\partial}{\partial\nu}|\nabla w|^p\,dx - \int_\Omega |\nabla w|^{p-2}\sum_{i,j=1}^n\left(\frac{\partial^2 w}{\partial x_i\partial x_j}\right)^2 dx \\
&\quad -(p-2)\sum_{j=1}^n\int_\Omega|\nabla w|^{p-4}\left(\nabla w\cdot\nabla\frac{\partial w}{\partial x_j}\right)^2 dx \\
&\leq\ -\int_\Omega|\nabla w|^{p-2}\sum_{i,j=1}^n\left(\frac{\partial^2 w}{\partial x_i\partial x_j}\right)^2 dx \\
&\quad -\frac{4(p-2)}{p^2}\int_\Omega\left|\nabla|\nabla w|^{p/2}\right|^2 dx. \qquad (7.44)
\end{aligned}$$

Operating ∇ to (7.43), multiplying $|\nabla(v_2-v_1)|^{p-2}\nabla(v_2-v_1)$, integrating over Ω, and applying (7.44) for $w = v_2 - v_1$, we have

$$\begin{aligned}
&\frac{1}{p}\frac{d}{dt}\|\nabla(v_2-v_2)\|_p^p + \int_\Omega|\nabla(v_2-v_1)|^{p-2}\sum_{i,j=1}^n\left(\frac{\partial^2(v_2-v_1)}{\partial x_i\partial x_j}\right)^2 dx \\
&\leq \int_\Omega |g(U_2,V_2)-g(U_1,V_1)| \\
&\quad\cdot\left|\nabla\cdot\left(|\nabla(v_2-v_1)|^{p-2}\nabla(v_2-v_1)\right)\right| dx. \qquad (7.45)
\end{aligned}$$

Here, in use of $C_{11} = \|g\|_{C^1(E_2)}$, we have

$$|g(U_2,V_2)-g(U_1,V_1)| \leq C_{11}\left(|U_2-U_1|+|V_2-V_1|\right). \qquad (7.46)$$

Next, we have

$$\begin{aligned}
&\left|\nabla\cdot\left(|\nabla(v_2-v_1)|^{p-2}\nabla(v_2-v_1)\right)\right| \\
&\leq |\nabla(v_2-v_1)|^{p-2}|\Delta(v_2-v_1)| \\
&+(p-2)|\nabla(v_2-v_1)|^{p-4}\left|\sum_{i,j=1}^n\frac{\partial^2(v_2-v_1)}{\partial x_i\partial x_j}\frac{\partial(v_2-v_1)}{\partial x_i}\frac{\partial(v_2-v_1)}{\partial x_j}\right| \\
&\leq |\nabla(v_2-v_1)|^{p-2}\left\{\sum_{i,j=1}^n\left(\frac{\partial^2(v_2-v_1)}{\partial x_i\partial x_j}\right)^2\right\}^{1/2}
\end{aligned}$$

$$+(p-2)\,|\nabla(v_2-v_1)|^{p-4}\left\{\sum_{i,j=1}^{n}\left(\frac{\partial^2(v_2-v_1)}{\partial x_i\partial x_j}\right)^2\right\}^{1/2}$$

$$\cdot\left\{\sum_{i,j=1}^{n}\left(\frac{\partial(v_2-v_1)}{\partial x_i}\right)^2\left(\frac{\partial(v_2-v_1)}{\partial x_j}\right)^2\right\}^{1/2}$$

$$=(p-1)\,|\nabla(v_2-v_1)|^{p-2}\left\{\sum_{i,j=1}^{n}\left(\frac{\partial^2(v_2-v_1)}{\partial x_i\partial x_j}\right)^2\right\}^{1/2}.$$

Therefore, it holds that

$$\begin{aligned}
&\int_\Omega |g(U_2,V_2)-g(U_1,V_1)| \\
&\qquad\cdot\left|\nabla\cdot\left(|\nabla(v_2-v_1)|^{p-2}\nabla(v_2-v_1)\right)\right|dx \\
&\le \frac{1}{2}\int_\Omega |\nabla(v_2-v_1)|^{p-2}\left\{\sum_{i,j=1}^{n}\left(\frac{\partial^2(v_2-v_1)}{\partial x_i\partial x_j}\right)^2\right\}dx \\
&\qquad+\frac{1}{p}(p-1)^pC_{11}^p\int_\Omega (|U_2-U_1|+|V_2-V_1|)^p\,dx \\
&\qquad+\frac{p-2}{2p}\int_\Omega |\nabla(v_2-v_1)|^p\,dx,
\end{aligned}$$

and hence it follows from (7.45) that

$$\begin{aligned}
&\frac{d}{dt}\,\|\nabla(v_2-v_1)\|_p^p \\
&\qquad+\frac{p}{2}\int_\Omega |\nabla(v_2-v_1)|^{p-2}\left\{\sum_{i,j=1}^{n}\left(\frac{\partial^2(v_2-v_1)}{\partial x_i\partial x_j}\right)^2\right\}dx \\
&\le (p-1)^pC_{11}^p\left(\|U_2-U_1\|_p+\|V_2-V_1\|_p\right)^p \\
&\qquad+\frac{p-2}{2}\,\|\nabla(v_2-v_1)\|_p^p.
\end{aligned}\tag{7.47}$$

Next, multiplying $|v_2-v_1|^{p-2}(v_2-v_1)$ to (7.43) and integrating over Ω, we get that

$$\frac{1}{p}\frac{d}{dt}\,\|v_2-v_1\|_p^p+(p-1)\int_\Omega |v_2-v_1|^{p-2}\,|\nabla(v_2-v_1)|^2\,dx$$

$$\leq \int_\Omega |g(U_2, V_2) - g(U_1, V_1)| \cdot |v_2 - v_1|^{p-1}\, dx$$
$$\leq \frac{1}{p} \|g(U_2, V_2) - g(U_1, V_1)\|_p^p + \frac{(p-1)}{p} \|v_2 - v_1\|_p^p.$$

Combining this with (7.46), we have

$$\frac{d}{dt}\|v_2 - v_1\|_p^p + p(p-1)\int_\Omega |v_2 - v_1|^{p-2} |\nabla(v_2 - v_1)|^2\, dx$$
$$\leq C_{11}^p \left(\|U_2 - U_1\|_p + \|V_2 - V_1\|_p\right)^p + (p-1)\|v_2 - v_1\|_p^p. \quad (7.48)$$

Next, multiplying $|u_2 - u_1|^{p-2}(u_2 - u_1)$ to (7.43) and integrating over Ω, we have

$$\frac{1}{p}\frac{d}{dt}\|u_2 - u_1\|_p^p + (p-1)\int_\Omega |u_2 - u_1|^{p-2} |\nabla(u_2 - u_1)|^2\, dx$$
$$= \int_\Omega (\chi_u(U_2, v_2)\nabla v_2 \cdot \nabla u_2 - \chi_u(U_1, v_1)\nabla v_1 \cdot \nabla u_1)$$
$$\cdot |u_2 - u_1|^{p-2}(u_2 - u_1)dx$$
$$+ \int_\Omega \left(\chi_v(U_2, v_2)|\nabla v_2|^2 - \chi_v(U_1, v_1)|\nabla v_1|^2\right)|u_2 - u_1|^{p-2}(u_2 - u_1)dx$$
$$+ \int_\Omega (\chi(U_2, v_2)\Delta v_2 - \chi(U_1, v_1)\Delta v_1)|u_2 - u_1|^{p-2}(u_2 - u_1)dx$$
$$+ \int_\Omega (f(U_2, V_2) - f(U_1, V_1))|u_2 - u_1|^{p-2}(u_2 - u_1)dx$$
$$= IV + V + VI + VII.$$

Here, we apply Lemmas 7.3, 7.4 and inequalities (7.34), (7.39), (7.38), and get the following:

$$|IV| \leq$$
$$\left|\int_\Omega (\chi_u(U_2, v_2) - \chi_u(U_1, v_1))(\nabla v_2 \cdot \nabla u_2)|u_2 - u_1|^{p-2}(u_2 - u_1)dx\right|$$
$$+ \left|\int_\Omega \chi_u(U_1, v_1)(\nabla(v_2 - v_1) \cdot \nabla u_2)|u_2 - u_1|^{p-2}(u_2 - u_1)dx\right|$$
$$+ \left|\int_\Omega \chi_u(U_1, v_1)(\nabla v_1 \cdot \nabla(u_2 - u_1))|u_2 - u_1|^{p-2}(u_2 - u_1)dx\right|$$
$$\leq C_5 \int_\Omega (|U_2 - U_1| + |v_2 - v_1|)(K_2 C_2^{1/p})(K_1 M)|u_2 - u_1|^{p-1}dx$$

$$+C_5\int_\Omega |\nabla(v_2-v_1)|\,(K_1M)\,|u_2-u_1|^{p-1}\,dx$$
$$+C_5\int_\Omega (K_1C_2^{1/p})\,|\nabla(u_2-u_1)|\cdot|u_2-u_1|^{p-1}\,dx$$
$$\le C_{12}(\|u_2-u_1\|_p^p+\|U_2-U_1\|_p^p+\|\nabla(v_2-v_1)\|_p^p+\|v_2-v_1\|_p^p)$$
$$+\frac{p-1}{2}\int_\Omega |u_2-u_1|^{p-2}\,|\nabla(u_2-u_1)|^2\,dx.$$

$$|V|\le\left|\int_\Omega (\chi_v(U_2,v_2)-\chi_v(U_1,v_1))\,|\nabla v_2|^2\,|u_2-u_1|^{p-2}\,(u_2-u_1)dx\right|$$
$$+\left|\int_\Omega \chi_v(U_1,v_1)\nabla(v_2+v_1)\cdot\nabla(v_2-v_1)\,|u_2-u_1|^{p-2}\,(u_2-u_1)dx\right|$$
$$\le C_5\int_\Omega (|U_2-U_1|+|v_2-v_1|)\left(K_2C_2^{1/p}\right)^2|u_2-u_1|^{p-1}\,dx$$
$$+C_5\int_\Omega (2K_2C_2^{1/p})\,|\nabla(v_2-v_1)|\cdot|u_2-u_1|^{p-1}\,dx$$
$$\le C_{13}(\|u_2-u_1\|_p^p+\|U_2-U_1\|_p^p+\|\nabla(v_2-v_1)\|_p^p+\|v_2-v_1\|_p^p).$$

$$|VI|\le\left|\int_\Omega (\chi(U_2,v_2)-\chi(U_1,v_1))\,\Delta v_2\,|u_2-u_1|^{p-2}\,(u_2-u_1)dx\right|$$
$$+\Bigg|\int_\Omega \nabla(v_2-v_1)\cdot\{[\chi_u(U_1,v_1)\nabla U_1+\chi_v(U_1,v_1)\nabla v_1]$$
$$\cdot|u_2-u_1|^{p-2}\,(u_2-u_1)$$
$$+(p-1)\chi(U_1,v_1)\,|u_2-u_1|^{p-1}\,\nabla(u_2-u_1)\}dx\Bigg|$$
$$\le C_5\int_\Omega (|U_2-U_1|+|v_2-v_1|)\left(K_2C_2^{1/p}\right)|u_2-u_1|^{p-1}\,dx$$
$$+C_5\int_\Omega |\nabla(v_2-v_1)|\cdot\Big\{\left(K_1M+K_2C_2^{1/p}\right)|u_2-u_1|^{p-1}$$
$$+(p-1)\,|u_2-u_1|^{p-2}\,|\nabla(u_2-u_1)|\Big\}\,dx$$
$$\le C_{14}(\|u_2-u_1\|_p^p+\|U_2-U_1\|_p^p+\|\nabla(v_2-v_1)\|_p^p+\|v_2-v_1\|_p^p)$$
$$+\frac{p-1}{2}\int_\Omega |u_2-u_1|^{p-2}\,|\nabla(u_2-u_1)|^2\,dx.$$

$$|VII|\le C_5\int_\Omega (|U_2-U_1|+|V_2-V_1|)\,|u_2-u_1|^{p-1}\,dx$$

$$\leq C_{15}(\|u_2 - u_1\|_p^p + \|U_2 - U_1\|_p^p + \|V_2 - V_1\|_p^p).$$

Those relations are summarized as

$$\frac{d}{dt}\|u_2 - u_1\|_p^p \leq C_{16}\|u_2 - u_1\|_p^p + C_{16}\|U_2 - U_1\|_p^p$$
$$+C_{16}\|V_2 - V_1\|_p^p + C_{16}\|\nabla(v_2 - v_1)\|_p^p + C_{16}\|v_2 - v_1\|_p^p. \quad (7.49)$$

From (7.47), (7.48), and (7.49) we get that

$$\frac{d}{dt}\left(\|u_2 - u_1\|_p^p + \|\nabla(v_2 - v_1)\|_p^p + \|v_2 - v_1\|_p^p\right)$$
$$\leq C_{17}\left(\|u_2 - u_1\|_p^p + \|\nabla(v_2 - v_1)\|_p^p + \|v_2 - v_1\|_p^p\right)$$
$$+C_{17}\left(\|U_2 - U_1\|_p^p + \|V_2 - V_1\|_p^p\right),$$

and hence it follows that

$$\|u_2(t) - u_1(t)\|_p^p + \|\nabla(v_2(t) - v_1(t))\|_p^p + \|v_2(t) - v_1(t)\|_p^p$$
$$\leq C_{17}\int_0^t e^{C_{17}(t-s)}\left(\|U_2(t) - U_1(t)\|_p^p + \|V_2(t) - V_1(t)\|_p^p\right) ds.$$

This implies

$$\|(u_2, v_2) - (u_1, v_1)\|_{C([0,T];L^p(\Omega)^2)}^p$$
$$\leq C_{18}\left(e^{C_{17}T} - 1\right)\|(U_2, V_2) - (U_1, V_1)\|_{C([0,T];L^p(\Omega)^2)}^p$$

and therefore, taking $T_1 \in (0, T_0]$ in

$$C_{18}\left(e^{C_{17}T_1} - 1\right) \leq \frac{1}{2^p},$$

we get the conclusion. □

The following lemma will assure the regularity of the fixed point.

Lemma 7.6 *Each $T \in (0, T_1]$ admits the estimate*

$$[\mathcal{F}(U, V)]_{\theta;Q_T} \leq C_{19}$$

for $(U, V) \in \mathcal{O}(M, T)$, where

$$[(u, v)]_{\theta;Q_T} \equiv [u]_{\theta;Q_T} + [v]_{\theta;Q_T}.$$

Proof. Let us recall that $(U,V) \in \mathcal{O}(M,T)$ and that $(u,v) = \mathcal{F}(U,V)$ denotes the classical solution to (IS). From (7.39), (7.19) and the definition of $\mathcal{O}(M,T)$, we have

$$\|v_t\|_p \leq \|\Delta v\|_p + \|g(U,V)\|_p \leq M + C_{11}\,|\Omega|^{1/p}$$

and

$$\|\nabla v\|_p \leq |\Omega|^{1/p}\,\|\nabla v\|_\infty \leq |\Omega|^{1/p}\,K_2 C_2^{1/p},$$

and hence we obtain

$$\|v\|_{W^{1,p}(Q_T)} \leq C_{20}.$$

Therefore, because

$$W^{1,p}(Q_T) \subset C^\theta(Q_T) \tag{7.50}$$

holds by $\theta < 1 - \frac{n+2}{p} < 1 - \frac{n+1}{p}$, we have

$$[v]_{\theta;Q_T} \leq C_{21}. \tag{7.51}$$

On the other hand, we have by (7.18) that

$$\begin{aligned}\|u_t\|_p &\leq \|\Delta u\|_p + \|\chi_u(U,v)\nabla v \cdot \nabla u\|_p \\ &\quad + \left\|\chi_v(U,v)\,|\nabla v|^2\right\|_p + \|\chi(U,v)\Delta v\|_p + \|f(U,V)\|_p \\ &\leq M + C_5(K_2C_2^{1/p})(K_1M)\,|\Omega|^{1/p} + C_5\left(K_2C_2^{1/p}\right)^2 |\Omega|^{1/p} \\ &\quad + C_5K_2C_2^{1/p}\,|\Omega|^{1/p} + C_5\,|\Omega|^{1/p}.\end{aligned}$$

We have also that

$$\|\nabla u\|_{L^p(\Omega)} \leq (K_1M)\,|\Omega|^{1/p},$$

and hence again by (7.50) it holds that

$$[v]_{\theta;Q_T} \leq C_{22}. \tag{7.52}$$

Thus, we get the conclusion by (7.51) and (7.52). □

7.2.4 Local Solvability

We will find a time-local solution to (CS) as a fix point of $\mathcal{F}$, taking $T \in (0, T_1]$. For this purpose, we put

$$\begin{aligned}\mathcal{K} &= \mathcal{K}(M,T)\\ &= \{(u,v) \in C^\theta(Q_T)^2 \cap C([0,T]; W^{2,p}(\Omega) \times W^{3,p}(\Omega))\\ &\mid \|(u,v)\|_{C([0,T];W^{2,p}(\Omega)\times W^{3,p}(\Omega))} \le M, \quad [(u,v)]_{\theta;Q_T} \le C_{19},\\ &\frac{\partial u}{\partial \nu} = \frac{\partial v}{\partial \nu} = 0 \quad \text{on} \quad \partial\Omega \times [0,T]\},\end{aligned}$$

where C_{19} is the constant prescribed in Lemma 7.6. This $\mathcal{K}$ is compact and convex in $C([0,T]; L^p(\Omega)^2)$, and we have $\mathcal{F}\mathcal{O}(M,T) \subset \mathcal{K}(M,T)$ by Lemma 7.6.

Now, we note the following.

Theorem 7.3 *If X is a Banach space with the norm $\|\cdot\|_X$ and F is a continuous mapping on X satisfying $\overline{O \cap K} \subset Dom(F)$, $F(O) \subset O \cap K$, and*

$$\|F(w_2) - F(w_1)\|_X \le k\,\|w_2 - w_1\|_X \qquad (w_2, w_1 \in O)\,, \tag{7.53}$$

where O, K are subsets in X and $k \in (0,1)$. Then, F has a unique fixed point in $\overline{O \cap K}$.

Proof. From the assumption, it holds that

$$F(\overline{O \cap K}) \subset F(\overline{O}) \subset \overline{F(O)} \subset \overline{O \cap K}$$

and that (7.53) for $w_2, w_1 \in \overline{O \cap K}$. Therefore, the conclusion follows from the *contraction mapping principle.* □

Now, we give the following.

Proof of Theorem 7.2: We have

$$\overline{\mathcal{O}(M,T) \cap \mathcal{K}(M,T)}^{C([0,T];L^p(\Omega)^2)} = \mathcal{K}(M,T)$$

and Lemma 7.5, and apply Theorem 7.3 for $X = C([0,T]; L^p(\Omega)^2)$, $O = \mathcal{O}(M,T)$, $K = \mathcal{K}(M,T)$, and $F = \mathcal{F}(\cdot, T)$. Then, we have a unique fix point denoted by $(u_*, v_*) \in C^{2,\theta}(Q_T)^2$, which becomes a classical solution to (CS). Conversely, if (u,v) is a classical solution to (CS) locally in time with the existence time $T_{\max} > 0$, then it is in $(u,v) \in \mathcal{O}(M,T') \cap \mathcal{K}(M,T')$ for some $T' \in (0, \min(T, T_{\max}))$ and is a fix point of $\mathcal{F}(\cdot, T')$. Therefore, it coincides with the above (u_*, v_*) for $t \in [0, T']$. Continuing the procedure, we see that $T_{\max} \geq T$ and $(u,v) = (u_*, v_*)$ for $t \in [0, T]$, and the proof is complete. □

Chapter 8

Appendix

This is the appendix. The first section is a catalogue of mathematical theories. Mathematical notions stated there are referred to in this monograph several times, and the reader will be able to get them easily. Detailed proofs of theorems are mostly written in Rudin [17] and Folland [9]. The second section, on the other hand, provides with some references to the theme treated in this monograph.

8.1 Catalogue of Mathematical Theories

8.1.1 *Basic Analysis*

Set of real numbers, denoted by $\mathbf{R}$, is provided with algebraic calculus, order, and continuity, which distinguishes that of rational numbers, denoted by $\mathbf{Q}$. Sets of integers and positive integers are denoted by $\mathbf{Z}$ and $\mathbf{N}$, respectively. Given $A \subset \mathbf{R}$, we say that it is bounded from above or below if there is $M \in \mathbf{R}$ or $m \in \mathbf{R}$ such that any $x \in A$ is in $x \leq M$ or $x \geq m$, respectively. It is said to be bounded if it is bounded from above and below. Such M or m is called the upper or lower bound of A, respectively. If A is bounded from above or below, then its least upper bound or largest lower bound is called the supremum or infimum, respectively. Then, the axiom of Weierstrass says that any set bounded from above and below has the supremum and infimum, respectively. Given a sequence in $\mathbf{R}$, denoted by $a_1, a_2, \cdots$ and $\alpha \in \mathbf{R}$, we mean $\lim_{n\to\infty} a_n = \alpha$ by $\lim_{n\to\infty} |a_n - \alpha| = 0$. This is expressed more precisely by the $\varepsilon - \delta$ argument that any $\varepsilon > 0$ admits $N \in \mathbf{N}$ such that $n \geq N$ implies $|a_n - \alpha| < \varepsilon$. A sequence $\{a_n\}$ is monotone increasing or decreasing if $a_n \leq a_{n+1}$ and $a_n \geq a_{n+1}$ for $n =$

$1, 2, \cdots$, respectively. Monotone increasing sequence bounded from above converges and so does monotone decreasing sequence bounded from below at the same time. Therefore, a bounded sequence always has a converging subsequence. If $\alpha \in \mathbf{R}$ is a limit of some subsequence of $\{a_n\}$, we say that it is an accumulating point of $\{a_n\}$. A sequence $\{a_n\}$ is said to be Cauchy if any $\varepsilon > 0$ takes $N \in \mathbf{N}$ such that $|a_n - a_m| < \varepsilon$ if $n, m \geq N$. Then, any Cauchy sequence converges. Those criteria of existence of supremum of the set bounded from above, convergence of the monotone increasing sequence bounded from above, and convergence of the Cauchy sequence are equivalent, and are indicated as the *continuity of real numbers* in short.

Given a sequence $\{a_n\}$, its supremum and infimum exist if $+\infty$ and $-\infty$ are admitted, and they are denoted by $\sup_n a_n$ and $\inf_n a_n$, respectively. Then, we can take the monotone decreasing sequence $\{b_n\}$ by

$$b_n = \sup_{k \geq n} a_k,$$

where $b_n = +\infty$ is admitted. In this case, we have $\lim_{n\to\infty} b_n$ with $\pm\infty$ admitted as its value, which is called the *limit supremum* of $\{a_n\}$ and is denoted by $\limsup_{n\to\infty} a_n$. The *limit infimum* of $\{a_n\}$, denoted by $\liminf_{n\to\infty} a_n$ is defined similarly by $\lim_{n\to+\infty} \inf_{k\geq n} a_k$. Any sequence $\{a_n\}$ admits subsequences converging to $\liminf_{n\to\infty} a_n$ and $\limsup_{n\to\infty} a_n$, respectively, and existence of $\lim_{n\to\infty} a_n$ is equivalent to $\liminf_{n\to\infty} a_n = \limsup_{n\to\infty} a_n$, where $\pm\infty$ is admitted as the value, and then, it holds that $\lim_{n\to\infty} a_n = \liminf_{n\to\infty} a_n = \limsup_{n\to\infty} a_n$. Convergence of a series $\sum_{n=1}^{\infty} a_n$ is discussed by the sequence $\{s_n\}$ made by its partial sum:

$$s_n = \sum_{k=1}^{n} a_k.$$

If $a_n \geq 0$ for any $n = 1, 2 \cdots$, then $\{s_n\}$ is monotone increasing so that $\sum_{n=1}^{\infty} a_n$ converges if and only if $\{s_n\}$ is bounded from above, in which case we refer to

$$\sum_{n=1}^{\infty} a_n < +\infty.$$

A series $\sum_{n=1}^{\infty} a_n$ is said to absolutely converge if $\sum_{n=1}^{\infty} |a_n| < +\infty$. In this case it is shown that $\{s_n\}$ forms a Cauchy sequence and hence the convergence of $\sum_{n=1}^{\infty} a_n$ follows.

Let $I = (a, b) \subset \mathbf{R}$ be an open interval with $a = -\infty$ and $b = +\infty$ admitted, and $f : I \to \mathbf{R}$ be given. Then its continuity at $x = x_0 \in I$ is indicated by $\lim_{x \to x_0} f(x) = f(x_0)$, so that we have to define $\ell = \lim_{x \to x_0} f(x)$ more precisely. Actually, this means that any $\varepsilon > 0$ admits $\delta > 0$ such that any x in $|x - x_0| < \delta$ takes the estimate $|f(x) - \ell| < \varepsilon$. Therefore, $f = f(x)$ is continuous at $x = x_0 \in I$ if and only if any $\varepsilon > 0$ admits $\delta > 0$ such that $|x - x_0| < \delta$ implies $|f(x) - f(x_0)| < \varepsilon$. Because I is an open interval and $x_0 \in I$, we have $x \in I$ for x close to x_0, and $f(x)$ is defined if above $\delta > 0$ is small enough. Then, we can show that $f = f(x)$ is continuous at $x = x_0$ if any sequence $\{x_n\}$ in $\lim_{n \to \infty} x_n = x_0$ admits $\lim_{n \to \infty} f(x_n) = f(x_0)$. We say that f is continuous in I if it is continuous at any $x_0 \in I$.

A set $A \subset \mathbf{R}$ is said to be open if any $x_0 \in A$ admits $\delta > 0$ such that $(x_0 - \delta, x_0 + \delta) \subset A$. It is said to be closed if $A^c = \mathbf{R} \setminus A$ is open. By definition, it is open even if $A = \emptyset, \mathbf{R}$. In particular, they are simultaneously closed. Subset $F \subset \mathbf{R}$ is closed if and only if $\{x_n\} \subset F$ and $\lim_{n \to \infty} x_n = x_0$ implies $x_0 \in F$. A real-valued function f defined on an open interval I is continuous there if and only if the inverse image $f^{-1}(A)$ of any open set $A \subset I$ is open. Given $X \subset \mathbf{R}$ not necessarily an open interval and $x_0 \in X$, we say that $f : X \to \mathbf{R}$ is continuous at $x = x_0$ if any $\varepsilon > 0$ admits $\delta > 0$ such that $x \in X$ and $|x - x_0| < \delta$ implies $|f(x) - f(x_0)| < \varepsilon$. It is continuous on X if it is so at any $x_0 \in X$. A subset $A \subset X$ is said to be open in X, if any $x_0 \in A$ admits $\delta > 0$ such that $(x_0 - \delta, x_0 + \delta) \cap X \subset A$. Then, f is continuous on X if and only if the inverse image $f^{-1}(A)$ of open $A \subset \mathbf{R}$ is open in X. Subset $X \subset \mathbf{R}$ is said to be connected if there is no pair of open sets U, V in $\mathbf{R}$ such that $U \cap X \neq \emptyset$, $V \cap X \neq \emptyset$, $U \cap V \cap X = \emptyset$, and $X \subset U \cup V$. It is equivalent for $X \subset \mathbf{R}$ to be an interval. If $f : X \to \mathbf{R}$ is continuous and $X \subset \mathbf{R}$ is connected, then $f(X)$ is connected. Then, the intermediate value theorem follows as if $f : [a, b] \to \mathbf{R}$ is continuous with $a < b$ and J denotes the closed interval with the endpoints composed of $f(a)$, $f(b)$, then any $\alpha \in J$ admits $x \in [a, b]$ such that $f(x) = \alpha$. Subset $X \subset \mathbf{R}$ is said to be compact if there is a family of open sets $\{U_\alpha\}_{\alpha \in \Lambda}$ in $\mathbf{R}$ such that $X \subset \bigcup_{\alpha \in \Lambda} U_\alpha$, then there is a finite $\alpha_1, \alpha_2, \cdots, \alpha_m \in \Lambda$ such that $X \subset \bigcup_{i=1}^{\alpha} U_{\alpha_i}$. If $X \subset \mathbf{R}$ is compact and $f : X \to \mathbf{R}$ is continuous, then $f(X)$ is compact in $\mathbf{R}$. On the other hand, Heine-Borel's theorem says that $X \subset \mathbf{R}$ is compact if and only if it is bounded and closed. From this we can show that if f is a real-valued continuous function defined on a bounded closed interval $I = [a, b]$ it takes minimum and maximum there.

A real-valued function f defined on $X \subset \mathbf{R}$ is said to be uniformly continuous if any $\varepsilon > 0$ admits $\delta > 0$ such that $x, x_0 \in X$ and $|x - x_0| < \delta$ imply $|f(x) - f(x_0)| < \varepsilon$. If X is compact and $f : X \to \mathbf{R}$ is continuous, then it is uniformly continuous. For a bounded closed interval $I = [a, b]$, its division is given by $\Delta : x_0 = a < x_1 < x_2 < \cdots < x_n = b$ and $\|\Delta\| = \max_{1 \leq i \leq n}(x_i - x_{i-1})$ is referred as the mesh size. For each small interval we take $\xi_i \in [x_{i-1}, x_i]$ arbitrarily and set

$$I_{\Delta,\xi} = \sum_{i=1}^{n} f(\xi_{i-1})(x_i - x_{i-1})$$

with $\xi = (\xi_1, \cdots, \xi_n)$. We say that f is Riemann integrable if there is I such that $\lim_{\|\Delta\| \to 0} I_{\Delta,\xi} = I$. We suppose that $f = f(x)$ is bounded, which means $\sup_{x \in I} |f(x)| \leq M$ with some $M > 0$, and let $M_i = \sup_{x \in [x_{i-1}, x_i]} f(x)$ and $m_i = \inf_{x \in [x_{i-1}, x_i]} f(x)$. Then, we can take that

$$S_\Delta = \sum_{i=1}^{n} M_i(x_i - x_{i-1}) \quad \text{and} \quad s_\Delta = \sum_{i=1}^{n} m_i(x_i - x_{i-1}),$$

which satisfies that $s_\Delta \leq I_{\Delta,\xi} \leq S_\Delta$ for any ξ. Then, Darboux's theorem assures $\lim_{\|\Delta\| \to 0} S_\Delta = S$ and $\lim_{\|\Delta\| \to 0} s_\Delta = s$ with $s \leq S$, where $S = \inf_\Delta S_\Delta$ and $s = \sup_\Delta s_\Delta$. Therefore, f is Riemann integrable on I if and only if $s = S$, or equivalently, any $\varepsilon > 0$ admits $\delta > 0$ such that $\|\Delta\| < \delta$ implies $S_\Delta - s_\Delta < \varepsilon$. Then, we can show that if $f : [a, b] \to \mathbf{R}$ is continuous, then it is Riemann integrable from its uniform continuity. Sequence of real-valued functions $\{f_n\}$ defined on $X \subset \mathbf{R}$ is said to converge f uniformly if any $\varepsilon > 0$ admits $N \in \mathbf{N}$ such that $|f_n(x) - f(x)| < \varepsilon$ for any $n \geq N$ and $x \in X$. In this case if each f_n is continuous on X, then so is the limit function f and furthermore,

$$\lim_{n \to \infty} \int_a^b f_n(x)dx = \int_a^b f(x)dx$$

if $X = [a, b]$.

8.1.2 *Topological Spaces*

First, *topological space* is a set provided with the family of *open* subsets, which satisfies the axioms that the whole space and the empty set are open and that any union of open subsets, and any finite intersection of open subsets, are again open. A subset of topological space is *closed* if its complement is open. A subset of topological space is *connected*, if it is not covered by any disjoint union of two non-empty open sets. A subset of a topological space is *compact* if any open covering of it is reduced to a sub-covering of finite number. A mapping from a topological space to another one is said to be *continuous* if its inverse image of any open set is open. If it is onto, one-to-one, and the inverse mapping is continuous, then it is said to be *homeomorphism*. Any subset A of a topological space L is regarded as a topological space under the agreement that a subset in A is open if and only if it is the intersection of an open set in L and A itself.

A set L is said to be a *metric space* if it is provided with the *distance* denoted by dist(,), which is a mapping from $L \times L$ to $[0, +\infty)$ satisfying the axioms that $\mathrm{dist}(u,v) = 0$ if and only if $u = v$, that $\mathrm{dist}(v,u) = \mathrm{dist}(u,v)$ for $u, v \in L$, and that

$$\mathrm{dist}(u,v) \leq \mathrm{dist}(u,w) + \mathrm{dist}(w,v) \tag{8.1}$$

for $u, v, w \in L$. Here, (8.1) is referred to as the *triangle inequality*.

If L is a metric space with the distance dist(,) and A is a subset of L, then $x_0 \in L$ is said to be an *interior* point of A if there is $r > 0$ satisfying $B(x_0, r) \subset A$, where

$$B(x_0, r) = \{x \in L \mid \mathrm{dist}(x, x_0) < r\}.$$

A subset of the metric space is said to be open if any element is an interior point. This notion agrees the axioms of open sets stated above and under this agreement the metric space is usually regarded as a topological space. Then, the set of interior points of A is called the interior of A and is denoted by $\mathrm{int}A$. Always, it is an open subset. On the other hand, an element x_0 of L is said to be on the *boundary* of A if $B(x_0, r) \cap A \neq \emptyset$ and $B(x_0, r) \cap A^c \neq \emptyset$ for any $r > 0$. The set of boundary points of A is denoted by ∂A and $\overline{A} = A \cup \partial A$ is said to be the *closure* of A. A subset of L is closed if and only if it is equal to its closure. A subset of L is said to be *dense* if its closure is equal to L itself. Also, L is said to be *separable* if there is a countable dense subset. Those notions of interior, boundary, and closure are extended to the general topological space.

A sequence $\{x_j\}$ in a metric space L with the distance dist(,) is said to converge to $x_0 \in L$ if any $r > 0$ admits k such that $x_j \in B(x_0, r)$ for

$j \geq k$. This is equivalent to

$$\lim_{j\to\infty} \mathrm{dist}(x_j, x_0) = 0,$$

and is denoted by $x_j \to x_0$ or $\lim_{j\to\infty} x_j = x_0$. Then, $x_0 \in \overline{A}$ if and only if there is a sequence $\{x_j\} \subset A$ converging to x_0. A continuous function f from the metric space (L, dist) to $\mathbf{R}$ is *uniformly continuous* on any compact subset $E \subset L$, which means that any $\varepsilon > 0$ admits $\delta > 0$ such that $\mathrm{dist}(x, y) < \delta$ and $x, y \in E$ imply $|f(x) - f(y)| < \varepsilon$. A sequence $\{u_j\}$ in the metric space (V, dist) is said to be a *Cauchy sequence* if it satisfies $\mathrm{dist}(u_j, u_k) \to 0$ as $j, k \to \infty$. A converging sequence is always a Cauchy sequence, and the metric space (V, dist) is said to be *complete* if any Cauchy sequence converges.

Euclidean space $\mathbf{R}^n$ is provided with the standard distance $\mathrm{dist}(\boldsymbol{x}, \boldsymbol{y}) = |\boldsymbol{x} - \boldsymbol{y}|$, where

$$|\boldsymbol{x}| = \sqrt{x_1^2 + x_2^2 + \cdots x_n^2} \qquad \text{for} \qquad \boldsymbol{x} = (x_1, x_2, \cdots, x_n).$$

Usually, *domain* indicates an open, connected subset of $\mathbf{R}^n$, and it is said to be *simply connected* if any Jordan curve can shrink to a point inside, where *Jordan curve* denotes a closed curve without self-intersections, and a closed curve in $\mathbf{R}^n$ is the image of a continuous mapping from the *unit circle* $S^1 = \{(\cos\theta, \sin\theta) \mid 0 \leq \theta < 2\pi\} \subset \mathbf{R}^2$ to $\mathbf{R}^n$. Finally, *closed region* indicates the closure of a domain.

Family of closed subsets satisfies the axioms that the whole space and the empty set are closed and that any intersection of closed subsets, and any finite union of closed subsets, are again closed. Topology can be introduced from this family by saying that a subset is open if its complement is closed. If $x \in L$ is given, its *neighborhood* indicates an open set containing it. A family $\mathcal{U}(x)$ is said to be a *fundamental neighborhood system* of $x \in L$ if any $U \in \mathcal{U}(x)$ admits an open V such that $x \in V \subset U$. It satisfies the axioms that any $V \in \mathcal{U}(x)$ satisfies $x \in V$, that any $V_1, V_2 \in \mathcal{U}(x)$ admits $V_3 \in \mathcal{U}(x)$ such that $V_3 \subset V_1 \cap V_2$, and that any $V \in \mathcal{U}(x)$ admits $W \in \mathcal{U}(x)$ such that for $y \in W$ there is $V_y \in \mathcal{U}(y)$ in $V_y \subset V$. In the case that the fundamental neighborhood system $\mathcal{U}(x)$ is given at each $x \in L$, we say that x is an interior point of $A \subset L$ if there is $U \in \mathcal{U}(x)$ such that $U \subset A$, and in this way the topology can be introduced from fundamental neighborhood systems instead of that of the family of open sets.

The notions of uniformly continuity and completeness are not extended to the general topological space, but can be to *uniform topological spaces.* In this connection, it should be noted that in the general topological space without *countability*, sequences are not enough to describe the full structure of topology and the notion of *net* takes place of. Thus, we say that the uniform topological space is *complete* if any Cauchy net converges, *sequentially complete* if any Cauchy sequence converges. Topological space L is said to satisfy the *first countability axiom* if each $x \in L$ admits a fundamental neighborhood system with countable members. In this case, its topology can be described by the notion of convergence of sequences. Topological space L is said to satisfy the *second countability axiom* if there is a family $\mathcal{O}$ consisting of countable open sets such that any open set is a union of some members in $\mathcal{O}$. The metric space (L, dist) always satisfies the first countability axiom. If it is separable, then it satisfies the second countability axiom.

Another important notion of topological space is the *separability*. First, topological space L is said to be a *Hausdorff space* if any $x, y \in L$ in $x \neq y$ admit open sets U and V such that $x \in U$, $y \in V$, and $U \cap V = \emptyset$. Hausdorff space L is said to be *normal* if any closed subsets F_1, F_2 with $F_1 \cap F_2 = \emptyset$ admits open sets U_1 and U_2 such that $F_1 \subset U_1$, $F_2 \subset U_2$, and $U_1 \cap U_2 = \emptyset$. Then, *Urysohn's theorem* guarantees that if F_0, F_1 are closed subsets in the normal Hausdorff space L satisfying $F_0 \cap F_1 = \emptyset$, then there is a continuous mapping $f : L \to [0, 1]$ with the values 0 and 1 on F_0 and F_1, respectively. This implies the *extension theorem of Tietze* that any continuous mapping $f : F \to \mathbf{R}$ defined on the closed set F in the normal Hausdorff space L has a continuous extension $\tilde{f} : L \to \mathbf{R}$.

A topological space is said to be *compact* if its any open covering is reduced to a finite subcovering. In the metric space it is equivalent to be *sequentially compact*, which means that any sequence there contains a converging subsequence. If X is a compact metric space, then the set of continuous functions on X, denoted by $C(X)$ forms a Banach space under the norm $\|f\| = \max_{x \in X} |f(x)|$. Then, *Ascoli-Arzelá's theorem* guarantees that $F \subset C(X)$ is compact if and only if it is *uniformly bounded* and *equicontinuous*. Here, the former means $\sup_{f \in F} \|f\| < +\infty$, while the latter says that any $\varepsilon > 0$ admits $\delta > 0$ such that $x, y \in X$ with $\text{dist}(x, y) < \delta$ implies that $\sup_{f \in F} |f(x) - f(y)| < \varepsilon$.

Exercise 8.1 Confirm that a metric space is a topological space and Euclidean space is a metric space under the distance given above.

Exercise 8.2 Confirm that if L and M are metric spaces, then $f : L \to M$ is continuous if and only if $x_j \to x_0$ in L implies $f(x_j) \to f(x_0)$ in M.

8.1.3 *Complex Function Theory*

Complex function theory is an interesting object by itself, and here we shall collect some basic facts.

First, a complex valued function $f(z)$ defined on a domain D in $\mathbf{R}^2 \cong \mathbf{C}$ is said to be *differentiable* at $z = c \in D$ if

$$\lim_{z\to c} \frac{f(z) - f(c)}{z - c} = f'(c)$$

exists. In terms of $z = x+\imath y$ and $f(z) = u(x,y)+\imath v(x,y)$ with real x, y, u, v, this condition is equivalent to saying that u, v are totally differentiable at c and satisfy *Cauchy-Riemann's relation*

$$u_x(c) = v_y(c), \qquad u_y(c) = -v_x(c).$$

We say that $f(z)$ is *holomorphic* in D and at c if it is differentiable at any point in D and in a neighborhood of c, respectively. The function $f(z) = u(x,y) + \imath v(x,y)$ with $z = x + \imath y$ is holomorphic in D if and only if u, v are totally differentiable at any point z in D and satisfy

$$u_x(z) = v_y(z), \qquad u_y(z) = -v_x(z).$$

We say that $w = f(z)$ is *conformal* at $z = c$ if γ is a curve crossing c with its tangential line there, then $f(\gamma)$ has the same property in w plane at $f(c)$, and the angle made by any of such two curves γ_1, γ_2 are equal to that made by $f(\gamma_1)$ and $f(\gamma_2)$. It is known that if $f'(c) \neq 0$, then $w = f(z)$ is conformal at c. *Men'shov's theorem* says that if $f : D \to \Omega$ is homeomorphic and conformal at any point in D, then it is holomorphic and satisfies $f'(z) \neq 0$ for any $z \in D$. A domain surrounded by a Jordan curve is called the *Jordan region*. Then, *Riemann's mapping theorem* guarantees that any Jordan region admits a conformal homeomorphism to the unit disc.

Cauchy-Hadamard's formula

$$R = \frac{1}{\limsup_{n\to\infty} |a_n|^{1/n}}$$

assures that the power series

$$\sum_{n=0}^{\infty} a_n z^n$$

converges absolutely and uniformly in $|z| < R$, while it does not converge for $|z| > R$. That is, R is the *convergence radius* and $|z| = R$ is the convergence circle. In this case,

$$f(z) = \sum_{n=0}^{\infty} a_n z^n$$

is holomorphic in $|z| < R$ with the relation that

$$f'(z) = \sum_{n=1}^{\infty} a_n n z^{n-1}$$

is also holomorphic in $|z| < R$. Actually, its convergence radius is again R.

The *integral theorem of Cauchy* guarantees that if $f(z)$ is holomorphic in a simply connected domain D and is continuous on $\overline{D}$, then it holds that

$$\int_C f(z)dz = 0,$$

where $C = \partial D$ is oriented counter-clockwisely. Conversely, *Morera's theorem* says that if $f(z)$ is continuous in D and satisfies

$$\int_C f(z)dz = 0$$

for any Jordan curve C in D, then it is holomorphic in D.

Cauchy's integral formula is indicated as

$$f(z) = \frac{1}{2\pi\imath} \int_C \frac{f(\zeta)}{\zeta - z} d\zeta \tag{8.2}$$

for $z \in D$, which guarantees that the holomorphic function $f(z)$ is differentiable arbitrary many times with the formula that

$$f^{(n)}(z) = \frac{n!}{2\pi\imath} \int_C \frac{f(\zeta)}{(\zeta - z)^{n+1}} d\zeta.$$

Furthermore, if $f(z)$ is holomorphic in $|z| < R$, then it has the *Taylor expansion* there as

$$f(z) = \sum_{n=0}^{\infty} c_n z^n \qquad \text{with} \qquad c_n = \frac{f^{(n)}(0)}{n!}.$$

This implies that if $f(z)$, $g(z)$ are holomorphic in D and there is a sequence $z_k \neq c$ converging to $c \in D$ with $f(z_k) = g(z_k)$ for $k = 1, 2, \cdots$, then it holds that $f(z) = g(z)$ for any $z \in D$.

Another application of (8.2) is the *maximum principle*, which says that if $f(z)$ is a non-constant holomorphic function in D, then $|f(z)|$ does not take the maximum in D. Therefore, if D is a bounded domain and $f(z)$ is continuous on $\overline{D}$, then the maximum of $|f(z)|$ is attained on ∂D. If $f(z)$ is holomorphic in $|z| < \infty$, then it is said to be *entire*. In this case, the maximum principle guarantees that

$$M(r) = \max_{|x|=r} |f(z)|$$

is non-decreasing in $r > 0$ and the value

$$\limsup_{r \to \infty} \frac{\log \log M(r)}{\log r} \tag{8.3}$$

is called the *order* of $f(z)$.

Schwarz' theorem says that if $f(z)$ is holomorphic in $|z| < R$, $|f(z)| \leq M$, and $f(0) = 0$, then it holds that

$$|f(z)| \leq \frac{M}{R} |z| \qquad (|z| < R).$$

If equality holds at some $z = z_1 \in D$ in the above inequality, then it follows that $f(z) = e^{\imath\theta} \cdot \frac{M}{R} z$ with some $\theta \in [0, 2\pi)$.

Cauchy's estimate means that if $f(z) = \sum a_n z^n$ is holomorphic in $|z| < R$, then it holds that

$$|a_n| \leq \frac{M(r)}{r^n} \qquad (r \in (0, R)),$$

with $M(r)$ defined by (8.3). This implies the *Liouville's theorem* that any bounded entire function must be a constant. This theorem makes it possible to give the proof of the fundamental theorem of algebra that any non-constant polynomial has a zero point in the complex plane $\mathbf{C}$.

The *Laurent series* is expressed as

$$\sum_{n=-\infty}^{+\infty} a_n(z-c)^n = \sum_{n=1}^{\infty} a_{-n}(z-c)^{-n} + \sum_{n=0}^{+\infty} a_n(z-c)^n, \tag{8.4}$$

where the first term of the right-hand side is called the *principal part*. If $f(z)$ is holomorphic in $\rho < |z-c| < R$ for $0 \le \rho < R \le \infty$, then it is expanded uniquely by the Laurent series. In the case that $f(z)$ is holomorphic in $0 < |z-c| < R$ and is not holomorphic at $z = c$, then $z = c$ is called the *isolated singular point* of $f(z)$. An isolated singular point is *removable*, *pole*, and *essential* if the principal part in (8.4) is composed of the terms of none, finite, and infinite, respectively. Then, *Riemann's theorem* says that if $f(z)$ is bounded and holomorphic in $0 < |z-c| < R$, then $z = c$ is removable so that it has a holomorphic extension in $|z-c| < R$. Conversely, *Weierstrass' theorem* says that if $f(z)$ is essentially singular at $z = c$, then any $\alpha \in \mathbf{C}$ admits $z_n \to c$ such that $f(z_n) \to \alpha$. If $z = c$ is an isolated singular point of $f(z)$, then the coefficient a_{-1} in (8.4) is called the *residue*. The isolated singularity of $f(z)$ at $z = \infty$ is defined by that of $g(w) = f(1/w)$ at $w = 0$. If $f(z)$ is holomorphic in D except for isolated poles, it is called a *meromorphic function*. A function meromorphic on $|z| \le \infty$ must be a rational function.

The *residue principle* says that if $f(z)$ is holomorphic in D except for finite isolated singular points, denoted by $\{a_1, \cdots, a_n\} \subset \mathbf{C}$, and C is a Jordan curve in $D \setminus \{a_1, \cdots, a_n\}$ containing those singular points inside, then it holds that

$$\frac{1}{2\pi\imath} \int_C f(z)dz = \sum_j \text{Res } (a_j),$$

where Res (a_j) indicates the residue of $f(z)$ at $z = a_j$.

The *argument principle* says that if $f(z)$ is meromorphic in D and holomorphic on C with $f(z) \neq 0$, then it holds that

$$\frac{1}{2\pi\imath} \int_C \frac{f'(z)}{f(z)} dz = N - P,$$

where C is a Jordan curve in D and N, P denote the number of zeros and poles of $f(z)$ inside C, respectively, with the multiplicities included. Then, *Rouche's theorem* says that if $f(z)$ is holomorphic in D, C is a Jordan curve

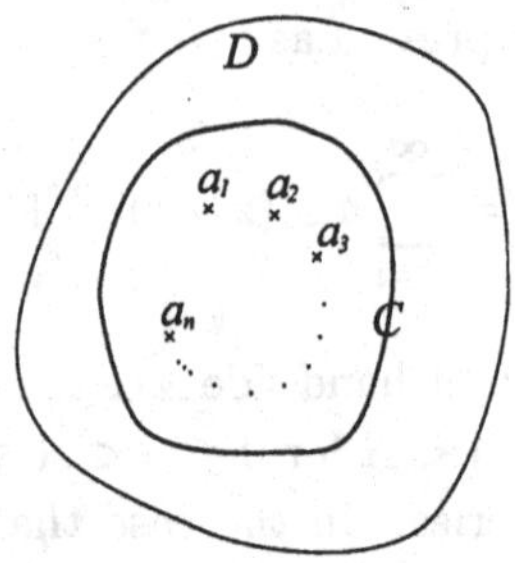

Fig. 8.1

in D, and $|f(z)| > |g(z)|$ holds on C, then the number of zeros of $f(z)$ and $f(z) + g(z)$ are equal in D.

Exercise 8.3 Show that if $f(z)$ is entire, then $M(r) = \max_{|z|=r} |f(z)|$ is a non-decreasing function of $r > 0$.

Exercise 8.4 Suppose that a polynomial $P(z)$ does not assume 0 in **C**, and take the entire function $f(z) = \frac{1}{P(z)}$. Then, apply Liouville's theorem to guarantee that $P(z)$ must be a constant.

Exercise 8.5 A function meromorphic on $|z| \leq \infty$ is holomorphic in $|z| \leq \infty$ except for finite number of poles, denoted by $c_1, \cdots, c_k$. Letting $P_i(z)$ be its principal part at $z = c_i$, take

$$\varphi(z) = f(z) - \sum_i P_i(z).$$

It is holomorphic in $|z| \leq \infty$ and therefore, must be a constant by Liouville's theorem. Thus, confirm that a function meromorphic on $|z| \leq \infty$ must be a rational function.

8.1.4 *Real Analysis*

The norm $\|\cdot\|_2$ introduced in §2.3.3 provides $L^2(0, \pi)$ with the complete metric. As is mentioned there, this fact is proven by the convergence theorems on Lebesgue integrals. In this connection, it may be worth noting that there are three important convergence theorems, *dominated convergence theorem*, *monotone convergence theorem*, and *Fatou's lemma*. The last one describes some kind of lower semi-continuity, described in § 2.3.4.

If X is a set, then 2^X denotes the set of all subsets of X. A subset $\mathcal{F}$ of 2^X is said to be *finitely additive* if it satisfies the axiom that $\emptyset \in \mathcal{F}$, $F^c \in \mathcal{F}$ if $F \in \mathcal{F}$, and $E \cup F \in \mathcal{F}$ if $E, F \in \mathcal{F}$. If $\mathcal{F}$ is such a family, then the mapping $m : \mathcal{F} \to [0, \infty]$ is said to be a *finitely additive measure* on $(X, \mathcal{F})$ if it satisfies the axiom that $m(\emptyset) = 0$ and $m(A \cup B) = m(A) + m(B)$ for $A, B \in \mathcal{F}$ with $A \cap B = \emptyset$. It is said to be a *pre-measure* if it satisfies the axiom that $m(A) = \sum_{k=1}^{\infty} m(A_k)$ if $\{A_k\}_{k=1}^{\infty} \in \mathcal{F}$ is a disjoint family and $A = \sum_{k=1}^{\infty} A_k (= \cup_k A_k) \in \mathcal{F}$.

If $\mathcal{F} \subset 2^X$ is a finitely additive family and m is a finitely additive measure on $(X, \mathcal{F})$, then

$$\Gamma(A) = \inf \left\{ \sum_{k=1}^{\infty} m(A_k) \mid A_k \in \mathcal{F},\ A \subset \cup_{k=1}^{\infty} A_k \right\}$$

defines the *outer measure* so that $\Gamma : 2^X \to [0, \infty]$ satisfies the axiom that $\Gamma(\emptyset) = 0$, $\Gamma(A) \leq \Gamma(B)$ if $A \subset B$, and $\Gamma\left(\cup_{k=1}^{\infty} A_k\right) \leq \sum_{k=1}^{\infty} \Gamma(A_k)$ for $\{A_k\}_{k=1}^{\infty} \subset 2^X$. Furthermore, if m is a pre-measure, then $\Gamma|_{\mathcal{F}} = m$ holds.

A family $\mathcal{D} \subset 2^X$ is said to be a *σ-algebra* if it satisfies the axiom that $\emptyset \in \mathcal{D}$, $A^c \in \mathcal{D}$ if $A \in \mathcal{D}$, and $\cup_{k=1}^{\infty} A_k \in \mathcal{D}$ if $\{A_k\}_{k=1}^{\infty} \subset \mathcal{D}$. If $\mathcal{D}$ is a σ-algebra and $\{A_k\}_{k=1}^{\infty} \subset \mathcal{D}$ is a disjoint family, then the latter is said to be a *division* of $A = \sum_{k=1}^{\infty} A_k (= \cup_{k=1}^{\infty} A_k) \in \mathcal{D}$. If $\mathcal{D}$ is a σ-algebra, then the mapping $\mu : \mathcal{D} \to [0, \infty]$ is said to be a *measure* if it satisfies the axiom that $\mu(\emptyset) = 0$ and $\mu(A) = \sum_{k=1}^{\infty} \mu(A_k)$ if $\{A_k\}_{k=1}^{\infty} \subset \mathcal{D}$ is a division of $A = \sum_{k=1}^{\infty} A_k \in \mathcal{D}$.

If $\mathcal{D} \subset 2^X$ is a σ-algebra, then $(X, \mathcal{D})$ is said to be a *measurable space*. If μ is a measure on $(X, \mathcal{D})$, then $(X, \mathcal{D}, \mu)$ is called the *measure space*. An assertion in the measure space is said to hold *almost everywhere* if it is valid except for a set of measure 0. A measure space $(X, \mathcal{D}, \mu)$ is said to be *σ-finite* if there are $X_k \in \mathcal{D}$ $(k = 1, 2, \cdots)$ such that $\mu(X_k) < +\infty$ and $\cup_k X_k = X$.

If Γ is an outer measure on X, then $E \in 2^X$ is said to be *Γ-measurable* if $\Gamma(A) = \Gamma(A \cap E) + \Gamma(A \cap E^c)$ holds for any $A \in 2^X$. Then, the family of Γ-measurable sets, denoted by m_Γ, forms a σ-algebra, and Γ restricted to m_Γ becomes a measure.

The *Jordan family* of $\mathbf{R}^n$ denotes the least finitely additive family in $2^{\mathbf{R}^n}$ containing n-dimensional rectangles, and the *Jordan measure* is the uniquely determined finitely additive measure defined on the Jordan family with the value of the n-dimensional rectangle equal to its n-dimensional volume. It is shown that the Jordan measure is a pre-measure, and from the above story, referred to as the *Carathéodory theory*, we get a measure space denoted by $(\mathbf{R}^n, \mathcal{L}_n, \mu_n(dx))$. Each element in $\mathcal{L}_n$ is said to be a (n-dimensional) *Lebesgue measurable set*, and $\mu_n(dx)$ is called the (n-dimensional) Lebesgue measure.

The minimum σ-algebra containing all open sets in $\mathbf{R}^n$ is said to be the *Borel family* in $\mathbf{R}^n$ and is denoted by $\mathcal{B}_n$. Each element in $\mathcal{B}_n$ is called the *Borel set*. Each Borel set is Lebesgue measurable, and in this way we get a smaller measure space by restricting $\mu_n(dx)$ to $\mathcal{B}_n$. Then, the original measure space $(\mathbf{R}^n, \mathcal{L}_n, \mu_n(dx))$ is regarded as the *completion* of $(\mathbf{R}^n, \mathcal{B}_n, \mu_n(dx))$ as a measure space. Actually, the measure space $(X, \mathcal{D}, \mu)$ is said to be *complete* if it satisfies the axiom that $A \in \mathcal{D}$, $\mu(A) = 0$, and $B \subset A$ imply $B \in \mathcal{D}$, and each measure space takes the least extended complete measure space, referred to as its *completion*. For $n = 1$, it is convenient to extend those notions of the Borel and the Lebesgue measurable sets to those in the two-point compactification of $\mathbf{R}$, denoted by $\overline{\mathbf{R}} = [-\infty, +\infty]$.

If $(X, \mathcal{D})$ is a measurable space, then the function $f : X \to [-\infty, \infty]$ is said to be *measurable* if any inverse image of the Borel set in $\overline{\mathbf{R}}$ is in $\mathcal{D}$. The measurability of functions is preserved under countably many limiting processes. Let $(X, \mathcal{D}, \mu)$ be a measure space with $\mu(X) < +\infty$, and let $f_k : X \to \mathbf{R}$ $(k = 1, 2, \cdots)$ be a family of measurable functions satisfying $\lim_{k\to\infty} f_k(x) = f(x)$ for almost every $x \in X$. Given $\varepsilon > 0$, then *Egorov's theorem* assures $E \in \mathcal{D}$ such that $\mu(E) < \varepsilon$ and $\lim_{k\to\infty} f_k(x) = f(x)$ uniformly in $x \in X \setminus E$. If $A \subset \mathbf{R}^n$ and $f : A \to \mathbf{R}$ are Lebesgue measurable with $\mu(A) < +\infty$ and $\varepsilon > 0$, then *Lusin's theorem* guarantees the existence of a compact set $K \subset A$ such that $\mu(A \setminus K) < \varepsilon$ and $f|_K$ is continuous. If $(X, \mathcal{D}, \mu)$ is a measure space and $\{f_k\}_{k=1}^{\infty}$ is a family of measurable functions, then we say that f_k *converges in measure* to f if $\lim_{k\to\infty} \mu\{x \in X \mid |f_k(x) - f(x)| > \varepsilon\} = 0$ for any $\varepsilon > 0$.

If $(X, \mathcal{D}, \mu)$ is a measure space, then $f : X \to [0, \infty]$ is said to be a (non-negative) *simple function* if it is written as a finite sum of $\alpha_j \chi_{A_j}$ with $\alpha_j \geq 0$ and $A_j \in \mathcal{D}$. The set of such functions is denoted by L_0^+. In this case, the quantity

$$\int f = \sum_j \alpha_j \mu(A_j)$$

is independent of the expression of $f = \sum_j \alpha_j \chi_{A_j}$ under the agreement of $0 \cdot \infty = 0$. This definition of $\int f$ for $f \in L_0^+$ is consistent with

$$\int f = \sup \left\{ \int \varphi \mid 0 \leq \varphi \leq f, \ \varphi \in L_0^+ \right\}$$

for the general measurable function $f : X \to [0, \infty]$, of which totality is denoted by L^+.

Then, the *monotone convergence theorem* guarantees that if $f_k \in L^+$ is monotone increasing pointwisely in $k = 1, 2, \cdots$, then

$$\lim_k \int f_k = \int f \tag{8.5}$$

follows for $f = \lim_k f_k$ with $f \in L^+$. On the other hand, *Fatou's lemma* assures that

$$\int \liminf f_k \leq \liminf_k \int f_k,$$

whenever $f_k \in L^+$ for $k = 1, 2, \cdots$, in which case it follows that $\liminf_k f_k \in L^+$.

In the general case that $f : X \to [-\infty, +\infty]$ is measurable, we take $f_\pm(x) = \max\{\pm f(x), 0\}$, which are again measurable. Then, we set

$$\int f = \int f_+ - \int f_-$$

at most one of $\int f_\pm$ is finite. If both of them are finite, then we say that f is *summable* and write that $f \in L^1(X, d\mu)$. Then, the *dominated convergence theorem of Lebesgue* assures (8.5) under the assumption that $f_k \to f$ and $|f_k| \leq g$ almost everywhere with some $g \in L^1(X, d\mu)$. Here, the latter assumption may be replaced by $|f_k| \leq g_k$, $g_k \to g$ almost everywhere, and

$$\lim_k \int g_k = \int g.$$

If $(X, \mathcal{M}, \mu)$ and $(Y, \mathcal{N}, \nu)$ are measure spaces, then $m(E \times F) = \mu(E) \cdot \nu(F)$ defined for $E \in \mathcal{M}$, $F \in \mathcal{N}$ generates the finitely additive family $\mathcal{G} \subset 2^{X \times Y}$ and the finitely additive measure $m : \mathcal{G} \to [0, \infty]$ under the agreement that $0 \cdot \infty = 0$. This m is a pre-measure, and from the Carathéodory theory it is extended to a measure on $\mathcal{M} \otimes \mathcal{N}$, the minimum σ-algebra containing $\mathcal{G}$, which is denoted by $\mu \otimes \nu$.

If $(X, \mathcal{M}, \mu)$ and $(Y, \mathcal{N}, \nu)$ are σ-finite, then so is $(X \times Y, \mathcal{M} \otimes \mathcal{N}, \mu \otimes \nu)$ and the extension is unique. In this case, it is called the *direct product* of $(X, \mathcal{M}, \mu)$ and $(Y, \mathcal{N}, \nu)$. Then, the *theorem of Tonelli* guarantees that if $f : X \times Y \to [0, \infty]$ is measurable, then

$$x \in X \mapsto g(x) = \int f_x d\nu \qquad \text{and} \qquad y \in Y \mapsto h(y) = \int f^y d\mu$$

are measurable and it holds that

$$\begin{aligned} \int f d(\mu \otimes \nu) &= \int \left[\int f(x,y) d\nu(y) \right] d\mu(x) \\ &= \int \left[\int f(x,y) d\mu(x) \right] d\nu(y), \end{aligned} \tag{8.6}$$

where $f_x(y) = f(x,y)$ and $f^y(x) = f(x,y)$. The *theorem of Fubini*, on the other hand, says that if $f : X \times Y \to [-\infty, \infty]$ is summable, then $y \in Y \mapsto f_x$ is ν-measurable for almost every $x \in X$ and $g(x) = \int f_x d\nu$ is μ-summable, that similarly, $x \in X \mapsto f^y$ is μ-measurable for almost every $y \in Y$ and $h(y) = \int f^y d\mu$ is ν-summable, and that (8.6) holds true.

In the case that $(X, \mathcal{M}, \mu)$ and $(Y, \mathcal{N}, \nu)$ are σ-finite complete measure spaces, we take the completion of $(X \times Y, \mathcal{M} \otimes \mathcal{N}, \mu \otimes \nu)$ denoted by $(X \times Y, \mathcal{L}, \lambda)$. Then, if $f : X \times Y \to [-\infty, \infty]$ is $\mathcal{L}$-measurable, and either $f \geq 0$ or $f \in L^1(X \times Y, \lambda(dxdy))$ is satisfied, then we have that f_x and f^y are $\mathcal{N}$- and $\mathcal{M}$-measurable for μ- and ν- almost every x and y, respectively, which are summable in the latter case, that $x \mapsto \int f_x d\nu$, $y \mapsto \int f^y d\mu$ are measurable and are summable in the latter case, and that

$$\int f d\lambda = \int \left[\int f(x,y) d\mu(x) \right] d\nu(y) = \int \left[\int f(x,y) d\nu(y) \right] d\mu(x).$$

If $(X, \mathcal{D})$ is a measurable space, the mapping $\mu : \mathcal{D} \to (-\infty, \infty)$ is said to be a *signed measure* if $\{A_k\} \subset \mathcal{D}$ is a division of $A = \sum_{k=1}^{\infty} A_k$, then $\mu(A) = \sum_k \mu(A_k)$ holds with the right-hand side converging absolutely. In this case, the *total variation* is given by

$$|\mu|(A) = \sup \left\{ \sum_{k=1}^{\infty} |\mu(A_k)| \mid \{A_k\}_{k=1}^{\infty} \subset \mathcal{D} \text{ is a division of } A \right\}$$

and is shown to be a measure on $(X, \mathcal{D})$ satisfying $|\mu|(X) < +\infty$. We call $\mu = \mu_+ - \mu_-$ and $|\mu| = \mu_+ + \mu_-$ the *Jordan decompositions*, where $\mu_\pm = \frac{1}{2}(|\mu| \pm \mu)$. A non-negative signed measure is called the *positive measure*, so that the total variation of a signed measure is a positive measure and so are $\mu_\pm$ given above.

If μ and λ are a measure and a (signed) measure on the measurable space $(X, \mathcal{D})$, we say that λ is *absolutely continuous* with respect to μ if $A \in \mathcal{D}$ and $\mu(A) = 0$ imply $\lambda(A) = 0$. It is written as $\lambda \ll \mu$. This condition is equivalent to that for any $\varepsilon > 0$ there is $\delta > 0$ such that $A \in \mathcal{D}$ and $\mu(A) < \delta$ imply $|\lambda(A)| < \varepsilon$. On the contrary, it is said that λ is concentrated on $E \in \mathcal{D}$ if $\lambda(A) = \lambda(E \cap A)$ holds for any $A \in \mathcal{D}$. Finally, two (signed) measures λ_1, λ_2 on $(X, \mathcal{D})$ is said to be *singular* to each other if there is a decomposition $A_1, A_2 \in \mathcal{D}$ of X such that λ_1, λ_2 are concentrated on A_1, A_2, respectively. This case is written as $\lambda_1 \perp \lambda_2$.

If $(X, \mathcal{D})$ is a measurable space and $\mu, \lambda : \mathcal{D} \to [0, \infty)$ are positive measures, then, the *theorem of Radon and Nikodym* says that there are unique positive measures λ_a and λ_s such that $\lambda_a \ll \mu$, $\lambda_s \perp \mu$, and $\lambda = \lambda_a + \lambda_s$, which is referred to as the *Lebesgue decomposition*, and that there is a μ-summable non-negative function h, which is unique up to μ-almost everywhere and is called the *Radon-Nikodym density* such that $\lambda_a(A) = \int_A h d\mu$ for any $A \in \mathcal{D}$. It is written as $d\lambda_a = h d\mu$ or $\frac{d\lambda_a}{d\mu} = h$. This is also the case that $\lambda : \mathcal{D} \to (-\infty, \infty)$ is a signed-measure if $(X, \mathcal{D}, \mu)$ is a σ-finite measure space and the density can change sign. The *differentiation theorem of Lebesgue* says that if $f(x)$ is *locally summable* in $\mathbf{R}^n$, which means that it is summable (with respect to the Lebesgue measure μ) on any compact set in $\mathbf{R}^n$, then it holds that

$$\lim_{r \downarrow 0} \frac{1}{\mu(B(x,r))} \int_{B(x,r)} |f(y) - f(x)| \, dx = 0$$

for almost every $x \in \mathbf{R}^n$.

A locally compact Hausdorff space X is said to be σ-compact if there is a family of compact subsets $\{X_k\}_{k=1}^\infty$ such that $X = \cup_{k=1}^\infty X_k$. Let X be such a space and $\mathcal{B}_X$ be the Borel family so that the minimum σ-algebra containing any open set. In this case, if μ is a measure on $(X, \mathcal{B}_X)$ such that $\mu(K) < +\infty$ for any compact set $K \subset X$, then it is a *Radon measure* so that $\mu(A) = \inf \{\mu(U) \mid A \subset U : \text{ open}\}$ for $A \in \mathcal{B}_X$ and $\mu(U) = \sup \{\mu(K) \mid U \subset K : \text{ compact}\}$ for $U \subset X$ open. If X is a locally compact Hausdorff space, provided with the property that any open set is σ-compact, and $\mathcal{B}_X$ denotes the Borel family, then a measure μ on $(X, \mathcal{B}_X)$ is Radon if and only if $\mu(K) < \infty$ for any compact K.

Here, we state some facts on the function of one variable used in §3.4.7.

First, a function f defined on the compact interval $[a, b]$ is said to be *absolutely continuous* if any $\varepsilon > 0$ admits $\delta > 0$ such that the division

$$\Delta : x_0 = a < x_1 < \cdots < x_n = b$$

of $[a, b]$ satisfies $\|\Delta\| = \max_{1 \leq i \leq n} (x_i - x_{i-1}) < \delta$ then it holds that

$$\sum_{i=1}^{n} |f(x_i) - f(x_{i-1})| < \varepsilon.$$

An absolutely continuous function $f(x)$ on $[a, b]$ is differentiable for almost every $x \in (a, b)$ with the derivative $f'(x)$ to be summable on $[a, b]$, which enjoys the property that

$$\int_a^x f'(y)dy = f(x) - f(a)$$

for each $x \in [a, b]$. A function $f(x)$ defined on $\mathbf{R}$ is said to be *locally absolutely continuous* if it is absolutely continuous on any compact interval. A *Lebesgue point* of a locally summable function $f(x)$ denotes x_0 satisfying

$$\lim_{h \downarrow 0} \frac{1}{h} \int_{x_0}^{x_0+h} |f(x) - f(x_0)|\, dx = \lim_{h \downarrow 0} \frac{1}{h} \int_{x_0-h}^{x_0} |f(x) - f(x_0)|\, dx = 0.$$

Then, *Lebesgue's differentiation theorem* guarantees that the complement of the set of Lebesgue points has the Lebesgue measure equal to 0.

8.1.5 *Abstract Analysis*

A complete normed space is called the *Banach space.* It is provided with three important properties, *Hahn-Banach's theorem*, the *uniformly bounded principle* of Banach-Steinhaus, and the *open mapping theorem* of Banach. The latter two are from *Baire's category theorem*, while the first one is based on *Zorn's lemma*, which is equivalent to the *axiom of selection.*

Hahn-Banach's theorem has several variations. In the analytic form it assures that any bounded linear operator $T_0 : H_0 \to \mathbf{R}$ defined on a subspace H_0 of a normed space H admits a bounded linear extension $T : H \to \mathbf{R}$ with $\|T_0\|_{H_0'} = \|T\|_{H'}$, where

$$\|T_0\|_{H_0'} = \sup \{ |T_0(f)| \mid f \in H_0, \ \|f\| \leq 1 \}$$

and

$$\|T\|_{H'} = \sup\{|T(f)| \mid f \in H,\ \|f\| \leq 1\}.$$

Those boundedness conditions may be unilateral in terms of the *semi-norm* defined on the *topological linear space*. The geometric version, on the other hand, is referred to as the *separation principle*, as disjoint convex sets A and B in a normed space are separated weakly, and strongly by a hyper-plane, if A is open, and if A is closed and B is compact, respectively.

One form of the theorem of Banach-Steinhaus says that if X and Y are Banach spaces and the bounded linear operators $T_n : X \to Y$ $(n = 1, 2, \cdots)$ satisfies $\sup_n \|T_n x\| < +\infty$ for each $x \in X$, then it holds that $\sup_n \|T_n\| < +\infty$. On the other hand, the open mapping theorem says that if X and Y are Banach spaces and $T : X \to Y$ is a surjective bounded operator, then it is open so that any image of open set is open. An equivalent form is the *closed graph theorem*. Namely, a linear operator T with the domain $D(T) \subset X$ and the range $R(T) \subset Y$ for Banach spaces X, Y is called *closed* if its graph $\mathcal{G} = \{(f, Tf) \mid f \in D(T)\}$ is closed in $X \times Y$. Then, it assures that a closed operator T with $D(T) = X$ is bounded. *Closed range theorem*, on the other hand, is concerned with the dual operator. Given a densely defined closed (linear) operator $T : D(T) \subset X \to Y$, we can define the dual operator $T' : Y' \to X'$ by $\langle T'y', x\rangle_{X',X} = \langle y', Tx\rangle_{Y',Y}$, where X and Y are Banach spaces. Then it says that $R(T)$ is closed in Y if and only if $R(T')$ is dense in X', and in this case it holds that $R(T) = N(T')^{\perp}$ and $R(T') = N(T)^{\perp}$.

If E is a closed non-empty subset of a Banach space X and a (nonlinear) mapping $f : E \to E$ is a *contraction* so that there is $\rho \in (0, 1)$ such that $\|f(x) - f(y)\| \leq \rho \|x - y\|$ for any $x, y \in E$, then it admits a unique *fixed point* x^* in E: $f(x^*) = x^*$. On the other hand *Schauder's theorem* guarantees the existence of a fixed point of $f : E \to E$ if $E \subset X$ is closed, convex, and non-empty, and f is *compact* so that image of any bounded set in E is relatively compact. Fundamentals of *nonlinear functional analysis* are composed of fixed point theorems, *topological degree*, and *variational methods*.

Important example of Banach space is the L^p space on measure space $(X, \mathcal{B}, \mu)$ for $p \in [1, \infty]$, with the norm given by

$$\|f\|_p = \begin{cases} \left\{\int_X |f(x)|^p \mu(dx)\right\}^{1/p} & (p \in [1, \infty)) \\ \text{ess.}\sup_{x \in X} |f(x)| & (p = \infty). \end{cases}$$

There, two functions equal to each other almost everywhere are identified. Then, convergence in L^p implies that in measure for $p \in [1,\infty)$, so that if a family converges in L^p for $p \in [1,\infty)$ there is a sub-family converging almost everywhere. It holds that $(L^p)' = L^{p'}$ for $p \in [1,\infty)$, where $\frac{1}{p} + \frac{1}{p'} = 1$. Another example is the set of continuous functions taking 0 at ∞ on a locally compact Hausdorff space X, denoted by $C_0(X)$, with the norm $\|f\|_\infty = \max_{x\in X}|f(x)|$. If $T \in C_0(X)'$ has the positivity, it is identified with the integration of a finite Radon measure, and $C_0(X)'$ is realized as the set of signed Radon measures in that sense. This fact is also referred to as *Riesz' representation theorem*. If $X = \Omega$ is a domain in $\mathbf{R}^n$, then the L^p space denoted by $L^p(\Omega)$ is introduced associated with the Lebesgue measure. *Fréchet-Kolmogorov's theorem* assures for a family $\mathcal{F} \subset L^p(\Omega)$ with $p \in [1,\infty)$ is relatively compact if and only if the following two conditions hold. Namely, first, any $\varepsilon > 0$ and sub-domain ω with $\overline{\omega} \subset \Omega$ admits $\delta > 0$ in $\delta < \mathrm{dist}(\omega, \Omega^c)$ such that $\|\tau_h f - f\|_{L^p(\omega)} < \varepsilon$ for any $h \in \mathbf{R}^n$ in $|h| < \delta$ and $f \in \mathcal{F}$, where τ_h denotes the *translation operator*: $(\tau_h f)(x) = f(x+h)$, and second, any $\varepsilon > 0$ takes a sub-domain ω in $\overline{\omega} \subset \Omega$ such that $\|f\|_{L^p(\Omega\setminus\omega)} < \varepsilon$. The *Sobolev space* $W^{m,p}(\Omega)$ denotes the set of p-integrable functions including their m-th order (distributional) derivatives. With its atural norm, it becomes a Banach space. On the other hand, $W_0^{m,p}(\Omega)$ indicates the closure of $C_0^\infty(\Omega)$, the set of C^∞ functions with compact supports in Ω, in $W^{m,p}(\Omega)$. If $\partial\Omega$ is Lipschitz continuous, $W_0^{1,p}(\Omega)$ with $p \in [1,\infty)$ is characterized as the kernel of the *trace operator* $\gamma : W^{1,p}(\Omega) \to W^{1,1-1/p}(\partial\Omega)$. The space $W^{m,2}(\Omega)$ becomes a Hilbert space, denoted by $H^m(\Omega)$. We also put $H_0^m(\Omega) = W_0^{m,2}(\Omega)$. *Sobolev's imbedding theorem* in the primitive form is described as $W_0^{1,p}(\Omega) \subset L^{p^*}(\Omega)$ for $p \in [1,n)$, where $\frac{1}{p^*} = \frac{1}{p} - \frac{1}{n}$. The best constant associated with this imbedding depends only on n and p. On the contrary, *Morrey's theorem* guarantees $W_0^{1,p}(\Omega) \subset C^\alpha(\overline{\Omega})$ for $p > n$, where $\alpha = 1 - \frac{n}{p}$. Those imbedding theorems extend to $W^{1,p}(\Omega)$ if $\partial\Omega$ is Lipschitz continuous, with the imbedding constant now depends on each Ω.

8.2 Commentary

8.2.1 *Elliptic and Parabolic Equations*

For the Strum-Liouville problem and the expansion theorem of Mercer, see Yosida [25] and Suzuki [20]. Concerning the method of separation of variables to seek eigenfunctions on symmetric domains described in §3.3.1, see Courant and Hilbert [6]. Justification of the eigenfunction expansion has been the fundamental theme of the *operator theory*. See Reed and Simon [16] and so forth. Concerning the eigenvalues for the general domains to (3.32) or (3.34), see Courant and Hilbert [6], Bandle [1], Chavel [4], and Suzuki [21]. Detailed justification of the Fourier transformation of distributions described in §5.2.6 including the theorem of Malgrange and Ehlenpreis is given in Yosida [24]. Layer potential is treated by Courant and Hilbert [6], Garabedian [11], and Folland [9]. For the modern treatment, see Fabes, Jodeit, and Riviére [33] and Verchota [91].

The standard text for the second order elliptic equation is Gilbarg and Trudinger [12]. For the parabolic equation, we refer to Ladyženskaja, Solonnikov, and Ural'ceva [15]. Hölder continuity of u_{jk} in §5.4.2 is described in [12]. Its L^p boundedness is established by [28], referred to as the theory of *singular integrals*. See also Stein [19] for this method of linear approach.

Theory of semilinear elliptic equations requires a different staff based on the topological consideration, and so forth, as is described in Suzuki [21]. As for more detailed justification of the method given in §§5.4.3 and 5.4.4 such as the trace operator, Sobolev's imbedding theorem, and so forth, see Brezis [3]. Among most important topics in recent study on nonlinear equations are the regularity and the blowup of the solution. The descriptions of §§5.4.4 and 5.4.5 are due to [63], [54], and [28]. For their extensions to the nonlinear problem, see Choe [5] and DiBenedetto [8].

The description of §6.1 follows Fujita [34]. The critical exponent $p = p_f$ is contained in the blowup case, which was proven by Hayakawa [42]. Fujita's triple law is obtained in Fujita [35] and is described in [21]. See also Kohda and Suzuki [59] for later developments. J.J.L. Velázquez' work on the best estimate of the dimension of the blowup set is done in [90]. The proof of Theorem 6.5 follows Ikehata and Suzuki [51]. The work by Y. Giga and R.V. Kohn characterizing the blowup point by the backward self-similar transformation is done in [39], [40], [41]. Theorem 6.7 is obtained by Otani [78]. The proof exposed here is based on Cazenave and Lions [29]. Theorems 6.9 and 6.10 are proven by Giga [38] and Merle [61], respectively. General theory of dynamical systems, particularly the omega-limit set of compact orbits in the presence of the Lyapunov function, is described in Henry [13]. The result by P.L. Lions on the semilinear parabolic dynamics is given in [60]. The unbounded global solution was introduced by W.-M. Ni, P.E. Sacks, and T. Tavantzis [73]. V.A. Galaktionov and J.L. Vazquez showed in

[37] that if Ω is a ball and $\frac{n+2}{n-2} < p < 1 + \frac{6}{(n-10)_+}$, $u_0 = u_0(|x|)$, and $u_{0r} < 0$ for $r = |x| > 0$, then it holds that $T_{\max} < +\infty$ for the unbounded global solution. Y. Naito and T. Suzuki studied the same problem for the case of $p = p_s$ in the general domain. Generally, this problem is concerned with the *post-blowup continuation*. In this connection, Sakaguchi and Suzuki [81] showed that if $u = u(x,t)$ is a super-solution to the linear heat equation and if its dead core $D(t) = \{x \mid u(x,t) = +\infty\}$ enclosed in a bounded domain, then it hold that $\liminf_{t \to t_0} L^n(D(t)) = 0$ for any t_0, where L^n denotes the n-dimensional Lebesgue measure. See also the references therein for the related work. The H^1 solution to (6.25) was constructed by Weissler [92], Hoshino and Yamada [50], and Ikehata and Suzuki [52]. Stable and unstable sets are introduced by Sattinger [82]. Fundamental properties of the Nehari manifold are described in Suzuki [21]. Theorem 6.11 is due to [51]. For the forward self-similar solution and its role to the asymptotic behavior of solutions to (6.1), see Kawanago [56] and the references therein. Theorem 6.12 is due to Naito and Suzuki [70]. See the references therein concerning the study on radially symmetric self-similar solutions. The results by M. Escobedo and O. Kavian are done in [32], [55].

For detailed studies on the numerical scheme to solve partial differential equations, see Fujita, Saito, and Suzuki [10]. For elements in nonlinear functional analysis, see Deimling [7].

8.2.2 *Systems of Self-interacting Particles*

Modelling of the motion of the mean field of self-interacting particles is treated by Samarskii and Mikhailov [18]. In Bensoussan and Frehse [2] the regularity of the solution to the (DD) model is described. More systematic study to the semiconductor device equation, modelling, simulation, and analysis, is done in Jüngel [14]. The statistical modelling described in §4.1 is based on Othmer and Stevens [79]. Semilinear elliptic equations with exponential nonlinearity arise also in the gauge field theory. See Yang [23] for this area.

T. Nagai's work on (7.1) is [65]. The original Keller-Segel model was proposed in [57] with the biological background and instability of the constant solution was studied. Then, V. Nanjundiah introduced a simplified system in [72], which now is called the Keller-Segel system, or the full system as in §7.1. Mathematical studies on this system, physical and biological motivations, and related references are described in Suzuki [22].

Problem (7.1) has several relatives and some of them were studied by [53], [65], [26], [43], [71]. Amang others are the existence and uniqueness of the classical solution [93], blowup of the solution and its singularity [53; 65; 85; 69; 46; 43], Lyapunov function and threshold for the blowup [68; 36; 26]), stationary solutions [84], and asymptotic behavior of the solution [75]. W. Jäger and J. Luckhaus introduced the simplified system whose second equation is slightly different, where the comparison theorem is valid to $U(r,t) = \int_{|x|<r} u(x,t)dx$. T. Nagai introduced (7.1) with the threshold expected by [30] for the blowup solution in the case of radially symmetric solutions. That method of using second moment $\int_\Omega |x|^2 u(x,t)dx$ is valid to non-radial case or the system

$$\begin{aligned} u_t &= \nabla\cdot(\nabla u - u\nabla\chi(v)), \\ 0 &= \Delta v - v + u, \end{aligned}$$

where $\chi = \chi(v)$ is a monotone increasing function. See [86; 66; 67]. Simplified system of chemotaxis has a remarkable structure in two space dimension. See [45; 44; 46] for the matched asymptotic expansion, [68; 36; 26] for the use of the Lyapunov function and the Trudinger-Moser type inequality, [22; 86] for the method of symmetrization, [85; 21; 49; 83] for stationary solutions and global dynamics. In three-dimensional case, there are self-similar blowup solutions and L^1 concentration blowup solutions. See [43] and the reference. The former case does not arise in two space dimensions. See [71]. Concerning with numerical results, see [47] and the references therein.

For the system whose second equation does not have the diffusion term, the structure of solutions is different from the systems with diffusion term. Yang, Chen and Liu treat the system

$$\begin{aligned} u_t &= \nabla\cdot(\nabla u - u\nabla\log(v)), \\ v_t &= F(u,v). \end{aligned}$$

They show results concerning with time-global existence and boundedness of the solution in the case where $F(u,v) = u - v$, and results concerning with growup and blowup of solutions in the case where $F(u,v) = u$ or uv. Then, the structure of solutions depends on the form $F(u,v)$. Othmer, Stevens, Levine and Sleeman introduce the system and numerically investigate. Concerning with background, mathematical and numerical results, see [94] and the references therein. Hillen and Stevens [48] introduced a hyperbolic model from the Keller-Segel system.

There are not so many mathematical results concerning the spatial pattern formulated by the solutions, but in the experiment and numerical computations a lot of systems of chemotaxis are reported to show the feature of aggregation, e.g., several patterns formed by bacterium such as the spot, ring, spike, swarm ring, spiral, and target patterns. One of the mathematical approach to the pattern formation is the dimension analysis of the attractor. See [75] for this. Profile or the speed of the traveling wave is an important information for the spatial pattern to understand, where the traveling wave is a solution determined by $x - ct$ with a constant c. Traveling bands was shown by Adler experimentally, which means that the gradient of chemical substance makes bacterium move toward higher density. E.F. Keller and L.A. Segel introduced a mathematical model for this phenomena with an explicit traveling wave solution. T. Nagai and others introduced a simplified system,

$$\begin{aligned} u_t &= (u_x - \alpha u(\log(v))_x)_x \\ v_t &= dv_{xx} - u \end{aligned}$$

in $\mathbf{R} \times (0, \infty)$, where α and d are positive constants, and studied the linearized instability of traveling waves. See [58; 31; 74] and the references therein.

Budrene and Berg's experiment showed that the bacterium exhibit complex two-dimensional spot or stripe patterns by the interplay of diffusion, growth, and chemotaxis. To analyze those patterns, M. Mimura and others proposed

$$\begin{aligned} u_t &= d_1 \nabla \cdot (\nabla u - ku \nabla \chi(v)) + \ell f(u), \\ v_t &= d_2 \Delta v - \alpha v + \gamma u, \end{aligned}$$

where d_1, d_2, α and γ are positive constants, $\chi(v)$ and $f(u)$ are smooth functions. T. Tsujikawa observed the existence of the explicit stationary solution in the limiting system. Then, in use of the singular perturbation method, he found one-dimensional and two-dimensional planar stationary solutions corresponding to the strip pattern, and investigated their stability. By a similar method, T. Tsujikawa and M. Mimura found a radially symmetric solution corresponding to the spot pattern and also two-dimensional stationary solutions different from the one previously found. See [27; 87] and the references therein.

After aggregation, cells of the cellular slime mold form a multicellular structure and show coherent motion such as vortices. For this phenomenon, T. Umeda and K. Inouye introduced a discrete model. In the numerical calculation, cells form some clusters, which merge to form larger clusters, and then the rotational cells movement can be seen. Also, they introduced the continuous model

$$\begin{aligned} \frac{1}{\rho}\nabla p &= -a\mathbf{v} + f\frac{\nabla c + b\mathbf{v}}{|\nabla c + b\mathbf{v}|}, \qquad \nabla\cdot\mathbf{v} = 0, \\ c_t &= D\Delta c - k_1 c + k_2\rho, \end{aligned}$$

where ρ is the density of cells, c is the chemical concentration, p is the pressure, $\mathbf{v}$ is the velocity of cells, and a, b, D, k_i $(i = 1, 2)$ are positive constants. They found a radial solution to the continuous model whose velocity has only azimuthal velocity, which corresponds to the rotational movement of cells. See [89] and the references therein.

Hildebrand and others introduced

$$\begin{aligned} u_t &= d_1\Delta u - \alpha\nabla\{u(1-u)\nabla\chi(v))\} + f(u,v), \\ v_t &= d_2\Delta v + \gamma v(v + u - 1)(1 - v), \end{aligned}$$

where micro-reactors with sub-micrometer and nanometer sizes are allowed to develop chemical reactions on surface by a non-equilibrium self-organization process. Here, d_i $(i = 1, 2)$, α and γ are positive constants, and $\chi(v)$ and $f(u, v)$ are smooth functions. This model has the chemotactic term, as $\alpha\nabla\{u(1-u)\nabla\chi(v)\}$. T. Tsujikawa and A. Yagi showed the existence of the time-global solution and the exponential attractor. See [88] and the reference.

The stream formation and spiral wave was studied by [77] to describe for finite amoebas from the numerical calculation. That model describes the case that each cell responses to one chemical substance. For the other case, Painter, Maini and Othmer proposed a system modelling bacterial chemotaxis or animal skin patterns. From the numerical calculation, solutions form spot, ring, and stream patterns. One of them is given by

$$\begin{aligned} u_t &= \nabla\cdot(\nabla u - u(\nabla(\chi_1(v_1) + \chi_2(v_2)))), \\ v_{it} &= d_1\Delta v_i + f_i(v_1, v_2) \quad (i = 1, 2). \end{aligned}$$

See [80]. Finally, let us recall that several experiments, mathematical models, and numerical calculations are exposed in [64].

Bibliography

1. Bandle, C. (1980) "Isoperimetric Inequalities and Applications", Pitman, London.
2. Bensoussan, A. and Frehse, J. (2002) "Regularity Results for Nonlinear Elliptic Systems and Applications", Springer, Berlin.
3. Brezis, H. (1983) "Analyse Fonctionnelle, Théorie et Applications", Masson, Paris.
4. Chavel, I. (1984) "Eigenvalues in Riemannian Geometry", Academic Press, New York.
5. Choe, H.J. (1992) "Degenerate Elliptic and Parabolic Equations and Variational Inequalities", Research Institute of Mathematics, Seoul National University.
6. Courant, R. and Hilbert, D. (1961) "Methods of Mathematical Physics", vol. 1 and vol. 2, Wiley, New York.
7. Deimling, K. (1985) "Nonlinear Functional Analysis", Springer, New York.
8. DiBenedetto, E. (1991) "Degenerate Parabolic Equations", Springer, New York.
9. Folland, G.B. (1995) "Introduction to Partial Differential Equations", second edition, Princeton, New Jersey.
10. Fujita, H., Saito, N. and Suzuki, T. (2001) "Operator Theory and Numerical Methods", Elsevier, Amsterdom.
11. Garabedian, P.R. (1964) "Partial Differential Equations", Chelsea, New York.
12. Gilbarg, D. and Trudinger, N.S. (1983) "Elliptic Partial Differential Equations of Second Order", second edition, Springer, Berlin.
13. Henry, D. (1981) "Geometric Theory of Semilinear Parabolic Equations", Springer, Berlin.
14. Jüngel, A. (2001) "Quasi-hydrodynamic Semiconductor Equations", Birkhäuser, Basel.
15. Ladyženskaja, A., Solonnikov, V.A. and Ural'ceva, N.N. (1968) "Linear and Quasilinear Equations of Parabolic Type", Amer. Math. Soc., Providence.

16. Reed, M., Simon, B.(1972-1978) "Methods of Modern Mathematical Physics", vol. 1-4, Academic Press, New York.
17. Rudin, W. (1987) "Real and Complex Analysis", third edition, McGraw-Hill, New York.
18. Samarskii, A.A. and Mikhailov, A.P. (2002) "Principle of Mathematical Modeling", Taylor and Francis, London.
19. Stein, E.M. (1970) "Singular Integrals and Differential Properties of Functions", Princeton, New Jersey.
20. Suzuki, T. (1991) "Mathematical Theory of Applied Inverse Problems", Lecture Notes, Sophia University, Tokyo.
21. Suzuki, T. (1994) "Semilinear Elliptic Equations", Gakkotosho, Tokyo.
22. Suzuki, T. (2004) "Free Energy and Self-interacting Particles", Birkhäuser, Basel.
23. Yang, Y. (2001) "Solitons in Field Theory and Nonlinear Analysis" Springer, New York.
24. Yosida, K. (1964) "Functional Analysis" Springer, Berlin.
25. Yosida, K. (1965) "Theory of Integral Equations" Interscience, New York.
26. Biler, P. (1998) "Local and global solvability of some parabolic systems modelling chemotaxis", *Adv. Math. Sci. Appl.* **8**, 715-743.
27. Budrene, E. O. and Berg, H. C. (1991) "Complex patterns formed by motile cells of Escherichia coli", *Biophys. J.* **58**, 919-930.
28. Calderón, A.P. and Zygmund, A. (1952) "On the existence of certain singular integrals", *Acta Math.* **88**, 85-129.
29. Cazenave, T. and Lions, P.L. (1984) "Solutions globales déquations de la chaleur semi lineaires", *Comm. Partial Differential Equations* **9**, (1984) 955-978.
30. Childress, S. and J.K. Percus, J.K. (1981) "Nonlinear aspects of chemotaxis", *Math. Biosci.* **56**, 217-237.
31. Ebihara, Y., Furusho, Y. and Nagai, T. (1992) "Singular solutions of traveling waves in a chemotactic model", *Bull. Kyushu Inst. Tech.* **39**, 29-38.
32. Escobedo, M. and Kavian, O. (1987) "Variational problems related to self-similar solutions for the heat equation" *Nonlinear Anal.* **33**, 51-69.
33. Fabes, E.B., Jodeit, Jr. M. and Riviére, N.M. (1978) "Potential techniques for boundary value problems on C^1-domains", *Acta Math.* **141**, 165-186.
34. Fujita, H. (1966) "On the blowing up of solutions of the Cauchy problem for $u_t = \Delta u + u^{1+\alpha}$", *J. Fac. Sci. Univ. Tokyo, Sec. IA* **13**, 109-124.
35. Fujita, H. (1969) "On the nonlinear equations $\Delta u + e^u = 0$ and $\partial v/\partial t = \Delta v + e^v$", *Bull. Amer. Math. Soc.* **75**, 132-135.
36. Gajewski, H and Zacharias, K. (1998) "Global behaviour of a reaction - diffusion system modelling chemotaxis", *Math. Nachr.* **195**, 77-114.
37. Galaktionov, V.A. and Vazquez, J.L. (1997) "Continuation of blow-up solutions of nonlinear heat equations in several space dimensions", *Comm. Pure Appl. Math.* **50**, 1-67.
38. Giga, Y. (1986) "A bound for global solutions of semilinear heat equations",

Comm. Math. Phys. **103**, 415-421.

39. Giga, Y. and Kohn, R.V. (1985) "Asymptotically self-similar blow-up of semilinear heat equations", *Comm. Pure Appl. Math.* **38**, 297-319.
40. Giga, Y. and Kohn, R.V. (1987) "Characterizing blowup using similarity variables", *Indiana Univ. Math. J.* **36**, 1-40.
41. Giga, Y. and Kohn, R.V. (1989) "Nondegeneracy of blowup for semilinear heat equations", *Comm. Pure Appl. Math.* **42**, 845-884.
42. Hayakawa, K. (1973) "On nonexistence of global solutions of some semilinear parabolic equations" *Proc. Japan Acad. Ser. A* **49**, 109-124.
43. Herrero, M.A., Medina, E. and Velázquez, J.J.L. (1998) "Self-similar blow-up for a reaction-diffusion system", *J. Comp. Appl. Math.* **97**, 99-119.
44. Herrero, M.A. and Velázquez, J.J.L. (1996) "Singularity patterns in a chemotaxis model", *Math. Ann.* **306**, 583-623.
45. Herrero, M.A. and Velázquez, J.J.L. (1997) "On the melting of ice balls", *SIAM J. Math. Anal.* **28**, 1-32.
46. Herrero, M.A. and Velázquez, J.J.L. (1997) "A blow-up mechanism for a chemotaxis model", *Ann. Scoula Norm. Sup. Pisa IV* **35**, 633-683.
47. Hillen, T. and Painter, K. (2001) "Global existence for a parabolic chemotaxis model with prevention of overcrowding", *Adv. in Appl. Math.* **26**, 280-301.
48. Hillen, T. and Stevens, A. (2000) "Hyperbolic models for chemotaxis in 1-D", *Nonlinear Anal. Real World Appl.* **1**, 409-433.
49. Horstmann, D. and Wang, G. (2001) "Blowup in a chemotaxis model without symmetry assumptions", *Euro. Jnl. of Applied Math.* **12**, 159-177.
50. Hoshino, H. and Yamada, Y. (1991) "Stability and smoothing effect for semilinear parabolic equations", Funkcial. Ekvac. **34**, 475-494.
51. Ikehata, R. and Suzuki, T. (1996) "Stable and unstable sets for evolution equations of parabolic and hyperbolic type", *Hiroshima Math. J.* **26**, 475-491.
52. Ikehata, R. and Suzuki, T. (2000) "Semilinear parabolic equations involving critical Sobolev exponent: local and asymptotic behavior of solutions", *Differential and Integral Equations* **13**, 869-901.
53. Jäger, W. and Luckhaus, S. (1992) "On explosions of solutions to a system of partial differential equations modelling chemotaxis", *Trans. Amer. Math. Soc.* **329**, 819-824.
54. John, F. and Nirenberg, L. (1961) "On functions of bounded mean oscillation", *Comm. Pure Appl. Math.* **14**, 415-426.
55. Kavian, O. (1987) "Remarks on the large time behavior of a nonlinear diffusion equation" *Ann. Inst. Henri Poincaré, Analyse non linéaire* **4**, 423-452.
56. Kawanago, T. (1996) "Asymptotic behavior of solutions of a semilinear heat equation with subcritical nonlinearity", *Ann. Inst. Henri Poincaré, Analyse non linéaire* **13**, 1-15.
57. Keller, E.F. and Segel, L.A. (1970) "Initiation of slime mold aggregation viewed as an instability", *J. Theor. Biol.* **26**, 399-415.
58. Keller, E.F. and Segel, L.A. (1970) "Traveling bands of chemotactic bacteria:

a theoretical analysis", *J. Theor. Biol.* **30**, 235-248.

59. Kohda, A. and Suzuki, T. (2000) "Blow-up criteria for semilinear parabolic equations", *J. Math. Anal. Appl.* **243**, 127-139.
60. Lions, P.-L. (1984) "Structure of the set of steady-state solutions and asymptotic behavior of semilinear heat equations", *J. Differential Equations* **53**, 362-386.
61. Merle, F. (1992) "Solution of a nonlinear heat equation with arbitrarily given blow-up points", *Comm. Pure Appl. Math.* **45**, 263-300.
62. Mimura, M. and Tsujikawa, T. (1996) "Aggregating pattern dynamics in a chemotaxis model including growth", *Physica A* **236**, 499-543.
63. Moser, J. (1961) "On Harnack's theorem for elliptic differential equations", *Comm. Pure Appl. Math.* **14**, 577-591.
64. Murray, J. D. (2003) "Mathematical Biology", thrid edition, Springer-Verlag, Berlin Heidelberg, 2003.
65. Nagai, T. (1995) "Blowup of radially symmetric solutions to a chemotaxis system", *Adv. Math. Sci. Appl.* **5**, 581-601.
66. Nagai, T. (2001) "Blowup of nonradial solutions to parabolic-elliptic systems modeling chemotaxis in two-dimensional domains", *J. Inequal. Appl.* **6**, 37-55.
67. Nagai, T. and Senba, T. (1997) "Global existence and blow-up of radial solutions to a parabolic-elliptic system of chemotaxis", *Adv. Math. Soc. Appl.* **8**,145-156.
68. Nagai, T., Senba, T and Yoshida, Y. (1997) "Application of the Trudinger-Moser inequality to a parabolic system of chemotaxis", *Funckcial. Ekvac.* **40**, 411-433.
69. Nagai, T., Senba, T. and Suzuki, T. (2000) "Chemotactic collapse in a parabolic system of chemotaxis", *Hiroshima Math. J.* **30**, 463-497.
70. Naito, Y. and Suzuki, T. (2000) "Radial symmetry of self-similar solutions for semilinear heat equations", *J. Differential Equations* **163**, 407-428.
71. Naito, Y., Suzuki, T. and Yoshida, K. (2002) "Existence of self-similar solutions to a parabolic system modelling chemotaxis", *J. Diff. Equations* **184**, 386-421.
72. Nanjundiah, V. (1973) "Chemotaxis, signal relaying, and aggregation morphology", *J. Theor. Biol.* **42**, 63-105.
73. Ni, W.-M., Sacks, P.E. and Tavantzis, J. (1984) "On the asymptotic behavior of solutions of certain quasilinear parabolic equations", *J. Differential Equations* **54**, 97-120.
74. Nagai, T. and Ikeda, T. (1991) "Traveling waves in a chemotactic model", *J. Math. Biol.* **30**, 169-184.
75. Osaki, K., Tsujikawa, T, Yagi, A. and Mimura, M. (2002) "Exponential attractor for a chemotaxis-growth system of equations", *Nonlinear Anal.* **51**, 119-144.
76. Osakai, K. and Yagi, A. (2002) "Global existence for a chemotxis - growth system in $\mathbf{R}^2$, *Adv. Math. Sci. Appl.* **12**, 587-606.

77. Oss, C., Panfilov, A. V., Hogeweg, P., Siegert, F. and Weijer, C. J. (1996) "Spatial pattern formation during aggregation of the slime mould Dictyostelium discoideum", *J. Theor. Biol.* **181**, 203-213.
78. Otani, M. (1980) "Existence and asymptotic stability of strong solutions of nonlinear evolution equations with a difference term of subdifferentials", *Colloq. Math. Soc. Janos Bolyai, Qualitative Theory of Differential Equations* **30**, North-Holland, Amsterdom.
79. Othmer, H.G. and Stevens, A. (1997) "Aggregation, blowup, and collapse: the ABC's of taxis in reinforced random walk", *SIAM J. Appl. Math.* **57**, 1044-1081.
80. Painter, K. J., Maini, P. K. and Othmer, H. G. (2000) "Development and applications of a model for cellular response to multiple chemotactic cues", *J. Math. Biol.* **40**, 285-314.
81. Sakaguchi, S. and Suzuki, T. (1998) "Interior imperfect ignition can not occur on a set of positive measure" *Arch. Rational Mech. Anal.* **142**, 143-153.
82. Sattinger, D. (1968) "On global solution of nonlinear hyperbolic equations", *Arch. Rational Mech. Anal.* **30**, 148-172.
83. Schaaf, R. (1985) "Stationary solutions of chemotaxis systems", *Trans. Amer. Math. Soc.* **292**, 531-556.
84. Senba, T. and Suzuki, T. (2000) "Some structures of the solution set for a stationary system of chemotaxis", *Adv. Math. Sci. Appl.* **10**, 191-224.
85. Senba, T. and Suzuki, T. (2001) "Chemotactic collapse in a parabolic-elliptic system of mathematical biology", *Adv. Differential Equations.* **30**, 463-497.
86. Senba, T. and Suzuki, T. (2001) "Parabolic system of chemotaxis: Blowup in a finite and the infinite time", *Meth. Appl. Anal.* **8**, 349-368.
87. Tsujikawa, T. (1996) "Singular limit analysis of planar equilibrium solutions to a chemotaxis model equations with growth" *Meth. Appl. Anal.* **3**, 401-431.
88. Tsujikawa, T. and Yagi, A. (2002) "Exponential attractor for an adsorbate-induced phase transition model", *Kyushu. J. Math* **56**, 313-336.
89. Umeda, T. and Inouye, K. (2002) "Possible role of contact following in the generation of coherent motion of Dictyostelium cells" *J. Theor. Biol.* **219**, 301-308.
90. Velázquez, J.J.L. (1993) "Estimates on the $(N-1)$-dimensional Hausdorff measure of the blow-up set for a semilinear heat equation", *Indiana Univ. Math. J.* **42**, 446-476.
91. Verchota, G. (1984) "Layer potentials and regularity for the Dirichlet problem for Laplace's equations in Lipschitz domains", *J. Funct. Anal.* **59**, 572-611.
92. Weissler, F. (1980) "Local existence and non-existence for semilinear parabolic equations in L^p", Indiana Univ. Math. J. **29**, 79-102.
93. Yagi, A. (1997) "Norm behavior of solutions to the parabolic system of chemotaxis", *Math. Japonica* **45**, 241-265.
94. Yang, Y., Chen, H. and Liu, W. (1997) "On existence of global solutions and blow-up to a system of reaction-diffusion equations modelling chemotaxis", *SIAM J. Math. Anal.* **33**, 763-785.

Index